Neural Control Engineering

Computational Neuroscience
Terence J. Sejnowski and Tomaso A. Poggio, editors

Neural Control Engineering

The Emerging Intersection between Control Theory and Neuroscience

Steven J. Schiff

The MIT Press
Cambridge, Massachusetts
London, England

This book was set in Times Roman by Westchester Book Composition.

Library of Congress Cataloging-in-Publication Data

Schiff, Steven J.
Neural control engineering : the emerging intersection between control theory and neuroscience / Steven J. Schiff.
    p. ;   cm. — (Computational neuroscience series)
Includes bibliographical references and index.
ISBN 978-0-262-01537-0 (hardcover : alk. paper), 978-0-262-54671-3 (paperback)
1. Computational neuroscience. 2. Nonlinear control theory. I. Title. II. Series: Computational neuroscience.
[DNLM: 1. Neural Networks (Computer) 2. Brian—physiology. 3. Models, Neurological. 4. Neuroscience.
5. Nonlinear Dynamics. 6. Robotics. WL 26.5]
QP357.5S35   2011
612.8—dc22

2010036051

to Eleanor

Contents

Series Foreword

Computational neuroscience is an approach to understanding the development and function of nervous systems at many different structural scales, including the biophysical, the circuit, and the systems levels. Methods include theoretical analysis and modeling of neurons, networks, and brain systems and are complementary to empirical techniques in neuroscience. Areas and topics of particular interest to this book series include computational mechanisms in neurons, analysis of signal processing in neural circuits, representation of sensory information, systems models of sensorimotor integration, computational approaches to biological motor control, and models of learning and memory. Further topics of interest include the intersection of computational neuroscience with engineering, from representation and dynamics, to observation and control.

Terrence J. Sejnowski
Tomaso A. Poggio

Preface

This book is a personal adventure. It reflects the present state of an effort on my part to fuse my experience in the medicine of dynamical diseases of the brain with an enduring fascination for the physics that underlies such disease dynamics, and the prospects for controlling such conditions.

The material covered in this book constitutes a one-semester graduate course, *Neural Control Engineering*, that I have taught at Penn State University for the past several years. The students who have taken this course have come from a variety of backgrounds: engineering science, mechanical engineering, aerospace engineering, kinesiology, bioengineering, mathematics, physics, and medicine. To teach such a course, I needed to develop the quantitative science for the students from biology and medicine, and the biology for the quantitative science and engineering students.

There are thirteen chapters in this book. I teach this material by devoting a week to each of the first twelve chapters. I have designed a series of exercises to accompany the first eight chapters. The exercises are designed to develop intuition in formulating computational neuroscience models in a control engineering framework. I have provided some fundamentals of the algorithms that are helpful in crossing the divide from intuition to effective application. I have tested these algorithms in Matlab as well as the open source Octave.

In the remaining weeks of the course, I present lectures on applications covered in chapters 9 through 12, while assigning a series of open problems to the students. The applications of the methods of control engineering presented in this book are, in general, new enough in neuroscience applications that students can produce a final project exploring at the edge of where scientific investigation lies. Some of these projects have reached the quality of publishable advances. Most have, I hope, been of lasting benefit to students who have come to this subject from a broad swath of disciplines.

But perhaps most important for the potential reader, this book reflects the journey of the author, a physician and physiologist by training, to understand this fusion of subjects. I have tried to make the text as self-contained as possible. It should constitute a means to acquire a working knowledge of the fundamentals of control theory and computational neuroscience sufficient to not just understand the literature in this transdisciplinary arena, but to begin working near the cutting edge of the research along these lines. It is my hope that this book will serve as a useful guide for the interested scientist, from either the biological or quantitative sciences or engineering, as well as for the interested physician who wishes to gain expertise in these areas.

I owe a special debt to my long-term collaborator and friend, Tim Sauer, who first showed me a copy of the pivotal paper by Voss and colleagues [VTK04] in 2006, and asked me what I thought of it. I suppose in a very real sense, this book is the result of my effort to answer that question. Tim also read the manuscript, and provided invaluable insights and suggestions.

Numerous colleagues have helped guide me toward a working knowledge of control theory. Richard Murray welcomed me to Caltech on several occasions, and his time, wisdom, and advice have been invaluable. My efforts to learn this subject also led me to get to know engineering scholars such as Alok Sinha, Sean Brennan, Jacob (Jack) Langelaan, Constantino Lagoa, Kenneth Jenkins, Raj Acharya, and John Guttag, whose insights into control engineering run deep. I also benefited greatly from attending the lectures on the Mathematical Theory of Control by Alberto Bressan (his book on this topic is superb [BP07]). Three colleagues already working within the interface of neural engineering—Dominique Durand, John White, and Brian Litt—have been of immeasurable help with frequent conversations. I have also, for many years now, benefited from the interest and insightful conversations of Partha Mitra on the potential for fusion of engineering principles with biology.

One of the deep privileges of my life was to have had the chance to encounter, in classrooms and scientific meetings, some of the pivotal people whose work is highlighted in this book. Kenneth Cole, a legendary figure in twentieth-century biophysics, was kind enough to read my first manuscript (with John Moore and Norman Stockbridge), and offer his comments and thoughts—I hope I have represented his work adequately in this book. An entire generation of students at MIT who were enthralled in the class taught by Jerome Y. Lettvin, were inspired to spend the rest of their lives thinking about brains and nervous systems. I benefitted greatly from spending a year studying the physics of the squid axon membrane with John Moore—John is one of the remaining links to the waning era of intense study of axonology. I had a chance to meet John (Jack) Eccles at the Neurosciences Institute in the 1980s, where Vernon Mountcastle, in a long walk on a snowy evening in New York City, convinced me to stay in my neurosurgical residency rather than refocusing my career on science alone. Wilfrid Rall graciously spent time listening to my efforts to better understand reflex variability's role in the spinal cord and cerebral palsy. And scientist-surgeons Arthur Ward, Dennis Spencer, and George Ojemann provided me with much inspiration to seek the science in surgery.

In computational neuroscience, I have had the privilege of long-term friendships with Jack Cowan, G. Bard Ermentrout, Jonathan Rubin, Willliam Troy, Eric Shea-Brown, Giorgio Ascoli, Eugene Izhikevich, Nancy Kopell, Maxim Bazhenov, and John Rinzel, whose collective wisdom has helped guide my professional work.

I have greatly benefited from my colleagues in Dynamical Systems, including Lou Pecora, Bill Ditto, Eric Kostelich, Edward Ott, Celso Grebogi, James Yorke, Michael Fisher, Predrag Cvitanovic, Mingzhou Ding, Alistair and Moira Steyn-Ross, Andrew Frasier, Rajarshi (Raj) Roy, Marty Golubitsky, Steven Strogatz, Holger Kantz, Thomas Schreiber, Danny Kaplan, Henry Abarbanel, John Beggs, Eshel Ben-Jacob, Raoul Kopelman, Michal Zochowski, Evelyn Sander, Leonard Sander, Michael Unser, Akram Aldroubi, David Collela, John Benedetto, James Collins, Paul Rapp, Frank Moss, Lenny Smith, André Longtin, James Theiler, Mark Spano, Paul So, David Walnut, Arnold Mandell, Alan Garfinkel, Walter Freeman, John Milton, Michael Mackey, Leon Glass, Paul So, Ernest Barreto, and J. Rob Cressman. George Benedek profoundly influenced my view of the possibilities of understanding the physics of biology and medicine—his class notes on biological physics are still available [BV73]. It was also inspirational to quietly sit in the back of H. Eugene Stanley's lectures on biological physics in the 1970s [Sta72]. Harold Morowitz gave me a chance to begin forming a group to fuse physics and neuroscience. Michael Shlessinger helped this physician over more scientific hurdles than I can remember, often over still lawnmowers or tea.

I also am very grateful to Steven Weinstein, an extraordinary children's epileptologist, and to my neuroscience collaborators Karen Gale and Jian-young Wu. A period of productive collaboration with Robert Burke led to a number of discussions of the spinal cord and motoneurons in this text about the brain.

In addition to the blessing of parents who encouraged me to pursue science and medicine, two mentors have had formative influence on my life and work. George Somjen—a deep scholar and superb mentor—taught me how to do science, and much about life and society outside of science. W. Jerry Oakes, the finest physician I have ever met, encouraged me to pursue Pediatric Neurosurgery as well as my scientific pursuits.

Two organizations have given me extraordinary support: the American Association of Neurological Surgeons, which provided a Research Fellowship that helped me pursue completion of my PhD while in the middle of my neurosurgical residency; and the American Physical Society, where my long association with the *Physical Review* has been a rare privilege for a physician.

At Penn State University, David Wormley and Judy Todd saw the value in having a physician join their College of Engineering and the Department of Engineering Science and Mechanics. I hope that I have finally learned enough about engineering to make this gamble seem less of a risk. Jayanth Banavar welcomed me to the Physics Department and provided endless thoughtful advice. Andrew Webb introduced me to the potentials of advanced imaging engineering, which will perhaps find a chapter if this book ever gets to a second edition. The profound support of Robert Harbaugh and Harold Paz is greatly appreciated—they have enabled me to keep my involvement in clinical medicine despite being at continual risk of being consumed by science and engineering. Our scientific institute directors at Penn State—Peter Hudson, Susan McHale, and Padma Raghavan—have been endlessly supportive of my efforts to cross disciplinary boundaries.

A special thanks goes to my long-term collaborator, friend, and now fellow "neural engineer" Bruce Gluckman, who helped keep our Center for Neural Engineering afloat while I buried myself in writing this book, and whose brilliant work with electric field control systems will provide a valuable path to implement many of the concepts developed in this book.

This book could never have been worth reading were it not for the students who have suffered through its early materials and drafts in my classes, and offered their suggestions for improvements. I am especially indebted to Andrew Whalen, Jason Mandell, Stephen van Wert, Balaji Shanmugasundaram, Yina Wei, Dana Andre, Guoliang Fang, Tucker Tomlinson, Eric Liauw, Bruce Langford, and Nick Chernyy, and to the long-distance Web attendees Francisco Zabala, John Hobbs, and Óscar Miranda-Dominguez. I extend particular thanks to Patrick Gorzelic, Sergey Gratiy, Madineh Sedigh-Sarvestani, and Nori Okita, for their thoughtful suggestions and corrections to the drafts.

Another special thanks goes to Dr. Ghanim Ullah, a brilliant postdoctoral scholar, who joined me for this adventure in control engineering for four years.

I am grateful to Mary Lee Carns for marvelous help organizing the Center, and to Rebecca Thomas for helping organize and transcribe early material prior to my descent into LaTeX.

Colleagues who have taken the time to read and comment on individual chapters include G. Bard Ermentrout, Henning Voss, Jens Timmer, Jack Cowan, Jürgen Kurths, Gregory Duane, Jonathan Rubin, Apostolos Georgopoulos, Dan Simon, John Rinzel, Richard Murray, Uri Eden, Liam Paninski, Marom Bikson, Lupjo Kocerev, Xiaorui Tang, Gregory Duane, Miguel Nicolelis, Bruce Gluckman, and I am grateful to David Atwill for his comments on the intersection of Jungian psychology and Taoism.

I am deeply grateful to Robert Prior and Susan Buckley at the MIT Press, who saw value in this book project, and helped advise and guide me all along the way.

I would like to warmly acknowledge the generosity of Harvey F. Brush, whose endowment fund supplied the resources necessary to support this adventure in neural control engineering at this early stage.

And finally, I need to express my endless thanks to my wife Eleanor, whose love and tireless support was able to tolerate the seemingly endless amount of time required to write this book.

1 Introduction

The Kalman filter is probably the single most useful piece of mathematics developed in this century.
—John L. Casti, 2000 [Cas00]

1.1 Overview

This morning you awoke to a not unreasonable weather forecast. The last airplane you flew on may well have landed using a computer controlled autolander. These are examples of the power of modern control theory to both observe (weather) and control (airframes).

Such control techniques have several features. First, we embody our knowledge of a natural or manmade system into a mathematical computer model. It's hard to keep track in your head of atmospheric dynamics throughout the planet, just as it is hard to keep track of all the airframe control surfaces and dynamics of a large airplane. Our recent successes exhibit the confluence of several factors: We have increasingly powerful computers, we have increasingly powerful numerical methods to perform calculations, and we have better models. Boeing spent a fortune to develop the airframe model of the 777—it was an airframe developed almost entirely on computers. Similarly, we have a very good grasp of the physics of the atmosphere—the Euler equations [Kal03]; these convection equations are hard to solve, and one has to apply them in many places simultaneously to model the weather.

But building great models of systems is not enough.

First, *all of our models are wrong*. We are not interested in fitting data from the past; who cares that you could have predicted yesterday's weather given what you know today. Your model, born of past data, needs to interact with your system in real-time moving forward. You don't get a lot of data to work with—most sensor systems give a sparse sampling, in space and time, of the things you wish to measure. Many parts of the globe have little coverage of temperature, pressure, and wind speed, just as there are many portions of a modern airframe with no sensors nearby. So, if your model represents the whole system, it will need to reconstruct the parts that are inaccessible to measurement.

Second, *all measurements are bad*. Sensors are noisy, imprecise, and deteriorate, and the amplifiers change their properties with age, temperature, and calibration. So we need a way of optimally feeding our bad data to our wrong models.

Last, our *computers are never fast enough*. Nature runs along in continuous time, while our digital circuitry and computers chug along in saltatory bursts. Even if your model and sensor data were perfectly aligned at the last reading, the model's forward iteration takes place while nature does its own thing. A random air gust blows a plane sideways. Thermal stochastic effects, or chaotic sensitivity to conditions, create an atmospheric state whose trajectory differs from the one your model is creating. In a robotic system such as an airplane, the models are simple enough, and the computations fast enough, that many iterations can take place per second. In numerical weather forecasting, 6 hours is a typical window for North American and European agencies. And in both cases, we never use our best models. Speed is traded for accuracy—it does no good to take two days to predict tomorrow's weather.

There are two key concepts here: *observability* and *controllability*. Rudolf Kalman [Kal60] demonstrated that these concepts are linked: If you can observe the state of a system, you have just determined the extent to which you can control it. For all complex systems, theory and modeling is not a luxury or the province of the socially impaired. It is the only way you can observe such a system. It is the *lens* through which your observation changes from vague subjectivity to a comprehensive estimation of what you are observing.[1]

Airframes are, if my colleagues in aerospace engineering will forgive me, simple. We have also poured an incredible amount of funding into building very good models of these structures. The weather is complex, but compared with biological systems, it is simple. Gas and fluid dynamics are characterized by neighbor-to-neighbor interactions only, the molecules are all indistinguishable, and were it not for nonlinear dynamics such as turbulence, forecasting would be a breeze.

Brains are the most complex structures in the known universe. So on the one hand, you would never want to observe raw data without the benefit of modeling. But if all models are bad, models of brains are terrible.

This book is a gamble—a race between our gathering knowledge of neuroscience and its embodiment in computational neuroscience, and our growing sophistication in computational engineering tools that can handle the complexity of brain dynamics. At some point, the convergence of these two areas of knowledge will make the fusion of computational neuroscience and control theory absolutely required just to record from a neuron, or decipher an epileptic electroencephalogram (EEG). Are we there yet? Probably. The rest of this book will try to justify this statement.

---

1. I was sure that this *lens* metaphor was suggested to me by Partha Mitra [MB08] in one of his talks, but neither he nor I are certain anymore. I will cite him more for inspiration if not fact.

1.2   A Motivational Example

The ultimate Figawi event.

In 1916, Ernest Shackleton was marooned with his crew in a place called Elephant Island after his ship sank in the Antarctic. They spent over a year there, and it looked like they were going to be trapped for another year as the winter again closed in. They had a photographer with them, and figure 1.1 shows the small lifeboat that they launched in an effort to traverse 800 miles across the worst ocean in the world to reach a tiny dot called South Georgia Island where there was a whaling station. By then the rest of the world had assumed that Shackleton, as had many other Antarctic explorers, been long dead. He took a sextant and a navigator, Frank Worsley, and a few strong men [Ale98].

Shackleton and Worsley had a near *perfect* model—a map of the position of land masses and oceans, and the navigational equations that could compute where Shackleton would be on a given day based on his previous position and his velocity. But perfect models, and deductive reckoning, can be terrible in implementation. Once he left Elephant Island his position grew increasingly uncertain. The only way to address this was to use additional data measurements to improve his estimation of position—his *state*. He needed to fuse new data, his position readings of the sun, with his existing model (map and estimates of position). He needed to *assimilate* data.

Shackleton had to take measurements at a time when the sky was clear and he could hold his sextant still. But in this part of the oceans, waves are often 40 feet high, and the wind

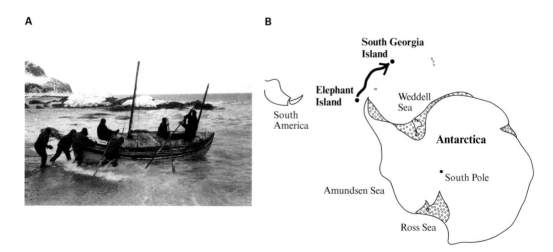

Figure 1.1
Voyage of the James Caird, 1916. Panel A reproduced from [Sha19], in public domain. Panel B courtesy of E. Schiff.

typically howls at 40 to 60 miles per hour. He was trying to grab a mast in the middle of the small boat to steady himself. The clouds broke only four times during the trip. There was enormous uncertainty about whether the position he measured in this turbulent world was accurate or not. Does he keep yesterday's prediction of where he expected to be that day, or use the measurement he just took? Compromising between model and data is a constant theme of this book. Flipping a coin to choose between estimate and data is not optimal—even bad measurements tell you *something*.

Data assimilation is a relatively new term forged out of two too common words. It is the fusion of data with your preexisting knowledge [WB07]. It is an act as old as navigation itself. Bayes's theorem (equation 1.73) tells you how to combine your previous guess with your new measurement, and this establishes a fundamental component of all data assimilation. But it would not be until the advent of the space program in the 1950s and 1960s that Kalman [Kal60] optimized how to apply Bayes's formula recursively to a dynamical process.

Let's simplify Shackleton's navigational nightmare a bit. Assume you are living in a one-dimensional world, and you only need to fix your position along a line (figure 1.2). Make the first measurement. I'm going to assume that when you made the first measurement, y, this is the first estimate of truth x. Not bad if you did this before pushing off Elephant Island—a good way to calibrate your instruments, or set your watch, because you really did know where you were and when the sun was at its noontime height. But perhaps your map of this rarely visited island did not fix its position well, or never captured its miserable coastline accurately. You might assume a probability distribution of your initial position x. Later in the book with nonlinear maps, you might pick a few initial conditions sprinkled

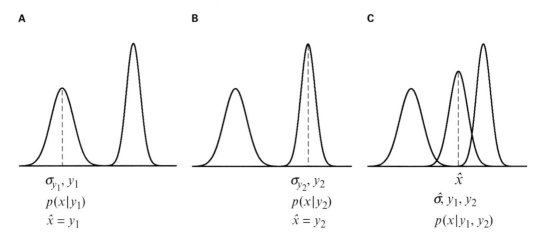

A   **B**   **C**

σ_{y_1}, y_1 σ_{y_2}, y_2 $\hat{x}$

$p(x|y_1)$ $p(x|y_2)$ $\hat{\sigma}, y_1, y_2$

$\hat{x} = y_1$ $\hat{x} = y_2$ $p(x|y_1, y_2)$

Figure 1.2
Bayes's rule tells you how to take sequential measurements and blend them. Two measurements, y_1 and y_2, and their respective uncertainties, σ_{y1} and σ_{y2}, are represented. The estimate of the true state, $\hat{x}$, is the mean of a probability distribution conditioned by y_1 or y_2, and then by y_1 and y_2.

about where you think you are, to determine what the consensus seems to be when you drop these initial guesses onto the rocky landscape of your topographical relief map.

Let's assume that you actually know something about the uncertainty, σ, for a given measurement, i, σ_i; that's quite a leap of faith,[2] but one that is often used in applications. Shackleton's first measurement, y_1, stinks, so as Caroline Alexander explained [Ale98], the rest of the crew helped to steady him and hold him up and reduce his weaving back and forth, for his second measurement, y_2. Assume that y_2 comes at a time very close to the first measurement. We don't think the boat has gone anywhere in the planet of significance during that short period of time, so we are estimating the same position on the map. Because two people were holding him up, the variance, the uncertainty of the second measurement, σ_{y_2}, is smaller than the first, σ_{y_1}. What is the optimal way to combine both measures to form an estimate, $\hat{x}$, of the true position, x?

There is a very critical, if subtle, truth here: *You never know the truth*. The truth is the true position x. You take measurements, y, and you use this information to *estimate* the truth, $\hat{x}$. It does not matter whether you are trying to estimate your position on a map, estimate a patient's pulse, asking what the voltage is within a neuron, or determining whether a seizure is going to occur soon. You have no direct access to the truth; you need to optimally estimate it given your knowledge (your model) and your data.

Shackleton's first measurement, y_1, had uncertainty σ_{y_1}, so if this were your only measurement, the conditional probability $p(x|y_1)$ of where you are, given what your measurement is (since you typically assume that your largest probability centers on your measurement), would lead you to pick $\hat{x}_1 = y_1$ as shown in figure 1.2A. Then you make the second measurement with its narrower uncertainty, as in figure 1.2B. If that were all you had, the conditional probability of x given only y_2 would be y_2. Now we combine these measurements into a single estimate (figure 1.2C).

There is no better intuitive explanation of this problem than that of Peter Maybeck, in his classic text [May82], and we will follow his framework here. Let's start with the answer (we will justify this later in the chapter):

$$\hat{x} = \frac{\sigma_{y_2}^2}{\sigma_{y_1}^2 + \sigma_{y_2}^2} y_1 + \frac{\sigma_{y_1}^2}{\sigma_{y_1}^2 + \sigma_{y_2}^2} y_2 \tag{1.1}$$

The estimated position, $\hat{x}$, is an average of the two position measurements, y_1 and y_2, weighted by the fraction of the variance from the *other* variances in equation (1.1). In other words, if the first measurement has a relatively large variance compared with the second, then pay attention to the second measurement more than the first (and vice versa). It also turns out that

$$\frac{1}{\hat{\sigma}_x^2} = \frac{1}{\sigma_{y_1}^2} + \frac{1}{\sigma_{y_2}^2} \tag{1.2}$$

---

2. Kierkegarrd's input will not formally be required until the last chapter in this book.

which implies that

$$\hat{\sigma}_x^2 < \sigma_{y_1}^2 \text{ and } \hat{\sigma}_x^2 < \sigma_{y_2}^2 \tag{1.3}$$

Equation (1.3) develops a fundamental philosophical point: *all measurements are of value*. No matter how bad the variances $\sigma_{y_1}^2$ and $\sigma_{y_2}^2$ are, they *always* make $\hat{\sigma}_x^2$ smaller if you know them. If a measurement had infinite uncertainty, then equation (1.1) tells you to throw out its measurement, and equation (1.2) tells you to set the estimated uncertainty equal to the uncertainty of the better measurement. If there was ever no uncertainty in a measure, you just use it and stop taking more measurements. Perhaps this all seems intuitive, but equation (1.1) is a statement of weighted least squares, and it was Carl Gauss who originally found that these weights y's were optimal [Str86].

What if the uncertainties in these measures were all equal to each other? The estimate of position would be the average of the measurements. And the uncertainty $\hat{\sigma}_x^2$ would be *half* of the individual uncertainties. This is the case of ordinary (unweighted) least squares.

Ronald Fisher pointed out that there was an inverse relationship of uncertainty to the information content [Fis34, FC96]. If uncertainty is infinite, you have no information from that measurement. If uncertainty goes to zero, you know everything; were such a thing possible, you would not need Kalman filtering, or this book. Equation (1.2) tells you that information is always useful. You are never worse off by taking a measurement—not in this worldview, at least—as long as the measurement carries some information about what you're trying to estimate.

So far, we have made these measurements all at the same time, when more measurements always makes certainty better. We will shortly propagate that state through time, and then time will start to pull certainty apart.

Let's return to equation (1.1). Let's recognize that y_1 came before y_2, and say that y_1 is the information we have *prior* to y_2—our a priori knowledge. We will use the following trick, adding two terms that sum to zero, to eliminate the coefficient that multiplies the prior repeatedly in this book

$$\hat{x}_2 = \frac{\sigma_{y_2}^2}{\sigma_{y_1}^2 + \sigma_{y_2}^2} y_1 + \left\{ \frac{\sigma_{y_1}^2}{\sigma_{y_1}^2 + \sigma_{y_2}^2} y_1 \right\} + \frac{\sigma_{y_1}^2}{\sigma_{y_1}^2 + \sigma_{y_2}^2} y_2 - \left\{ \frac{\sigma_{y_1}^2}{\sigma_{y_1}^2 + \sigma_{y_2}^2} y_1 \right\} \tag{1.4a}$$

$$= y_1 + \frac{\sigma_{y_1}^2}{\sigma_{y_1}^2 + \sigma_{y_2}^2} (y_2 - y_1) \tag{1.4b}$$

$$= \hat{x}_1 + \frac{\sigma_{y_1}^2}{\sigma_{y_1}^2 + \sigma_{y_2}^2} (y_2 - \hat{x}_1) \tag{1.4c}$$

$$\hat{x}_2 = \hat{x}_1 + K(y_2 - \hat{x}_1) \tag{1.4d}$$

where

$$K = \frac{\sigma_{y_1}^2}{\sigma_{y_1}^2 + \sigma_{y_2}^2} = \frac{\sigma_{\hat{x}_1}^2}{\sigma_{\hat{x}_1}^2 + \sigma_{y_2}^2} \tag{1.5}$$

In (1.4b) the weighting is now applied to the difference $y_2 - y_1$. We assume in (1.4c) that when all you had was y_1, this was your best estimate of state $\hat{x}_1$. The weighting in (1.4d) is assigned the variable K, and in calculating K we can replace the a priori uncertainties $\sigma_{y_1}^2$ with $\sigma_{\hat{x}_1}^2$ as in equation (1.5).

K will shortly become Kalman's gain function [Kal60]. Kalman's gain, in this static case, is the ratio of the previous uncertainty to the total uncertainty. This ratio will tell you whether to weigh your model strongly using what you calculated a priori, or pay more attention to your new measurement, y_2. If $\sigma_{y_1}^2$ (or $\sigma_{\hat{x}_1}^2$) is small compared with the new uncertainty, ignore the new measure and "fly" the model. If your new measure is much more precise than the old one, you might want to forget your previous calculation and look out the window.

In *prediction-corrector* systems, you make a prediction, $\hat{x}_1$, and then correct it based on new measurements. The new prediction, $\hat{x}_2$, for linear systems with Gaussian errors, is the mean, the median, the mode, the maximum likelihood estimator, the weighted least squares error, the best linear unbiased estimator. All of this drops out of the simple formulation in equations (1.4).

K has been a guiding principle in water, air, and space navigation for over half a century. Let's take a look at how K relates to propagating the uncertainty.

From (1.2) we write

$$\frac{1}{\sigma_{\hat{x}_2}^2} = \frac{1}{\sigma_{y_1}^2} + \frac{1}{\sigma_{y_2}^2} = \frac{\sigma_{y_1}^2 + \sigma_{y_2}^2}{\sigma_{y_1}^2 \sigma_{y_2}^2} \tag{1.6}$$

so that

$$\sigma_{\hat{x}_2}^2 = \frac{\sigma_{y_1}^2 \sigma_{y_2}^2}{\sigma_{y_1}^2 + \sigma_{y_2}^2} \tag{1.7}$$

We'll use our adding zero trick from equation (1.4a) again

$$\sigma_{\hat{x}_2}^2 = \frac{\sigma_{y_1}^2 \sigma_{y_2}^2}{\sigma_{y_1}^2 + \sigma_{y_2}^2} - \left\{ \frac{\sigma_{y_1}^2 \sigma_{y_1}^2}{\sigma_{y_1}^2 + \sigma_{y_2}^2} \right\} + \left\{ \frac{\sigma_{y_1}^2 \sigma_{y_1}^2}{\sigma_{y_1}^2 + \sigma_{y_2}^2} \right\} \tag{1.8}$$

which resolves to

$$\sigma_{\hat{x}_2}^2 = \sigma_{y_1}^2 - K\sigma_{y_1}^2 \tag{1.9}$$

or

$$\sigma_{\hat{x}_2}^2 = \sigma_{\hat{x}_1}^2 - K\sigma_{\hat{x}_1}^2 \tag{1.10}$$

What happens to uncertainty when you make a second measurement? Uncertainty *always* goes down with a new measurement. You knew that from before, but in this framework, the Kalman gain tells you how to do a weighted average of not just position—it also lets you do a weighted adjustment of your ongoing knowledge of uncertainty. It's not obvious or trivial.

Now add dynamics. In the simplest case, the boat, after measurement y_2, moves. In Shackleton's case, the wind probably blows with a very nice constant mean velocity, v, say 50 miles per hour all day and all night long, but there are some random fluctuations, q, which we will add to our constant velocity

$$\frac{dx}{dt} = velocity + noise = v + q, \qquad q = N(0, \sigma_q^2) \tag{1.11}$$

where $N(0, \sigma_q^2)$ indicates that q is drawn from a normal (Gaussian) distribution, with mean $= 0$ and variance $= \sigma_q^2$.

Integrate this equation to make your next position prediction. We integrate the position, velocity, and uncertainty from time $t = t_2$ to $t = t_3$

$$\int_{t_2}^{t_3} dx = \int_{t_2}^{t_3} v dt + \int_{t_2}^{t_3} q dt \tag{1.12}$$

which gives

$$x_3 - x_2 = v(t_3 - t_2) + \sigma_q^2(t_3 - t_2) \tag{1.13}$$

The integration of position and velocity gives the trivial results $x_3 - x_2$ and $v(t_3 - t_2)$. The integration of the random q is extremely nontrivial: $\sigma_q^2(t_3 - t_2)$. Intuitively, you can sense why this is true. If you integrate a Brownian noise process, that is integrate values drawn from a Gaussian distribution throughout a time interval, the result happens to be the variance times that time interval. This is known in physics as a Langevin equation (a particle diffusing with the wind also blowing), and in economics it is related to the Black-Scholes equation (a way to price options). Klebaner's textbook [Kle05] is an excellent resource to seek further understanding of such stochastic integrals.

In the previous case, from t_1 to t_2, there was no mean velocity, and if we had waited a given amount of time, the boat's position would diffuse on the map from random gusts. The uncertainty in position increases uniformly with time for the case of static estimation.

We will need to introduce some new notation to clarify two very different time scales that will be referred to throughout this book. There is a slow time scale that represents the natural system's dynamics over an interval, say t_2 to t_3, or the analogous time computed from the predictive model such as equation (1.12). In contrast, there is the much faster time scale that represents the state of a system just before you make a new measurement, $\hat{x}_3^-$, and the advance in your knowledge just after you take in a new measurement, $\hat{x}_3^+$. Indeed, even the computation required to absorb this new piece of data, as when you use equation (1.4d), is faster than model propagation as with the integration in equation (1.12). And in our boat example, the boat has not gone any appreciable distance just as you make a measurement—but you have learned something afterwards.

So we take our best estimated position at time t_2, $\hat{x}_2$, and the estimated uncertainty, $\hat{\sigma}_q^2$, and propagate both forward with equation (1.13). You now have a best predicted position, $\hat{x}_3^-$. There is the additive integrative uncertainty, $\sigma_q^2(t_3 - t_2)$, which if we permitted too long a time interval, would generate planetary scale uncertainty. So we don't want to take measurements too far apart.

Now take y_3, the next measure. Shackleton could complete only four measurements during the journey [Ale98]. What is the new position and uncertainty?

If the uncertainty of the new measurement is too high, the Kalman gain goes to zero, and by equation (1.4d), you ignore y_3. If the uncertainty is very small, then just use y_3 and ignore the model. Kalman filtering gives you a prescription for what data to use and what data to ignore based on the uncertainties of either of your measurements or the underlying process. But we have considerable ground yet to cover before discussing the Kalman filter in detail.

If we focus on the mean from the example in equation (1.13) (the mean of σ_q^2 is zero), the mean propagates as

$$\hat{x}_3^- = \hat{x}_2^+ + v\,(t_3 - t_2) \tag{1.14}$$

and the variance as

$$\sigma_{\hat{x}_3^-}^2 = \sigma_{\hat{x}_2^+}^2 + \sigma_q^2\,(t_3 - t_2) \tag{1.15}$$

all before the measurement at $t = t_3$ is taken.

Now make measurement y_3 with variance σ_{y3}^2. To assimilate this new measurement, we write

$$\hat{x}_3^+ = \hat{x}_3^- + K_3(y_3 - \hat{x}_3^-) \tag{1.16}$$

where we now index K acknowledging our new time scale as

$$K_3 = \frac{\sigma_{\hat{x}_3^-}^2}{\sigma_{\hat{x}_3^-}^2 + \sigma_{y3}^2} \tag{1.17}$$

Figure 1.3
The minor planet Ceres—the largest asteroid. Image from Hubble Space Telescope.

and

$$\sigma^2_{\hat{x}^+_3} = \sigma^2_{\hat{x}^-_3} - K_3 \sigma^2_{\hat{x}^-_3} \tag{1.18}$$

Note carefully the subtleties of how increasing or decreasing $\sigma^2_{y_3}$ or $\sigma^2_{\hat{x}^-_3}$ affects K_3, and the implications of $\sigma^2_{y_3} \ll \sigma^2_{\hat{x}^-_3}$, or $\sigma^2_{\hat{x}^-_3} \ll \sigma^2_{y_3}$.

1.3   Least Squares

But since all our measurements and observations are nothing more than approximations to the truth, the same must be true of all calculations resting on them, and the highest aim of all computations made concerning concrete phenomena must be to approximate, as nearly as practicable, to the truth.
—Carl Frederic Gauss, 1809

At the end of the eighteenth century, astronomers had sighted the minor planet Ceres (now relegated to largest asteroid status, figure 1.3). In his 1809 book [Gau09] on tracking asteroids and comets,[3] Gauss detailed that he had been using his least squares method since 1795 (when he was 18), much to the distress of his contemporary Legendre [Sor70]. Gauss worked out how to solve least squares and weighted least squares solutions to deal with the measurement of celestial objects. His was the only technique that seemed capable of

---

3. Henry Davis's very readable translation [GD57] of Gauss's 1809 book, *Theory of the Motion of Heavenly Bodies*, is now readily available through the Google book project on the Internet, and well worth examining.

completing the elliptical orbit and estimating where Ceres was going to be after it transited behind the sun for a period of weeks, using only the incomplete measurements of previous sightings before the transit [TW99].

Gauss, by 1809, had laid down most of the foundations for later measurement theory, Kalman filtering, and this book. His insights included conceptualizing that one needed at least an approximate knowledge of the dynamics of a system in order to assimilate data, and that errors in measurements required measuring more than the minimum number of observations needed in order to solve the equations in your possession (i.e., your problem should be *overdetermined*) [Sor70]. He realized that one needed to minimize the errors between what you predicted and what you observed—what will later be called the *innovations* in modern control theory. And he invented and understood the far-reaching properties of least squares:

[T]he most probable system of values of the unknown quantities . . . in which the sum of the squares of the differences between the observed and computed values of the functions is a minimum. . . . This principle, which promises to be of most frequent use in all applications of the mathematics to natural philosophy, must, everywhere, be considered an axiom with the same propriety as the arithmetical mean of several overserved values of the same quantity is adopted as the most probable value. [GD57]

Let's introduce the matrix formalism of Gauss's least squares. Throughout this book, we wish to solve the basic problem

$$Ax = y \tag{1.19}$$

where y is our observation and, if y is a vector, then A is a matrix.[4] We need matrices because we assume that all of our natural (and neural) systems have more than one variable that describes their states x, and that there will be more than one variable measured in any observation y. Even in this world of multielectrode arrays, one still might record from, for instance, a single deep brain microelectrode, but you would always consider that univariate y to be a condensation of a multivariable (multiple neuron) state. The matrix A tells us how to *remix* x, that is form a linear combination of the actual underlying variables that make up the vector or matrix x, in order to form the measurement y. We always assume that the true state x of any system, neural or otherwise, is hidden from our direct observation. Our task is to find the best (most probable) estimate of x given observations y, and we call this best estimate $\hat{x}$.

As a simple example, let's assume that y is measured twice, with values 1 and 3, and that x is one-dimensional with a coefficient $a = 1$ at each measurement time, then[5]

---

4. We will use capital letters to indicate matrices in the text.

5. If you are unfamiliar with matrix mathematics and linear algebra, I would strongly recommend putting this book down and spending a few weeks with one of Gilbert Strang's introductory linear algebra textbooks [Str06]. These are well suited for self-instruction for the neuroscientist or physician who did not realize that such skills would be needed later in life. Note, also, that MIT has made the video recordings of Professor Strang's course on this subject openly available on the MIT Open Courseware Web site.

$$\begin{bmatrix} 1 \\ 1 \end{bmatrix} x = \begin{bmatrix} 1 \\ 3 \end{bmatrix} \tag{1.20}$$

You do this every day in a laboratory or clinic—make up for imprecision by repeating measurements. Such systems are termed *overdetermined*. But if the measurements differ, and the system is assumed to remain the same, *none* of the individual measurements is correct. Some overdetermined systems are famous examples of great science; Millikan reported measuring the electron charge on his oil drop 58 times on 60 consecutive days— the result was fundamental and permanent [McG71].[6] We define the vector of errors, r, as

$$\text{error} = r = Ax - y = \begin{bmatrix} x - 1 \\ x - 3 \end{bmatrix} \tag{1.21}$$

The length of a vector is indicated by the *norm* $\|\cdot\|$, and following the suggestion of Gauss, we wish to minimize the square of the norm of r

$$\|r\|^2 = r_1^2 + r_2^2 + \ldots = [Ax - y]^T [Ax - y] \tag{1.22}$$

where $[\cdot]^T$ indicates matrix transpose of rows and columns. The square of the norm is found by taking the inner product of the transpose of the error vector with itself. To minimize it, take the derivative and set it equal to zero

$$\frac{d}{dx}\left[(x-1)^2 + (x-3)^2\right] = 2(x-1) + 2(x-3) = 0 \tag{1.23}$$

which tells you that the best estimate of x is $\hat{x} = 2$. In matrix notation, this would be

$$\frac{d}{dx}\left[x^T A^T Ax - y^T Ax - x^T A^T y + y^T y\right] = 0 \tag{1.24}$$

Taking derivatives of each term with respect to x yields[7]

$$2A^T Ax - 2A^T y = 0 \tag{1.25}$$

so we reduce to one equation with one unknown

---

6. Some considerable controversy has arisen in recent years over these measurement reports. See [Goo00], for instance.

7. The derivative of the quadratic form $x^T A^T Ax$ is a bit confusing at first sight. For vectors that are functions of x, $u(x)$ and $v(x)$, the derivative with respect to x of $u^T v$ is

$$\frac{d}{dx}\left[u^T v\right] = \left[\frac{du}{dx}\right]^T v + \left[\frac{dv}{dx}\right]^T u$$

So write $Ax = u = v$, and express $x^T A^T Ax$ as $(Ax)^T (Ax) \equiv u^T v$, and the derivative is seen to readily be $2A^T Ax$.

$$A^T A x = A^T y \tag{1.26}$$

Using our example from equation (1.20), we get

$$[1\ 1] \begin{bmatrix} 1 \\ 1 \end{bmatrix} x = [1\ 1] \begin{bmatrix} 1 \\ 3 \end{bmatrix} \tag{1.27}$$

where $2x = 4$, and we find our best estimate, $\hat{x} = 2$, as we already knew. So in matrix formalism, the solution to least squares for such problems is

$$\hat{x} = (A^T A)^{-1} A^T y \tag{1.28}$$

Our discussion above was for the case where each measurement had equal uncertainty. Now let's introduce *weighted* least squares, where we assume, as in our Shackleton story, that the measurements have different uncertainties.

Strang [Str86], provides an excellent discussion of weighted least squares, and how they lead to the Kalman filter, and we will follow that discussion closely here. The weights will be indicated by matrix W, so that

$$W A x = W y \tag{1.29}$$

and using our example from equation (1.20)

$$\begin{bmatrix} \sigma_{11} & 0 \\ 0 & \sigma_{22} \end{bmatrix} \begin{bmatrix} 1 \\ 1 \end{bmatrix} x = \begin{bmatrix} \sigma_{11} & 0 \\ 0 & \sigma_{22} \end{bmatrix} \begin{bmatrix} 1 \\ 3 \end{bmatrix} \tag{1.30}$$

The σ's are standard deviations. Our weighted error vector is now

$$\text{error} = W r = W A x - W y \tag{1.31}$$

and we need to minimize

$$\|W r\|^2 = \sigma_{11}^2 r_1^2 + \sigma_{22}^2 r_2^2 + \dots \tag{1.32}$$

which leads to

$$A^T W^T W A x = A^T W^T W y \tag{1.33}$$

The best estimator, $\hat{x}$, is

$$\hat{x} = (A^T W^T W A)^{-1} A^T W^T W y \tag{1.34}$$

Equation (1.34) is unpleasant, but using our previous example, we write

$$[1\ 1] \begin{bmatrix} \sigma_{11} & 0 \\ 0 & \sigma_{22} \end{bmatrix} \begin{bmatrix} \sigma_{11} & 0 \\ 0 & \sigma_{22} \end{bmatrix} \begin{bmatrix} 1 \\ 1 \end{bmatrix} x = [1\ 1] \begin{bmatrix} \sigma_{11} & 0 \\ 0 & \sigma_{22} \end{bmatrix} \begin{bmatrix} \sigma_{11} & 0 \\ 0 & \sigma_{22} \end{bmatrix} \begin{bmatrix} 1 \\ 3 \end{bmatrix}$$

which yields

$$(\sigma_{11}^2 + \sigma_{22}^2)x = (\sigma_{11}^2 \cdot 1 + \sigma_{22}^2 \cdot 3)$$

so our best estimate, $\hat{x}$, is

$$\hat{x} = (\sigma_{11}^2 \cdot 1 + \sigma_{22}^2 \cdot 3)/(\sigma_{11}^2 + \sigma_{22}^2) \tag{1.35}$$

If the weightings are all equal, we can set all the σ_{ii}'s to 1, and $W = I$, the *identity* matrix with all 1's on the diagonal and zeros elsewhere. This would bring us back to ordinary least squares. Otherwise,

$$\hat{x} = \frac{\sigma_{11}^2}{(\sigma_{11}^2 + \sigma_{22}^2)}y_1 + \frac{\sigma_{22}^2}{(\sigma_{11}^2 + \sigma_{22}^2)}y_2 \tag{1.36}$$

which explains the form of our first lifeboat-inspired example[8] in equation (1.1).

1.4   Expectation and Covariance

We need to develop a few more essential concepts. The *central limit theorem* states[9] that the sum of many random processes gives a Gaussian probability distribution $p(x)$,

$$p(x) = \frac{1}{\sqrt{2\pi\sigma^2}} \exp\left[-x^2/2\sigma^2\right] \tag{1.37}$$

where we define that the total probability must equal 1

$$\int_{-\infty}^{\infty} p(x)dx = 1 \tag{1.38}$$

The *expectation* of x, the mean μ_x of what you would expect after a large (infinite) number of samples, is defined as

$$E[x] = \sum_x xp(x) = \mu_x \tag{1.39}$$

for discrete, and

---

8. Note that the indices used in W, σ_{ij}, are used to label the rows i and columns j in the matrix, and are not the indices of the weightings from equation (1.1), where σ_{11} was σ_{y_2}.

9. The proof of the central limit theorem is not at all trivial [KF70], but most readers will be content if it is just stated as done here.

$$E[x] = \int_{-\infty}^{\infty} x p(x) dx \tag{1.40}$$

for continuous processes. For our Gaussian probability distribution with zero mean

$$E[x] = 0 \tag{1.41}$$

Variance is the expectation of the squared deviation from the mean (for discrete)

$$\text{var}[x] = E[(x - E[x])^2] = \sum_x (x - \mu_x)^2 p(x) \equiv \sigma_x^2 \tag{1.42}$$

The variance for the continuous Gaussian distribution is

$$E[x^2] = \int_{-\infty}^{\infty} x^2 p(x) dx \tag{1.43}$$

The *covariance*[10] is the expectation of the squared deviation from the mean for multiple variables

$$\text{cov}[x, y] = E[(x - \mu_x)(y - \mu_y)] = \sum_x \sum_y (x - \mu_x)(y - \mu_y) p(x, y) \tag{1.44}$$

where $p(x, y)$ is the joint probability distribution of x and y. If x and y are vectors, then covariance is the expectation of the outer product

$$\text{cov}[x, y] = E[(x - \mu_x)(y - \mu_y)^T] \tag{1.45}$$

So for an error vector

$$r = \begin{bmatrix} r_1 \\ r_2 \\ \vdots \end{bmatrix} \tag{1.46}$$

the covariance matrix is

$$E[rr^T] = R \tag{1.47}$$

Gauss showed [Str86] that the inverse of the measurement covariance matrix, R^{-1}, gives the *best linear unbiased estimator* for the least squares solution, so we replace $W^T W$ in equation (1.34) with R^{-1}

10. A superb introduction to covariance and multivariable statistics in general can be found in Bernhard Flury's textbook [Flu97].

$$\hat{x} = (A^T R^{-1} A)^{-1} A^T R^{-1} y \qquad (1.48)$$

There are now two sets of errors to keep track of for the rest of the book. R is the measurement error covariance matrix, the measurement errors being $Ax - y$. The other set of errors that concern us are the errors in the estimation of x, which are $x - \hat{x}$. We define P as the covariance in the errors in the estimation of x

$$P = E[(x - \hat{x})(x - \hat{x})^T] \qquad (1.49)$$

Once we know the result in equation (1.48), it follows that the best estimate of P is

$$P = (A^T R^{-1} A)^{-1} \qquad (1.50)$$

The proof is nontrivial, and the least unpleasant description of it is in Strang (p. 144 in [Str86]).

If one refers back to figure 1.2c, R describes the errors in measurements y, and P describes the errors in the estimate of $\hat{x}$.

Let's assume that you take several measurements of a neuron firing rate. Assume for now that the uncertainty in measurements, σ^2, are all the same and independent from each other. Then the covariance matrix is just the set of individual variances

$$R = \begin{pmatrix} \sigma^2 & 0 \\ 0 & \sigma^2 \end{pmatrix} \qquad (1.51)$$

and the inverse of a diagonal matrix is just the reciprocal of the diagonal values

$$R^{-1} = \begin{pmatrix} 1/\sigma^2 & 0 \\ 0 & 1/\sigma^2 \end{pmatrix} \qquad (1.52)$$

Then from $P = (A^T R^{-1} A)^{-1}$,

$$P^{-1} = \begin{bmatrix} 1 & 1 \end{bmatrix} \begin{pmatrix} 1/\sigma^2 & 0 \\ 0 & 1/\sigma^2 \end{pmatrix} \begin{bmatrix} 1 \\ 1 \end{bmatrix} \qquad (1.53)$$

which is just

$$P = \frac{\sigma^2}{2} \qquad (1.54)$$

and for n measurements

$$P = \frac{\sigma^2}{n} \qquad (1.55)$$

P decreases with every measurement. It does not matter how uncertain the measurements are—all measurements tell you something about the truth x.

1.5 Recursive Least Squares

Solving for $\hat{x}$ or P using equation (1.48) or (1.50) gets more complex as the number of measurements and the size of matrices A and R enlarge. But each new measurement changes the covariances R and P only incrementally. If the errors

$$\begin{bmatrix} r_1 \\ r_2 \\ \vdots \end{bmatrix} \tag{1.56}$$

are independent, then for two measurements at t_0 and t_1 (again, following [Str86])

$$R = \begin{bmatrix} R_0 & \\ & R_1 \end{bmatrix} \tag{1.57}$$

and

$$P^{-1} = \begin{bmatrix} A_0^T & A_1^T \end{bmatrix} \begin{bmatrix} R_0^{-1} & \\ & R_1^{-1} \end{bmatrix} \begin{bmatrix} A_0 \\ A_1 \end{bmatrix} \tag{1.58}$$

Multiply this out and you get

$$P^{-1} = A_0^T R_0^{-1} A_0 + A_1^T R_1^{-1} A_1 \tag{1.59}$$

which is the key to making a recursive formula. Using equation (1.50) we can write

$$P_1^{-1} = P_0^{-1} + A_1^T R_1^{-1} A_1 \tag{1.60}$$

and this holds for any sequential estimations.

Substituting equation (1.50) in (1.48)

$$\hat{x}_1 = P_1 A^T R^{-1} y \tag{1.61}$$

which, for our simple two-measurement example, yields

$$\hat{x}_1 = P_1 \begin{bmatrix} A_0^T & A_1^T \end{bmatrix} \begin{bmatrix} R_0^{-1} & \\ & R_1^{-1} \end{bmatrix} \begin{bmatrix} y_0 \\ y_1 \end{bmatrix} \tag{1.62}$$

which multiplied out gives

$$\hat{x}_1 = P_1 \begin{bmatrix} P_0^{-1} A_0^{-1} y_0 + A_1^T R_1^{-1} y_1 \end{bmatrix} \tag{1.63}$$

Substituting x_0 for $A_0^{-1} y_0$

$$\hat{x}_1 = P_1 \left[P_0^{-1} x_0 + A_1^T R_1^{-1} y_1 \right] \tag{1.64}$$

and finally substituting P_0^{-1} with equation (1.60)

$$\hat{x}_1 = P_1 \left[P_1^{-1} x_0 - A_1^T R_1^{-1} A_1 x_0 + A_1^T R_1^{-1} y_1 \right] \tag{1.65}$$

gives

$$\hat{x}_1 = x_0 + K_1(y_1 - A_1 x_0) \tag{1.66}$$

with

$$K_1 = P_1 A_1^T R_1^{-1} \tag{1.67}$$

Equations (1.66) and (1.67) are the beginning steps of *recursive least squares*. Note several things. If $y_1 = A_1 x_0$, then $\hat{x}_1 = x_0$. If $y_1 = A_1 x_0 + e$, where e is the unexpected part of y_1, the *innovation*, then $\hat{x}_1 = x_0 + K_1(e)$, where $K_1(e)$ is the correction to the previous x_0.

The fundamental equations of recursive least squares are therefore

$$P_i^{-1} = P_{i-1}^{-1} + A_i^T R_i^{-1} A_i$$

$$K_i = P_i A_i^T R_i^{-1} \tag{1.68}$$

$$\hat{x}_i = \hat{x}_{i-1} + K_i(y_i - A_i \hat{x}_{i-1})$$

Let's measure the firing rate of a neuron in spikes per minute by counting spikes within a 10-second window.[11] You get one, two, and four spikes, with calculated $y_0 = 6$, $y_1 = 12$, and $y_2 = 24$ spikes per minute. We assume $A = 1$. Then

$$A^T A \hat{x} = A^T y$$

$$\begin{bmatrix} 1 & 1 & 1 \end{bmatrix} \begin{bmatrix} 1 \\ 1 \\ 1 \end{bmatrix} \hat{x} = \begin{bmatrix} 1 & 1 & 1 \end{bmatrix} \begin{bmatrix} 6 \\ 12 \\ 24 \end{bmatrix} \tag{1.69}$$

$$3\hat{x} = 42$$

$$\hat{x} = 14 \text{ spikes per minute}$$

and

11. Strang [Str86] shows a similar example for estimating heart rate.

$$P^{-1} = \begin{bmatrix} A_0^T & A_1^T & A_2^T \end{bmatrix} \begin{bmatrix} R^{-1} & & \\ & R^{-1} & \\ & & R^{-1} \end{bmatrix} \begin{bmatrix} A_0 \\ A_1 \\ A_2 \end{bmatrix}$$

$$P^{-1} = \begin{bmatrix} 1 & 1 & 1 \end{bmatrix} \begin{bmatrix} 1/\sigma^2 & & \\ & 1/\sigma^2 & \\ & & 1/\sigma^2 \end{bmatrix} \begin{bmatrix} 1 \\ 1 \\ 1 \end{bmatrix} \qquad (1.70)$$

$$P = \frac{\sigma^2}{3}$$

The best estimate of this static problem (we assume that this pacemaker neuron always has the same rate) is 14 spikes per minute.

Now let's estimate this firing rate recursively:

$$P_0^{-1} = A_0^T R^{-1} A_0 = [1][1/\sigma^2][1] = 1/\sigma^2$$

$$P_1^{-1} = P_0^{-1} + A_1^T R^{-1} A_1 = 1/\sigma^2 + 1/\sigma^2 = 2/\sigma^2 \qquad (1.71)$$

$$P_2^{-1} = P_1^{-1} + A_2^T R^{-1} A_2 = 3/\sigma^2$$

with

$$K_n = P_n A_n^T R_n^{-1} = \frac{\sigma^2}{n}[1]\frac{1}{\sigma^2} = \frac{1}{n}$$

$$\hat{x}_1 = \hat{x}_0 + K_n(y_1 - \hat{x}_0) = 6 + \frac{1}{2}(12 - 6) = 9 \qquad (1.72)$$

$$\hat{x}_2 = 9 + \frac{1}{3}(24 - 9) = 14$$

Note that in the recursive formulation that you throw away all of the history with each step—there is no need to store steadily increasing amounts of data with such recursion. Three measurements are not so bad to calculate. But later, we will have data sets with many thousands of sequential samples, and if you could find a computer with enough memory to solve equation (1.69), it would not be able to keep up with tracking a process in real time. Equations (1.71) and (1.72) never get more complex than what you see.

1.6 It's a Bayesian World

If our data were y, and our true underlying states x, then one could describe probability distributions of y independent of x, $p(y)$, and distributions of x independent of y, $p(x)$.

But if x and y are related, from knowledge of $p(x)$, we could refine the distribution of y given $p(x)$, the *conditional distribution* $p(y|x)$. But we already have y—this book is about estimating the underlying truth x, and often about estimating the most likely x, the mean of x, from the data observed. The conditional distribution $p(x|y)$ is the *posterior* or a posteriori distribution of x given data y, and $p(x|y)$ is our primary concern.

In taking conditional expectations, one takes a slice of a *joint* probability distribution, $p(x, y)$, and since all probability distributions must add to one as in (1.38), we need to normalize things. The rule for this comes from *Bayes's rule* [Flu97]

$$p(y|x) = \frac{p(x, y)}{p(x)} \tag{1.73}$$

So instead of integrating over $p(x)$ as in (1.40), the conditional expectation $E[x|y]$ is

$$E[x|y] = \sum_x x p(x|y) = \sum_x x \frac{p(x, y)}{p(y)} = \mu_{x|y} \tag{1.74}$$

and with similar structure for the continuous version. Multiply two (independent) Gaussian probability distributions as in equation (1.37) to get the joint Gaussian distribution

$$p(x, y) = \frac{1}{\sqrt{2\pi \sigma_x^2 \sigma_y^2}} \exp\left[-x^2/2\sigma_x^2 - y^2/2\sigma_y^2\right] \tag{1.75}$$

and the conditional distribution $p(y|x)$ is constructed as [Flu97]

$$p(y|x) = \frac{1}{\sqrt{2\pi \sigma_{y|x}^2}} \exp\left[-(y - \mu_{y|x})^2/2\sigma_{y|x}^2\right] \tag{1.76}$$

So let's assume that the truth is normally distributed with a mean of μ_x and variance σ_x^2, and that the observations y are normally distributed with a mean of $\mu_{y|x}$ and variance $\sigma_{y|x}^2$. Then, following [WB07],

$$p(y|x) = \frac{1}{\sqrt{2\pi \sigma_{y|x}^2}} \exp\left[-(y_1 - \mu_{y|x})^2/2\sigma_{y|x}^2\right] \cdot \frac{1}{\sqrt{2\pi \sigma_{y|x}^2}} \exp\left[-(y_2 - \mu_{y|x})^2/2\sigma_{y|x}^2\right] \tag{1.77}$$

which is proportional to

$$\exp - \left[(y_1 - \mu_{y|x})^2/2\sigma_{y|x}^2 + (y_2 - \mu_{y|x})^2/2\sigma_{y|x}^2\right] \tag{1.78}$$

To simplify this a bit, we can assume that the real position x is known, and the y's are distributed about the true position x, which serves as the y mean. Bayes's rule tells you that $p(x|y)$ is proportional to $p(y|x)p(x)$, so we multiply (1.78) by $p(x)$ to get

$$\exp - \left[(y_1 - x)^2/2\sigma_{y|x}^2 + (y_2 - x)^2/2\sigma_{y|x}^2 + (x - \mu_x)^2/2\sigma_x^2 \right] \tag{1.79}$$

which, because the sum of y^2 terms can be factored out (they are the variance of y plus the mean of y, all constants), simplifies to

$$\exp - \left[x^2 \left(\frac{2}{\sigma_{y|x}^2} + \frac{1}{\sigma_x^2} \right) - 2x \left(\frac{y_1 + y_2}{\sigma_{y|x}^2} + \frac{\mu_x}{\sigma_x^2} \right) \right] \tag{1.80}$$

and after completing the square,[12] this new Gaussian distribution has a mean of

$$E[x|y] \equiv \hat{x} = \frac{\sigma_x^2}{2\sigma_x^2 + \sigma_{y|x}^2}(y_1 + y_2) + \frac{\sigma_{y|x}^2}{2\sigma_x^2 + \sigma_{y|x}^2}\mu_x \tag{1.83}$$

or for n measurements, letting $\bar{y} = \frac{1}{n} \sum_{i=1}^{n} y_i$,

$$E[x|y] \equiv \hat{x} = \frac{n\sigma_x^2}{n\sigma_x^2 + \sigma_{y|x}^2}\bar{y} + \frac{\sigma_{y|x}^2}{n\sigma_x^2 + \sigma_{y|x}^2}\mu_x \tag{1.84}$$

Notice that these fractions, weighting the means of the distributions by the fraction of the *other* distribution's variance, is the reason why such weighting was used in equation (1.1) at the start of this chapter. It was based on Bayes's rule.

Note some important subtleties of equation (1.84). If the uncertainty in the model, σ_x^2, grows large, you can ignore the model—it's a bad map. If the number of measurements, y_i, becomes large, the data overwhelms your prior information, μ_x and σ_x^2, and you can again throw out the model. If $\sigma_x^2 \to 0$, ignore the measurements. Last, note that if we rewrite equation (1.84) as

$$\hat{x} = \mu_x + \frac{n\sigma_x^2}{n\sigma_x^2 + \sigma_{y|x}^2}(\bar{y} - \mu_x) \tag{1.85}$$

12. If $ax^2 + bx = 0$, then

$$x^2 + \frac{b}{a}x + \left(\frac{b}{2a} \right)^2 = \left(\frac{b}{2a} \right)^2 = \left(x + \frac{b}{2a} \right)^2 \tag{1.81}$$

where

$$\frac{b}{2a} = -\frac{\sigma_x^2 \sigma_{y|x}^2}{n\sigma_x^2 + \sigma_{y|x}^2} \left[\frac{y_1 + y_2}{\sigma_{y|x}^2} + \frac{\mu_x}{\sigma_x^2} \right] \tag{1.82}$$

for n measurements.

and let

$$K = \frac{n\sigma_x^2}{n\sigma_x^2 + \sigma_{y|x}^2} \tag{1.86}$$

we have the more general form of equation (1.4d).[13]

Kalman filtering is a subset of Bayesian analysis. But Kalman added dynamics to this static data assimilation framework. And that's the subject of the next chapter.

Exercises

1.1. The inverse of a matrix can be calculated in several ways [Str06]. One approach is to use the determinant of a matrix and calculate a set of matrix cofactors, where

$$(A^{-1})_{ij} = \frac{C_{ji}}{\det(A)}$$

and the cofactors, C_{ij}, are

$$C_{ji} = (-1)^{i+j} M_{ij}$$

where, for a 2×2 matrix,

$$\begin{bmatrix} a_{11} & a_{12} \\ a_{21} & a_{22} \end{bmatrix}$$

the M's are matrix minors,[14] which for this 2×2 case are

$$M_{11} = a_{22}, M_{22} = a_{11}, M_{12} = a_{12}, M_{21} = a_{21}$$

and the determinant is $a_{11}a_{22} - a_{12}a_{21}$. The inverse is

$$\frac{1}{a_{11}a_{22} - a_{12}a_{21}} \begin{bmatrix} a_{22} & -a_{12} \\ -a_{21} & a_{11} \end{bmatrix}$$

Suppose you are recording from two neurons, and measure their firing rates as $y_1 = 20$, and $y_2 = 30$. Given a linear system of equations

$$Ax = y$$

---

13. Unresolved in this chapter, or history in general, is just how Shackleton and Worsley actually found South Georgia Island. Armed with the techniques discussed in this book, you have a better chance. Armed with the technology of the turn of the twentieth century, their feat stands as a remarkable human achievement.

14. A matrix minor is the determinant after deleting the row and column containing a_{ij}.

where

$$A = \begin{bmatrix} 1 & 1 \\ 1 & 2 \end{bmatrix}$$

solve by hand the best least squares solution for

$$\hat{x} = (A^T A)^{-1} A^T y \tag{1.87}$$

1.2. Using the same A and y from Exercise (1.1), solve the least squares solution to (1.87) using a computer.[15] Note that, although you can solve this in Matlab-Octave as xhat = inv(A'*A)*A'*y, that a set of more numerically stable algorithms is accessed using the notation xhat = (A'*A)\A'*y. If the inverse of (A'*A) does not exist, the pseudoinverse can be tried as xhat = pinv(A'*A)*A'*y.

1.3. Another computer exercise. Assume the measured firing rates of a neuron are

$$y = \begin{bmatrix} 3 \\ 5 \\ 4 \\ 8 \end{bmatrix}$$

spikes per second. If, as in equation (1.69), we assume that

$$A = \begin{bmatrix} 1 \\ 1 \\ 1 \\ 1 \end{bmatrix}$$

solve for $\hat{x}$ using

$$\hat{x} = (A^T A)^{-1} A^T y$$

Assuming that

$$R = \begin{bmatrix} \sigma^2 & & & \\ & \sigma^2 & & \\ & & \sigma^2 & \\ & & & \sigma^2 \end{bmatrix}$$

---

15. Throughout this text, I will sketch out algorithms compatible with Matlab and Octave. Although Matlab is a commonly used language for scientific computing, it is expensive. Octave is an open source equivalent, and I will strive to ensure that any algorithmic examples are code compatible between these languages.

calculate P from

$$P = (A^T R^{-1} A)^{-1}$$

Now repeat the calculation using the fundamental equations (1.68) of recursive least squares

$$P_i^{-1} = P_{i-1}^{-1} + A_i^T R_i^{-1} A_i$$

$$K_i = P_i A_i^T R_i^{-1}$$

$$\hat{x}_i = \hat{x}_{i-1} + K_i (y_i - A_i \hat{x}_{i-1})$$

2 Kalman Filtering

2.1 Linear Kalman Filtering

Most books on the Kalman filter are rather discouraging. [Str86]
—Gilbert Strang, 1986

In the previous chapter, we presented the fundamental equations of recursive least squares in (1.68)

$$P_i^{-1} = P_{i-1}^{-1} + A_i^T R_i^{-1} A_i$$

$$K_i = P_i A_i^T R_i^{-1}$$

$$\hat{x}_i = \hat{x}_{i-1} + K_i(y_i - A_i \hat{x}_{i-1})$$

Such equations, in essence, assume that the model of x is *static*

$$x_i = x_{i-1} \tag{2.1}$$

Kalman [Kal60] added dynamics to this static picture

$$x_i = F_{i-1} x_{i-1} + G_{i-1} u_{i-1} + q_{i-1}$$

$$y_i = A_i x_i + r_i \tag{2.2}$$

where F_{i-1} is a matrix linearly combining the variables in the vector x_{i-1} to generate x_i, u_{i-1} will be a control vector (either a function of time, open loop, or of state x, closed loop) acted on by G_{i-1} (not necessarily the same size as F), q_{i-1} is random noise in the model, r_i is noise in the measurements y_i, and A_i is our matrix combining the variables in state x_i to produce measurement y_i.[1]

---

1. The q really involves more than random noise. Although there is a stochastic element to these process disturbances, q also represents the unknown and unmodelable aspects of the process. We will address this in much greater detail in chapter 8 on model inadequacy.

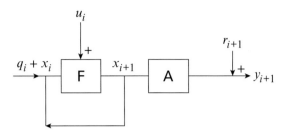

Figure 2.1
Schematic of Kalman filtering equations.

A frequently used schematic to illustrate such a system is shown in figure 2.1, which illustrates the dynamical mapping of x_i to x_{i+1} through F, the filtering through A, the addition of u to the dynamics, and the explicit addition of noise to dynamics, q, and measurement, r, where q_i and r_i are individual values drawn from Gaussian distributions with zero mean and variances Q and R, respectively

$$q_i \sim N(0, Q_i)$$

$$r_i \sim N(0, R_i)$$
$$(2.3)$$

Let's propagate the mean of x. We will follow the logic and notation of Simon [Sim06a][2]

$$E(x_i) = E(F_{i-1}x_{i-1} + G_{i-1}u_{i-1} + q_{i-1})$$

$$= F_{i-1}E(x_{i-1}) + E(G_{i-1}u_{i-1}) + E(q_{i-1})$$

$$\bar{x}_i = F_{i-1}\bar{x}_{i-1} + G_{i-1}u_{i-1} + 0 \qquad \{\text{treat } G_{i-1}u_{i-1} \text{ as const}\}$$
$$(2.4)$$

$$\bar{x}_i = F_{i-1}\bar{x}_{i-1} + G_{i-1}u_{i-1}$$

And now propagate the covariance P, the expectation of

$$(x_i - \bar{x}_i)(\cdots)^T = (F_{i-1}x_{i-1} + G_{i-1}u_{i-1} + q_{i-1} - \bar{x}_i)(\cdots)^T$$

$$= (F_{i-1}x_{i-1} + G_{i-1}u_{i-1} + q_{i-1} - F_{i-1}\bar{x}_{i-1} - G_{i-1}u_{i-1})(\cdots)^T$$

$$= (F_{i-1}(x_{i-1} - \bar{x}_{i-1}) + q_{i-1})(\cdots)^T$$
$$(2.5)$$

$$= F_{i-1}(x_{i-1} - \bar{x}_{i-1})(x_{i-1} - \bar{x}_{i-1})^T F_{i-1}^T + q_{i-1}q_{i-1}^T$$

$$+ F_{i-1}(x_{i-1} - \bar{x}_{i-1})q_{i-1}^T + q_{i-1}(x_{i-1} - \bar{x}_{i-1})^T F_{i-1}^T$$

2. Dan Simon's clear and self-contained book on Optimal State Estimation [Sim06a] is highly recommended as a complete and in-depth exposition on the details and logic of linear and nonlinear Kalman filtering.

where the use of $(\cdots)$ indicates replication of the previous bracket contents. If we now take expectations, recalling that q_{i-1} is, by definition, uncorrelated with x_{i-1}

$$P_i = F_{i-1} P_{i-1} F_{i-1}^T + Q_{i-1} \tag{2.6}$$

Since we are working with the two time scales introduced in chapter 1, we again indicate the separation of time scales. A priori is the time, at $(k-1)^-$ or k^-, just before a new measurement, at $(k-1)^+$ or k^+, is made, producing an a posteriori value. Figure 2.2 schematizes this separation of time scales.

It is important to note now that in equation (2.4), the propagation of the mean $\bar{x}$ due to system dynamics, and the propagation of the covariance P in equation (2.5), both take place within the relatively long time scale from a posteriori to a priori in figure 2.2. We now recognize that the mean of x, $\bar{x}$, is now a moving target, the best estimate of x, $\hat{x}$ is

$$\hat{x}_i^- = F_{i-1} \hat{x}_{i-1}^+ + G_{i-1} u_{i-1} \tag{2.7}$$

and in the same spirit, a best estimate for the propagation of P is

$$P_i^- = F_{i-1} P_{i-1}^+ F_{i-1}^T + Q_{i-1} \tag{2.8}$$

Note, critically, why the variance of the noise on the dynamics of x drops out of the estimate of $\hat{x}$, but stays with the propagation of P. In equation (2.4), the expectation is taken over q_i, which is zero. In equations (2.5), the covariance produces a product of qq^T, whose expectation is the definition of variance Q. *The estimate of state error always worsens between measures. Wait long enough, and your uncertainty in a noisy process becomes global.*

This brings us to the relatively fast transition from a priori to a posteriori, that occurs after a new measurement is made (e.g., at time k^+ in figure 2.2). We first need to drag P_i^- across the threshold from a priori to a posteriori. And we return to our equations (1.68) for

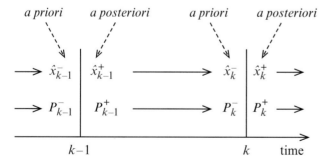

Figure 2.2
Separation of time scales. A priori is the time, at $(k-1)^-$ or k^-, just before a new measurement, at $(k-1)^+$ or k^+, is made, producing an a posteriori value. Modified after Simon [Sim06a] with permission.

recursive least squares, which were intended for instantaneous use after measurements, to write[3]

$$P_i^+ = \left[(P_i^-)^{-1} + A_i^T R_i^{-1} A_i \right]^{-1} \tag{2.9}$$

and use this value of P_i^+ to calculate the next K_i

$$K_i = P_i^+ A_i^T R_i^{-1} \tag{2.10}$$

We use this K_i to bring $\hat{x}_i^-$ across the measurement threshold

$$\hat{x}_i^+ = \hat{x}_i^- + K_i(y_i - A_i\hat{x}_i^-) \tag{2.11}$$

Note that during the time without measurement, P_i^- was propagated in proportion to the dynamical noise covariance Q. This drives P_i larger. At the time of measurement, look at equation (2.9). A small measurement error covariance, R_i, will drive the inverse R_i^{-1} large. This tends to overwhelm $(P_i^-)^{-1}$, and the inverse of this sum will become small. In (2.10), a small P_i^+ will be countered by the large R_i^{-1}, and the sizable K_i will pay credence to the new measure y_i. Note that if R_i were huge relative to P_i, then the term $A_i^T R_i^{-1} A_i$ can be neglected, and in equation (2.10) K_i is driven small. Ignore y_i when the measure is unreliable. *When the situation is fluid, and the balance of uncertainties changes, Kalman's filter iteratively adjusts for these shifting priorities.*

To summarize the equations for linear Kalman filtering:

$$x_i = F_{i-1}x_{i-1} + G_{i-1}u_{i-1} + q_{i-1}$$

$$y_i = A_i x_i + r_i$$

$$q_i \sim N(0, Q_i)$$

$$r_i \sim N(0, R_i)$$

$$\hat{x}_i^- = F_{i-1}\hat{x}_{i-1}^+ + G_{i-1}u_{i-1} \tag{2.12}$$

$$P_i^- = F_{i-1}P_{i-1}^+ F_{i-1}^T + Q_{i-1}$$

$$P_i^+ = \left[(P_i^-)^{-1} + A_i^T R_i^{-1} A_i \right]^{-1}$$

$$K_i = P_i^+ A_i^T R_i^{-1}$$

$$\hat{x}_i^+ = \hat{x}_i^- + K_i(y_i - A_i\hat{x}_i^-)$$

---

3. There are a variety of ways of writing what are mathematically equivalent formulas for the Kalman filter. One often seen in the literature is $P_i^+ = (I - K_i A_i)P_i^-$, whose product can be less numerically stable [Sim06a] than the sum in (2.9). Later, we will introduce other formulations that are more convenient for the nonlinear ensemble methods of Kalman filtering.

The balance between Q and R has some other considerations as well. From equation (2.6), P grows with propagation in proportion to the process (model) noise Q. Your intuition tells you that noise is bad, but intuition needs some more experience here. Note that if Q were very small, then P_i is minimized through the propagation step. In equation (2.9), a small P_i^- generates a term that tends to overwhelm the other term when the inverse $(P_i^-)^{-1}$ is taken. The final inverse in equation (2.9) makes P_i^+ small, which in (2.10) drives K_i small. Over time, the iterations of very small process noise tend to drive P small, and the filter stops paying attention to new measurements. So adding additional uncertainty Q to the filter can actually improve its performance. There is something deeper here—it involves what we will later term *covariance inflation*, as well as the broader issue of *model inadequacy*, to which chapter 8 is devoted. But for now, use enough Q.

If the Kalman filter in equations (2.12) is an accurate one, without significant modeling errors, the innovations $(y_i - A_i \hat{x}_i^-)$ become a Gaussian random variable. These *residuals* are what is left after extracting the maximal amount of information from a signal. *The Kalman filter really is a filter*. It extracts information out of a noisy signal.

One unfortunate aspect of the original Kalman filter equations (2.12) is that they describe linear systems.[4] It is unfortunate because there are so few close-to-linear systems in biology.[5] And I'll just be blunt in saying that there are no linear neuronal dynamics. Nevertheless, the power of linear Kalman filtering has intrigued neuroscientists. In a technical report, Bousquet and colleagues laid out a linear Kalman filter framework to account for hippocampal probabilistic spatial localization [BBH98]. Just as in spatial learning in the hippocampus, there appears to be explicit predictive coding in the visual system, and extended (linearized) Kalman filtering has been placed in this framework [RB97]. In addition, the issue of content-addressable memory and Kalman filtering was discussed by Ballard [Bal97]. Limitations of these linear approaches were discussed in some detail by Eliasmith and Anderson [EA03]. While these researchers in the 1990s were struggling with linear Kalman frameworks, the robotics and numerical weather prediction communities were developing explicit nonlinear adaptations of the Kalman filter that were computationally tractable. These newer methods provided the motivation for this book.

2.2 Nonlinear Kalman Filtering

Unfortunately, linear systems do not exist...
—Dan Simon, 2006 [Sim06b]

---

4. If you are engineering a man-made system, then you can constrain the nonlinearities and let Kalman filtering revolutionize your control capabilities.

5. A thoughtful discussion of control of *bilinear systems* in biology can be found in [Moh91a, Moh91b].

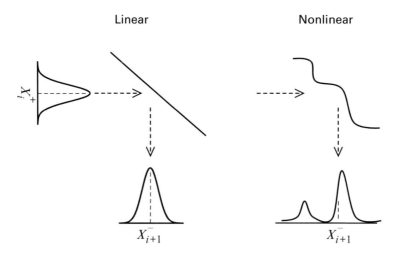

Figure 2.3
Comparison of the propagation of the mean and uncertainty of $\hat{x}_i^+$ to $\hat{x}_{i+1}^-$. This propagation corresponds to the time from a posteriori to a priori in figure 2.2. The black dashed lines within the distributions indicate the true mean values of the state x. Modified after Thrun et al. [TBF05].

As illustrated in figure 2.2, Kalman filtering has two parts: a propagation of state and uncertainty, and a faster Bayesian fusion of state prediction with data. Bayes' rule is immutable for our purposes. The propagation is not.

Thrun and colleagues [TBF05] provide a detailed analysis of comparative propagation schemes. We will follow their logic here. First, in linear Kalman filtering, one passes the mean (your best estimate of state $\hat{x}$) and uncertainty from Gaussian distributions through linear transformations—what comes out of linear transformation of a Gaussian distribution is another Gaussian (figure 2.3).

Propagating the same Gaussian distribution through a nonlinear function distorts it (see figure 2.2). If you had an analytic solution to this propagation, and the subsequent propagation of the no-longer-Gaussian distributions that result, you could write the general nonlinear version of Kalman's filter and use it. But there are no such general solutions to this problem.

The extended Kalman filter (EKF) is where one linearizes a nonlinear function at a particular point. Literally, draw a straight-line tangent to this point, and convert the nonlinear propagation problem into the linear one. Figure 2.4 illustrates the linearized solution.

For a wide range of problems this works very well. But in many circumstances, there is no adequate linearization that will capture the propagation of mean and uncertainty with reasonable fidelity to the underlying problem. Figure 2.4 illustrates why there can often be significant error in the linearized propagation of the mean $\hat{x}_{i+1}^-$.

Although the EKF (in its many forms) is a widely used filtering strategy, over thirty years of experience with it has led to a general consensus within the tracking and control community that it is difficult to

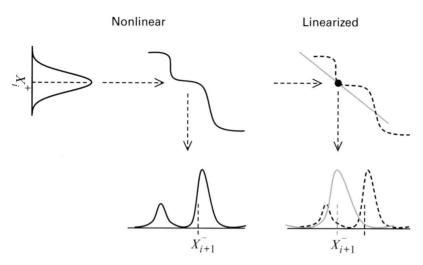

Figure 2.4
Comparison of nonlinear versus linearized propagation of the mean and uncertainty of $\hat{x}_i^+$ to $\hat{x}_{i+1}^-$. There is significant distortion of the mean (gray dashed line) from the true mean (black dashed line). Modified after Thrun et al. [TBF05].

implement, difficult to tune, and only reliable for systems which are almost linear on the time scale of the update intervals. [JU97a]

In range finding, one might wish to convert sines and cosines from individual sonar or radar systems to a common Cartesian coordinate. Linearization gives inaccurate results. This simple yet intractable problem motivated the push toward explicitly nonlinear methods [JU97b], [JU97a].[6]

The most accurate way to numerically solve the nonlinear propagation problem is to approximate the distribution by many individual states, and solve the passage of each state through the nonlinear propagation equations. Such methods fall under the general rubric of Monte Carlo methods, although in control engineering the terminology is generally that of particle filters. If you use enough particles (generally picked at random), you can faithfully approximate any distribution, and then process each particle one-by-one as shown in figure 2.5. The mean value of x is more faithfully propagated in this illustration (left side of figure 2.5). Nevertheless, the computational load of such Monte Carlo methods can be high, and this often prohibits real-time use.

In 1997, Julier and Uhlmann published two proceedings papers. In the first [JU97a], they demonstrated that by parameterizing a Gaussian distribution (such distributions are fully characterized by their means and standard deviations, as in equation (2.3)), that they could

---

6. A lucid and expanded discussion of Julier and Uhmann's more difficult discussion can be found in [Sim06a]. Note that were a Taylor's series expansion used to keep the second order terms of the sines and cosines, the inaccuracies of this simple range-finding problem would be better addressed. Nevertheless, many of the problems in this book would not be well served by higher-order attempts at expansion.

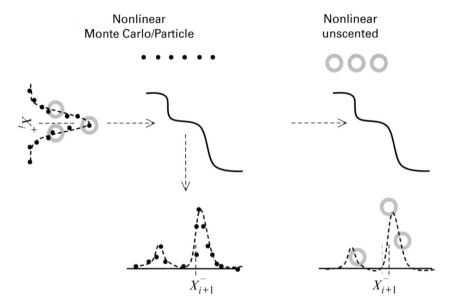

Figure 2.5
Comparison of Monte Carlo (particle) propagation versus unscented Kalman filter using the parameterization of the mean and uncertainty of $\hat{x}_i^+$ by individual *sigma* points. There is less distortion of the mean (gray dashed line) from the true mean (black dashed line) in comparison with linearization (cf. figure 2.4). Modified after Thrun et al. [TBF05].

dramatically cut down on the number of particles required, yet preserve accuracy. They called their parameterized points *sigma points*. Their method was "founded on the intuition that it is easier to approximate a Gaussian distribution than it is to approximate an arbitrary nonlinear function or transformation" [JU97a].

Compared with the extended Kalman filter, their *unscented* Kalman filter required no computation of derivatives in linearizing nonlinear systems (Jacobians in higher dimensions), and proved more accurate. There is no truncation of the nonlinear function from linearization (recall that linearization is a truncation of the Taylor's series expansion of the nonlinear function at a particular point). And they preserved the statistics of the moments of the distribution they were propagating to second order (mean and variance). Figure 2.5 illustrates this method in comparison with a full Monte Carlo simulation.

Significantly, in their second 1997 paper [JU97b], less well read by the community, Julier and Uhlmann pointed out that one can add more points to the parameterization, such as kurtosis of the distribution, to bring the accuracy of the representation to fourth order. This substantially improved the accuracy of their range finding position calculations.[7]

---

7. It is a remarkable testament to the peer review system that the peer review of the findings in these two papers would take four submissions over six years. They graciously wrote: "The authors would like to thank ... the referees for their many constructive remarks" [JUDW00].

Sigma points are generated deterministically. In the second order unscented Kalman filter, for an n-dimensional state variable x, there is an n-dimensional uncertainty distribution. The $2n + 1$ sigma points, X, are generated as

$$X_0 = \bar{x}$$

$$X_i = \bar{x} + \left(\sqrt{(n+\kappa)P}\right)_i \qquad\qquad (2.13)$$

$$X_{i+n} = \bar{x} - \left(\sqrt{(n+\kappa)P}\right)_i$$

where $i = 1\ldots n$ and $\bar{\ }$ indicates average. The mean of the distribution is chosen as the first sigma point $\bar{x}$. P is the covariance matrix for the errors in x. The square root is the matrix square root, whose computation is greatly abetted by the availability of efficient decompositions such as Cholesky's.[8] The matrix square root of M is defined as

$$M \equiv \left(\sqrt{M}\right)^T \sqrt{M}$$

and in practice, one chooses the ith row (or column) from $\sqrt{M}$ to sum with the average of $\bar{x}$ in equation (2.13).

As in other forms of particle filtering, one weights the sigma points. Julier and Uhlman used for their weights, q,

$$q_0 = \kappa/(n+\kappa)$$

$$q_i = 1/2(n+\kappa) \qquad\qquad (2.14)$$

$$q_{i+n} = 1/2(n+\kappa)$$

but in practice, one can often drop the q_0 weight (it is only one point), and then the weights are all equal and can be ignored. Later in the chapter, we will just use $2n$ sigma points.

Kappa (κ) is fascinating. Julier and Uhlman used this as an extra degree of freedom to "fine tune" the approximation and reduce errors [JU97a]. What κ does in practice is *increase* the size of the uncertainty "ball" that the sigma points represent. In recent years, numerical meteorology has investigated this type of *covariance inflation* in considerable detail [AA99], [HKS07], and we will deal with this in more detail later.

In higher dimensions, the unscented Kalman filter propagation appears as in figure 2.6.

The propagation of sigma points provides a ready route to calculate the mean and covariation of the propagated points in order to realize the new a priori values of $\hat{x}_i^-$ and P_i^-. But Julier and Uhlmann [JU97a] used a different formulation for calculating P_i^+ and K when using the unscented Kalman filter to make it computationally simpler.

Recall the formula (2.11) for the final update of $\hat{x}_i^+$

---

8. Cholesky's work on this decomposition far predated the advent of digital computers.

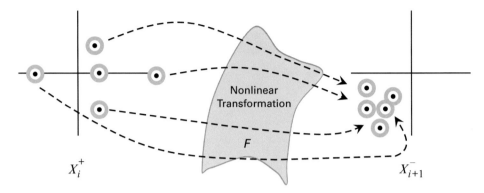

Figure 2.6
Schematic of unscented Kalman filter propagation of sigma points (circles). The transformed points are used to calculate the new a priori estimate of x and its covariance P. Modified after [JU97a].

$$\hat{x}_i^+ = \hat{x}_i^- + K_i(y_i - A_i\hat{x}_i^-)$$

One way to view K_i is as a "derivative" that evaluates the rate of change of x in relation to y

$$K_i = \frac{\hat{x}_i^+ - \hat{x}_i^-}{y_i - \hat{y}_i^-} \tag{2.15}$$

Using this notion of linearity in K, a *best linear unbiased estimate* for K can be alternatively derived by setting $z = x - \hat{x}_i$, and noting that the covariance of z, P_z, is

$$P_z = E\left[(x - \hat{x} - (\bar{x} - \hat{\bar{x}}))(\cdots)^T\right] \tag{2.16}$$

Minimizing the trace of the matrix P_z minimizes the sum of the variances of the individual components of x, and this will generate the best K. A somewhat intricate derivation in Simon [Sim06a] demonstrates that this trace is minimized when

$$K = P_{xy}P_{yy}^{-1} \tag{2.17}$$

and further consideration [Sim06a] yields

$$P_{xx}^+ = P_{xx}^- - KP_{xy}^T \tag{2.18}$$

This best linear statistical derivation makes use of some new notation on the covariances

$$P_{xx} = E\left[(x - \hat{x})(x - \hat{x})^T\right]$$

$$P_{xy} = E\left[(x - \hat{x})(y - \bar{y})^T\right] \tag{2.19}$$

$$P_{yy} = E\left[(y - \bar{y})(y - \bar{y})^T\right]$$

Despite the very different appearance of these equations (2.17) through (2.19) from (2.9) through (2.11), they are mathematically equivalent [Sim06a]. We will now see why this formulation is so transparent for the unscented Kalman filter.

We now write our equations (2.2) as

$$x_i = F_i(x_{i-1}, u_{i-1}) + q_{i-1}$$
$$y_i = A_i(x_i) + r_i \tag{2.20}$$

where F, G, and A are no longer just matrices multiplying x and u, but are now possibly nonlinear functions $F(\cdot)$ and $A(\cdot)$ of x and u.

In figure 2.7, we show a schematic of the full propagation of values through the unscented Kalman filter. Sigma points are generated from the covariance P_{xx}^+ of $\hat{x}_i^+$, which are then propagated through the dynamical equation $F(\hat{x}_i^+) + q_i$ from equation (2.20). This generates a new ensemble of points whose average is the new $\hat{x}_{i+1}^-$. One can read the propagated P_{xx}^- off of the ensemble of points by taking their covariance with respect to the average $\hat{x}_{i+1}^-$. Then you pass these points through the observation function $\hat{y}_{i+1} = A_{i+1}(\hat{x}_{i+1}) + r_{i+1}$, which lets you estimate both the expected $\hat{y}$, and the covariance in the expected $\hat{y}$, P_{yy}.

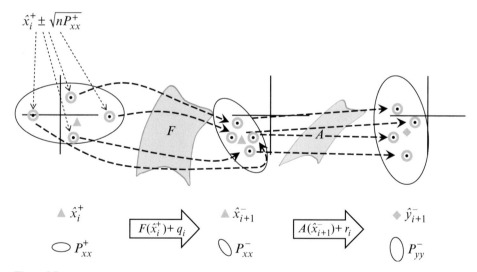

Figure 2.7
Schematic of the propagation components of the unscented Kalman filter. The notation in this figure is matched to the notation in the summary equations (2.22). At the left, using the present estimate of $\hat{x}_i^+$, the weighted matrix square root of the covariance P_{xx}^+ is used to generate sigma points (circles). These sigma points are iterated forward using the nonlinear equations F. The mean of the new ensemble is $\hat{x}_{i+1}^-$, and the new covariance P_{xx}^- generated from the ensemble about this mean. The points are further transformed by the observation function A, to generate the estimate of the next measurement, $\hat{y}_{i+1}$, and calculate P_{yy}. All of this takes place *before* the new measurement y is taken, after which K is used to propagate P_{xx}^- and $\hat{x}_{i+1}^-$ to P_{xx}^+ and $\hat{x}_{i+1}^+$.

Using P_{xy} calculated as in (2.19), one then calculates K from (2.17), and finally everything is in order to calculate P_{xx}^+ from (2.18). The final update of x is now written as

$$\hat{x}_{i+1}^+ = \hat{x}_{i+1}^- + K_{i+1}(y_{i+1} - \hat{y}_{i+1}^-) \tag{2.21}$$

Let's summarize the equations for the unscented Kalman filter using $2n$ sigma points:

$$x_i = F_i(x_{i-1}, u_{i-1}) + q_{i-1}$$

$$y_i = A_i(x_i) + r_i$$

$$q_i \sim N(0, Q_i)$$

$$r_i \sim N(0, R_i)$$

$$X_j = \hat{x}_i^+ \pm \left(\sqrt{n P_{xx}}\right)_j, \quad j = 1 \ldots 2n$$

$$\tilde{X}_j = F(X_j)$$

$$\tilde{Y}_j = A(\tilde{X}_j)$$

$$\hat{x}_{i+1}^- = \frac{1}{2n} \sum_{j=1}^{2n} \tilde{X}_j$$

$$\hat{y}_{i+1}^- = \frac{1}{2n} \sum_{j=1}^{2n} \tilde{Y}_j \tag{2.22}$$

$$P_{xx}^- = \frac{1}{2n} \sum_{j=1}^{2n} (\tilde{X}_j - \hat{x}_{i+1}^-)(\tilde{X}_j - \hat{x}_{i+1}^-)^T + Q_i$$

$$P_{xy} = \frac{1}{2n} \sum_{j=1}^{2n} (\tilde{X}_j - \hat{x}_{i+1}^-)(\tilde{Y}_j - \hat{y}_{i+1}^-)^T$$

$$P_{yy} = \frac{1}{2n} \sum_{j=1}^{2n} (\tilde{Y}_j - \hat{y}_{i+1}^-)(\tilde{Y}_j - \hat{y}_{i+1}^-)^T + R_{i+1}$$

$$K = P_{xy} P_{yy}^{-1}$$

$$P_{xx}^+ = P_{xx}^- - K P_{xy}^T$$

$$\hat{x}_{i+1}^+ = \hat{x}_{i+1}^- + K(y_{i+1} - \hat{y}_{i+1}^-)$$

A few comments are in order for what appears complex. The notation here is unified with figure 2.7. We have indexed the sigma points X and Y by j, which reflects that, for dimension n (the number of elements in the state vector x), we have $2n$ sigma points. Don't confuse j with the time indices i. The difference between Kalman's original filter and the unscented Kalman filter is all in the ensemble updating for the propagation steps. These propagations bring us to the a priori condition for the final update when the new measurement y_{i+1} arrives.

I list in equations (2.22) adding the process noise Q to the estimation of P_{xx}, and observation noise R to the estimation of P_{yy}. There are many variants in how to handle adding these quantities, and we will deal with many of these intricacies later in the book. For now, I have dropped the extra parameter κ that Julier and Uhlmann used in equations (2.13)—but it is all the same issue. Embracing uncertainty, whether random noise, or model inadequacy, is a major topic to be dealt with.

The last three equations of (2.22) are identical to the original Kalman filter. Bayes' rule still rules the *assimilation* of the data point. On the other hand, there is no longer any need to be concerned with approximating the nonlinear functions F and A by something inaccurate such as a linearization. The sigma points are each numerically passed one-by-one through these intact nonlinear equations, regardless of how unpleasant or difficult to solve analytically such equations might be. The only issue remaining is how to numerically integrate those equations, and we will deal with this later in the book.

While Julier and Uhlmann were working out their method, they were not alone. One of the strange facts of academic life, is that we can spend a lifetime in one building, and never meet, discuss, or perish the thought read the literature of our colleagues in another department. In 1994, Evensen [Eve94] published an article seeking to adapt the Kalman filter to the tracking and prediction of ocean dynamics. His ensemble Kalman filter, has had a profound impact on numerical meteorology and oceanography (see, e.g., [EvL00], [OHS+04], [BHK+06], [HKS07]). In the geophysical literature, ensemble Kalman filtering is the more commonly used term, whereas in robotics, the term unscented Kalman filter is more common. The particular method of deterministically choosing the sigma points is associated with unscented Kalman filtering. But this is a fascinating example of parallel development of scientific ideas.

2.3 Why Not Neuroscience?

It is interesting to consider why the techniques discussed this chapter that have revolutionized robotics and weather prediction have not had a significant impact on neuroscience. There are several good reasons for this.

First, the subject matter, despite my attempts at an approachable exposition, is unfriendly. No neuroscientist or physician has in the past come to their profession with a strong grounding in the appropriate background.

Second, the older theory, linear and linearized approximations to dynamics, was not broadly applicable to the nonlinearities of neuronal dynamics.

Third, we needed neuronal models suitable to the tasks we would ask of them. Computational neuroscience models have been maturing for nearly as long as modern control theory—both had their origins in the 1950s. For the past several decades, the fidelity of these neural models has grown ripe for fusing with control theory.

Fourth, we needed suitable nonlinear control algorithms that were suitable for nonlinear dynamics. Once Evensen, or Julier and Uhlmann, had done their work, it needed to percolate through and across disciplinary boundaries and creep into neuroscience settings. The first point of entry seems clear. In a relatively obscure physics journal, Voss, Timmer, and Kurths [VTK04] made the leap in 2004.

And finally, one might ask why I chose to focus on the unscented Kalman filter for this introduction? Because this is what Voss and colleagues first used in their seminal paper bridging from nonlinear control theory to neuronal dynamics. Their algorithmic framework has subsequently proven instrumental in applications to several neuronal systems that we will discuss in detail in this book. Nevertheless, a considerable amount of cutting-edge research in ensemble tracking and prediction of nonlinear systems is coming out of meteorology, and we will cross over between these literature streams frequently.

But before we can get to the results of Voss et al. [VTK04], we must first discuss the Hodgkin-Huxley equations in the next chapter, and then the reduction of Hodgkin-Huxley dynamics in the Fitzhugh-Nagumo equations in the subsequent chapter, before the application of equations (2.22) will be transparent.

Exercises

2.1. For the first (and likely last) time in your life, let's work the steps of the Kalman filters by hand. Return to the simple example from equation (1.20)

$$\begin{bmatrix} 1 \\ 1 \end{bmatrix} x = \begin{bmatrix} 1 \\ 3 \end{bmatrix}$$

where we calculated from least squares in (1.27) that $\hat{x} = 2$, and from (1.54) that $P = \sigma^2/2$, and let's use the linear Kalman filter equations from (2.12).

Let's assume a static model, where the dynamics are trivial

$$F = I, \text{ so that } F_{i+1} = F_i$$

and let's assume again that the variable x is directly observable, so that

$$A = I, \text{ and } A_{i+1} = A_i$$

We will assume that the uncertainty in both model, F, and observation, A, are both equal to σ^2. The initial conditions are therefore

$$y_1 = 1$$

$$\hat{x}_1^+ = y_1$$

- Calculate $\hat{x}_2^-$
- What is P_1^+?
- What is P_2^-?
- What is P_2^+?
- What is K_2?
- Calculate $\hat{x}_2^+$.

Does your answer differ from the least squares solution to this problem? It should.

2.2. To clarify the results in exercise 2.1, reverse the order of the measurements, so that $y_1 = 3$ and $y_2 = 1$. Calculate $\hat{x}_2^+$.

Hopefully your answer again differs from the least square solution, and also differs from the result in exercise 2.1. Kalman filtering tends to (exponentially) *forget* past data, and weights the most recent data more heavily.

2.3. Let's now calculate the same results for the static model in exercise 2.1, but now use the unscented Kalman filter using equations (2.22). Assume that the variance P_{xx} to start with is 1, and that for this simple one-dimensional problem, $n = 1$. We will start again setting $\hat{x}_1 = y - 1 = 1$, and the two sigma points will simply be 2 and 0.

- What are P_{xx}^-, P_{xy}, and P_{yy}?
- What is K?
- Calculate $\hat{x}_2^+$.
[Hint: The answer should be the same as the least squares solution.]
- Now calculate P_{xx}^+.

- Assuming that you were running your unscented filter for awhile, the newest P_{xx} would be the P_{xx} for calculating the next set of sigma points. So repeat the preceding calculation for $y_1 = 1$ and $y_2 = 3$, except start by using the last P_{xx} you had calculated above to generate the sigma points. What is $\hat{x}_2^+$ now? It should now be very close to the value from the linear Kalman filtering in 2.1, weighted in favor of the most recent measurement. Sometimes you just have to let a dynamical system settle down from its initial transient response. You should think of Kalman filters as dynamical systems.

3 The Hodgkin-Huxley Equations

This book is about model-based tracking and control. All of the neuronal models we will employ are related, directly or indirectly, to the work of Hodgkin and Huxley in the 1940s and 1950s. Although research into *axonology* is no longer viewed as the cutting edge of neuroscience research, the potential pitfalls raised without a deep understanding of these dynamics are very much at the forefront of current work [MSY07]. Later, in chapter 12, we will reexamine the Hodgkin-Huxley equations in a control theory framework. These equations are the foundation of biophysically based models of neuron dynamics.

This will be a long chapter. Few readers will likely ever have much experience in testing in the laboratory the fundamental features of excitable neuronal membranes. With squid axonology largely over, the present-day experience is most often with mammalian neurons of extraordinary spatial complexity. These neurons are typically populated with a *zoo* of ion channels [GK02], whose temporal and spatial dynamics are very complex. At the same time, some of our best textbooks give only brief descriptions of the foundational works that are insufficient to develop intuition. In this chapter, I will take some time to develop a description of the foundational membrane phenomenology enough for the reader to develop a sense for the dynamical intuition behind the Hodgkin-Huxley formalism. All of our biophysically based neuronal models developed in recent decades is commentary and complexity. And understanding these more complex biophysical models is key to our later discussion of Parkinson's disease and epilepsy in chapters 10 and 12.

3.1 Pre-Hodgkin and Huxley

By the turn of the twentieth century, Julius Bernstein had formulated the hypothesis that all living cells were surrounded by a membrane with low permeability to ions [Ber12]. This was a remarkable speculation well before there was direct proof of the existence of such

A   B

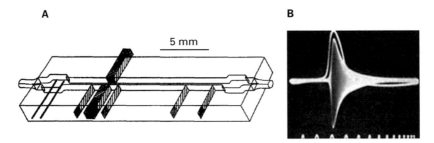

Figure 3.1
Cole and Curtis's apparatus, A, for measuring the impedance change during passage of an action potential, B. In B the action potential and bridge imbalance are shown superimposed. Modified from [CC39] with permission.

A   B

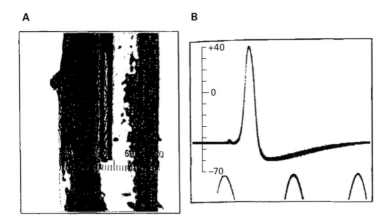

Figure 3.2
The first measurement of overshoot of an action potential. A shows the impalement of the squid giant axon by a glass micropipette. B shows the measured action potential overshoot to near +50 mV with respect to the external bath solution. Reproduced from [HH39] with permission.

structures.[1] Bernstein further hypothesized that there was a resting transmembrane potential gradient, and that action potentials (recorded extracellularly from peripheral nerve) were the result of the breakdown of permeability barriers in the neuronal membrane [Ber12]. This hypothesis predicted that during the action potential, the transmembrane voltage gradient would collapse to near zero.[2]

---

1. The indirect evidence was strong. Biophysical work with living tissues, especially suspensions of blood cells, demonstrated impedance and reactance characteristics that firmly pointed to electrolyte-containing structures surrounded by a dielectric capacitance [Col68].

2. It would actually collapse to the Donnan equilibrium potential, if impermeant anions remained within the cell [NRB04]

Cole and Curtis [CC39] performed an ingenious series of transverse measurements of the impedance of a squid giant axon,[3] and demonstrated through imbalance of an alternating-current Wheatstone bridge that during the action potential, a large decrease in impedance occurred consistent with Bernstein's hypothesis (figure 3.1). Unexplained in figure 3.1B was the unusual temporal interplay between the course of the action potential (thin line), and the impedance change (broad envelope). It will take the rest of this chapter to clear this up.

Then came the first direct measurements of transmembrane voltage [HH39]. A significant overshoot of transmembrane potential was observed (figure 3.2). Bernstein's hypothesis was dead.

3.2 Hodgkin and Huxley and Colleagues

As the concept of feedback control electronics matured during the 1940s [Wie48], the ability to devise voltage feedback control systems in biological experiments became feasible. Cole [Col49] and Marmont [Mar49] devised a *voltage clamp*. A key problem was that the voltage of an action potential varied in time as well as space. To eliminate spatial variation, a *space clamp* was employed, whereby the longitudinal (axial) electrode shown in figure 3.3 was used to eliminate potential variations along the length of the axon. To ensure that there were no voltage gradients except across the central region of the axon, they used guard electrodes polarized to the same potential as the central electrode [Mar49] (figure 3.3).

An additional technical need, as Cole[4] introduced [Col68], was to separate the electrode pairs that pass current from the ones that sense the potential changes. Hodgkin, Huxley, and Katz [HHK52] varied Cole's design in order to measure both transmembrane voltage and current, as shown in figure 3.4.

The electronic schematic for this four-electrode technique is illustrated in figure 3.5. We will incorporate this latter principle in the large-scale voltage clamps that we use in neuronal network control experiments described in chapter 11.

The notation used by Hodgkin and Huxley (and Katz) is as follows [HHK52]. V is the displacement of membrane potential, E, from rest, E_r

$$V = E - E_r$$

where E is taken with respect to a distant reference, and sign is by convention outside minus inside. The resting membrane potential (with respect to the fluid outside of a cell) is

---

3. The description by J. Z. Young of the giant axon of the squid [You36] greatly aided the avalanche of neurophysiological advances over the next sixteen years. The axon was large enough to permit not just impalement with an electrode, but insertion of space clamp electrodes, as well as squeezing out and replacement of the axoplasm.

4. The author is grateful to K. C. Cole for informally reviewing his first manuscript when working with John Moore and Normal Stockbridge, later published as [MSS82]. Such formative encounters inspire preservational chapters.

A

B

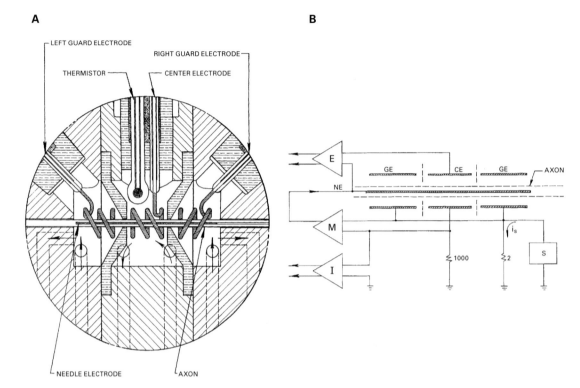

Figure 3.3
Marmont's schematic of his voltage clamp apparatus, A, and electronics, B, that he shared with Cole. Reproduced from [Mar49] with permission.

considered 0 mV. The current sign is positive for positive charge flowing into a cell. They had a strong measurement for the cellular membrane capacitance, C_m, measured from alternating currents passed through suspensions of blood cells measured by Fricke in the 1920s to be 1 μF/cm^2 [Fri24, Fri25]. The *unit membrane* [Rob57], the lipid bilayer of all cells, vertebrate and invertebrate, would turn out to have a universal physical capacitance constant.

We assume that the total membrane current, I_m, is a sum of membrane capacitive current, $C_m \partial V/\partial t$ and membrane ionic current, I_i.

$$I_m = C_m \frac{\partial V}{\partial t} + I_i \tag{3.1}$$

Using voltage clamp feedback to hold the membrane voltage constant, which in equation (3.1) would force the capacitive current to zero, would leave $I_m = I_i$. This was the key to studying the ionic membrane currents, and sorting them out one by one.

The first thing that Hodgkin, Huxley, and Katz [HHK52] observed was the response to membrane potential from brief shocks (figure 3.6). For hyperpolarizing charges, driving

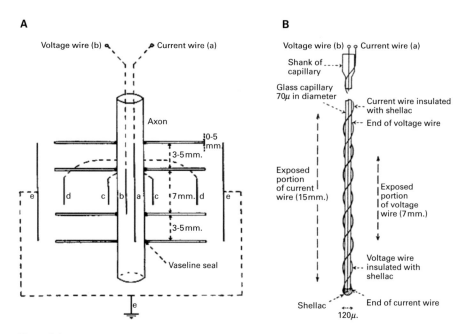

Figure 3.4
Voltage clamp schematic, A, and axial electrode configuration, B, from Hodgkin, Huxley, and Katz [HHK52]. Reproduced from [HHK52] with permission.

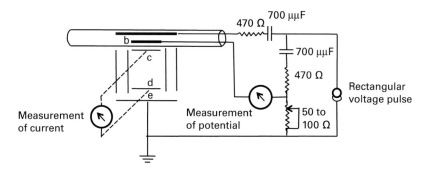

Figure 3.5
Four-electrode technique used to separate current driving and sensing electrode pairs. Reproduced from [HHK52] with permission.

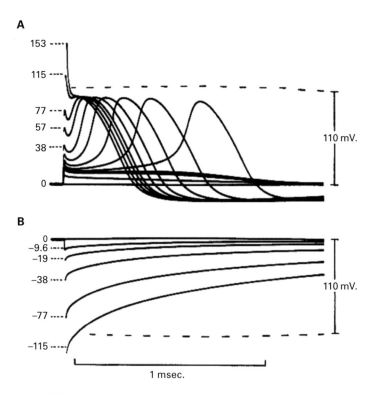

Figure 3.6
Membrane action potentials in response to brief shocks using the apparatus in figure 3.4. Reproduced from [HHK52] with permission.

the inside of the axon more negative than that of rest potential (normally about -70 mV with respect to the outside of the axon), the membrane is charged with the pulse, and then decays like a leaky capacitor (figure 3.6B). For small depolarizing positive shocks, there is similar decay with sign inverted (figure 3.6A). But for slightly larger depolarizing shocks, there is a much larger excursion (figure 3.6A). These *membrane action potentials* were uniform within the central chamber between the guard chambers. They are not propagating, and uniform within the restricted region of the space clamp. Note that, for depolarizing shocks just over the apparent threshold, the delay to the regenerative action potential is longer (figure 3.6A).

Exploring the region of the threshold gave amazing responses (figure 3.7). *If the potential was displaced to the threshold level it might remain in a state of unstable equilibrium for considerable periods of time* [HHK52]. Neural fibers close to threshold are indeterminate in that you can't tell whether they will or won't fire an action potential or when that action potential will occur. They are much more reliable with larger stimuli.

I must say when I first saw these pictures, this indeterminacy appeared to have philosophical implications. It still does. Much has been written about the origins of such concepts of

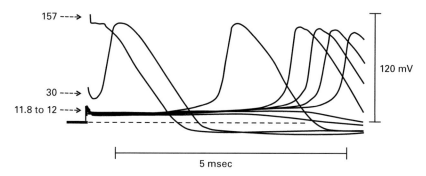

157 ---→

30 ---→

11.8 to 12 ---→

120 mV

5 msec

Figure 3.7
Indeterminacy in action potential behavior near threshold. Reproduced from [HHK52] with permission.

free will, which appeared difficult to reconcile with the deterministic formulation of brain dynamics, such as the Hodgkin-Huxley equations we are about to examine. There is a considerable literature, starting with Schrodinger [Sch44b], and promulgated by Eccles [Ecc53] and later Penrose [Pen94], exploring whether quantum indeterminacy needs to be invoked to render a nervous system free from automatism. Missing from the following description of the Hodgkin-Huxley equations is the fundamental stochastic molecular conformational changes that the ion channels exhibit. In later decades, such molecular fluctuations, based on quantum mechanics but only requiring a classical probability description, would be used to explain, first, hemoglobin's binding of oxygen [MWC65] and, later, to account for the indeterminacy as seen here [Fit65] [SFS98].

There was also something deeply wrong with figure 3.6. Look at the plot of voltage delivered and the current that developed 0.29 msec after the shock was delivered (figure 3.8). From rest, a positive (hyperpolarizing), or small negative (depolarizing) deflection yields a linear response. This is Ohm's law. Give a larger negative response, and the response is not just nonlinear. Ohm's law reverses, and, were you to measure between the starting point and the point on the curve for a given depolarizing pulse, the chord conductance [Tho86] would be negative.

Now introduce a voltage clamp. Under voltage clamp, neurons do not fire action potentials (because their voltages are kept from changing). But the electronics in figure 3.5 permit us to change this voltage quickly. In figure 3.9, the upper trace was the current that evolves in response to a +65-mV step hyperpolarization, while the lower trace was in response to a −65-mV depolarization. There is relatively little current that flows for the hyperpolarizing pulse. But for the depolarizing pulse, there is a transient inward current, followed by a prolonged outward current.

Now plot the differences between the current-voltage (I-V) curves for short (0.6 msec) and long (steady state) times as in figure 3.10. The change in current that develops is a function of both voltage and time.

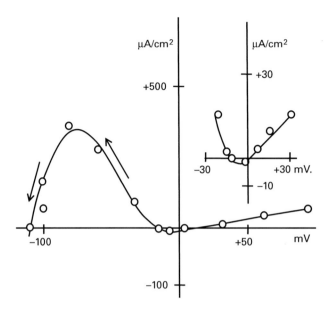

Figure 3.8
Current-voltage (I-V) curve at 0.29 msec after shock is delivered. Note that they flipped their voltage convention. Rest is still at 0 mV. Hyperpolarization is positive, and depolarization is negative. Reproduced from [HHK52] with permission.

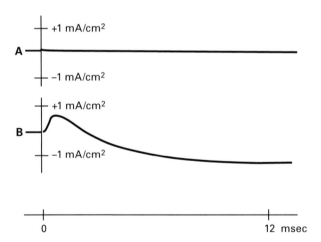

Figure 3.9
Current responses to symmetric positive A and negative B step polarizations. Up in these figures is for inward current, and down from baseline is for outward current. Note the discontinuity at the step onset at 0, which reflects the capacitive current surge not shown on this time frame. Reproduced from [HHK52] with permission.

A   **B**

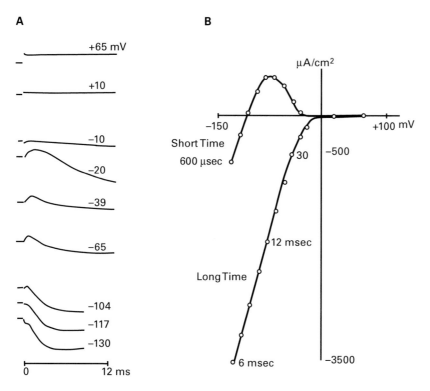

Figure 3.10
A shows currents as a function of time and voltage. B shows current-voltage curves at short versus longer time periods after voltage step. Modified from [HHK52] with permission.

There was a small gap in the records in figure 3.9 at the onset of the change in voltage. If one examines such traces at higher speed, this capacitive current *surge* is seen (figure 3.11). It's important to note that this capacitive current is symmetric, very much unlike the currents in figure 3.9. It is also much faster than the more biologically based mechanisms behind the ionic currents.

3.3 Hodgkin and Huxley

We now move on to the sequence of four dual-authored papers typically referred to as the Hodgkin-Huxley papers.[5]

---

5. The reader need not be too concerned for the missing colleague Bernard Katz. He would receive the 1970 Nobel Prize for Medicine, seven years after Hodgkin and Huxley.

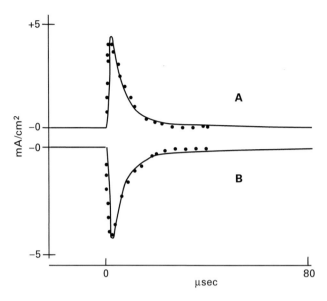

Figure 3.11
Very short time symmetric capacitive currents. Reproduced from [HHK52] with permission.

Hodgkin-Huxley Paper 1

The first task was to separate the ions. The first experiment was to replace, in their artificial seawater, sodium chloride with choline chloride. Choline is a big enough molecule so it's not going to go through the membrane as easily as sodium. Choline rendered squid axons inexcitable, but did not change the resting membrane potential substantially. In choline seawater, the early inward current disappeared (figure 3.12). Replace the sodium and the early inward current reappears. Note the early outward current bump (gray arrow) in figure 3.12. It is due to sodium current flowing from inside to outside as the electrical gradient holding it in is decreased under the depolarizing pulse.

A more complete sequence of hyperpolarizing (positive) and depolarizing (negative) pulses are shown in normal versus choline seawater in figure 3.13. Note that the transient inward sodium current decreases as depolarization becomes more intense. It reverses at about −98 mV, the sodium reversal potential E_{Na} (arrow in figure 3.13). In choline seawater the small outward sodium current is seen at all depolarizations. Replacing sodium restores the original findings.

The origin of the negative resistance (conductance) in figures 3.8 and 3.10 is now apparent. At modest depolarizations, the sodium gradient caused sufficient inward current to overpower the expected outward current from a polarization that decreased the transmembrane potential. The reversal of sodium current at strong depolarizations (figure 3.13) now

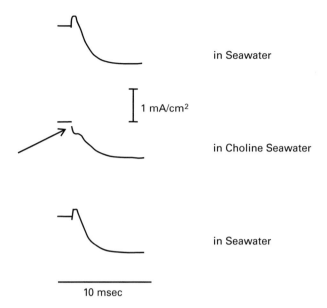

Figure 3.12
Response to 65-mV depolarizations. Modified from [HH52a] with permission.

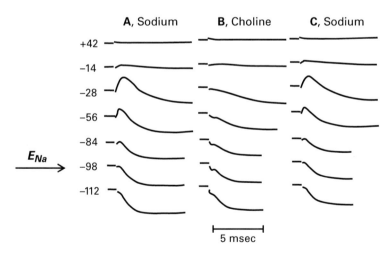

Figure 3.13
Voltage clamp responses in (A) seawater, (B) choline seawater, and (C) seawater. Modified from [HH52a] with permission.

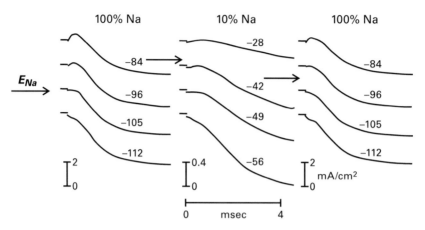

Figure 3.14
Changing external sodium concentration to test whether the Nernst potential accounted for the early outward current's reversal potential. Modified from [HH52a] with permission.

gives you the ability to estimate what the internal sodium concentration is. From the Nernst potential

$$E_{Na} = \frac{RT}{nF} \ln \frac{[Na]_o}{[Na]_i} \approx 26mV \ln \frac{[Na]_o}{[Na]_i} \tag{3.2}$$

where R is the gas constant, n the valence, T temperature in degrees Kelvin, and F the Faraday constant, so that RT/nF is about 26 mV at 25°C. Given the outside sodium concentration $[Na]_o$ of 460 mM in seawater, they estimated the internal $[Na]_i$ to be 60 to 70 mM.[6]

Hodgkin and Huxley did a series of experiments (figure 3.14) to test whether sodium was the early current carrier by varying the sodium concentration, and asking if the change in reversal potential tracked the Nernst potential in equation (3.2). The comparison of theory with experiment was nearly exact. The Nernst potential could be used to calculate at what value to set the voltage clamp in order to reverse the early current.

The early current was therefore a sodium current. It flowed in response to both concentration difference of sodium inside and outside the nerve, as well as the electrical potential gradient. One gradient could overcome the other.

The late current appeared Ohmic (figure 3.10), but it is huge, and much greater (50 to 100 times) than what would be expected from just the voltage change alone. It seemed unaffected by substituting choline for sodium in the external bath. So assume that the total ionic current, $I_i = I_{Na} + I_K$, and in low sodium, the total current, I_i', will be $I_i' = I_{Na}' + I_K'$. If we assume

6. We will explicitly examine mammalian $[Na]_i$ and $[Na]_o$ dynamics in chapter 12.

that $I_K = I'_K$, then $I_i - I'_i = I_{Na} - I'_{Na}$. It follows that $I_K = I'_K = I_i - I'_{Na}$. Last, assume that I_i and I'_i have the same time course, so that they are related as $k = I'_{Na}/I_{Na}$.

So measure the total current in normal I_i and low sodium I'_i conditions, and one backs out a whole set of relationships to calculate both I_K and I_{Na}

$$I_i - I'_i = I_{Na} - I'_{Na} = I_{Na} - I_{Na}\left(\frac{I'_{Na}}{I_{Na}}\right) = I_{Na}(1-k)$$

$$I_{Na} = \left(I_i - I'_i\right)\Big/(1-k)$$

$$I'_{Na} = k\left(I_i - I'_i\right)\Big/(1-k)$$

$$I_K = I'_K = I_i - I_{Na} = \left(I'_i - kI_i\right)\Big/(1-k)$$

(3.3)

Using these relationships, they can now estimate the individual currents as shown in figure 3.15.

Their next assumption was that the driving force for these ion currents was equal to the voltage applied by their voltage clamp, V, minus the Nernst potential for a given ion species, $E_{Na/K}$. $I_{Na/K}$ would then be proportional to the driving force $(V - E_{Na/K})$. If you put the command potential equal to the Nernst potential, no current would flow. The

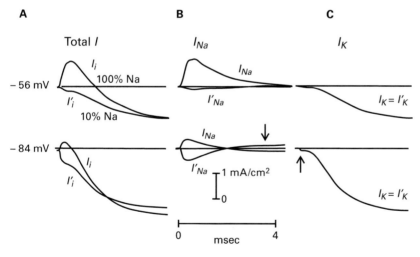

Figure 3.15
Separating currents. Note that near the sodium equilibrium potential in the lower plots, the tails in calculated I_{Na} and I'_{Na} cross (down gray arrow), showing that the assumptions in equations (3.3) were not perfect. Another imperfection is evident in the small offset (up gray arrow) at the beginning of the calculated I_K at -84 mV. This is due to other ions, such as chloride, not included in these assumptions, but which will form the *leak* current to be discussed shortly. Modified from [HH52a] with permission.

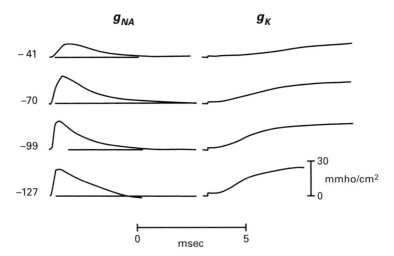

Figure 3.16
Families of conductance curves. Values of E_K are calculated from the subsequent Hodgkin-Huxley paper. Note that mho, Ohm spelled backwards, is an obsolete term for Seimens (1/Ohm). Modified from [HH52a] with permission.

magnitude of the current would then be proportional to the permeability or conductance, g, for a given ion species, or

$$g = I_{Na/K}/(V - E_{Na/K})$$

Calculating families of conductance curves, they obtain figure 3.16. Note that both the maximal conductances and the rate of change in conductance are voltage dependent. These two features are important in what follows.

The increase in conductances from rest were tremendous. Putting together data from the experiments that were sufficient to piece together curves,[7] they obtained the data in figure 3.17. The logarithmic plots demonstrate up to 1,000-fold increases in conductance when depolarizing from rest. These figures were fully consistent with what Cole and Curtis [CC39] found from their experiment in figure 3.1, measuring total conductance increase at the peak of the action potential.

Last, they proposed in this first paper that each ion species crossed the membrane independently of each other. This *independence principle* was a key to working out their model. It will also prove the likely explanation for why, in chapter 12, we will be able to fully reconstruct Hodgkin-Huxley dynamics using a control theoretic approach.

---

7. There were a whole series of experiments that Hodgkin and Huxley had performed previously. Throughout their four 1952 papers they carefully pick their examples. Many of these axons were also deteriorating, some oscillated, others didn't, but they picked the ones that seemed to give them data that they could use. One gets the impression that if they actually used their data in an unbiased fashion, perhaps with the benefit of a statistician looking over their shoulder reminding them not to throw any of the data away or be biased about what they picked, they would never have actually gotten to the conclusions that they came to.

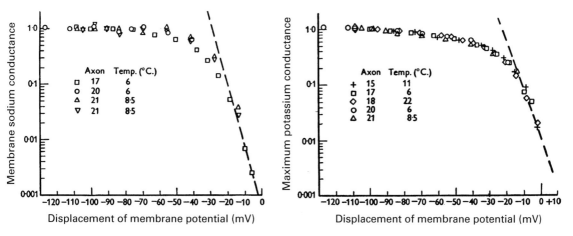

Figure 3.17
Semilogarithmic plots of peak conductances for sodium and potassium from selected axons. Reproduced from
[HH52a] with permission.

Hodgkin-Huxley Paper 2

The second concerns general changes in voltage and time. The first experiments used pulses
of different time lengths. If they hit the nerve with a short enough pulse—5, 50, maybe
80 μsec—they got nothing. The large inward shift in sodium required that they hold the
nerve membrane depolarized for a sufficient length of time for some process to occur. They
could then stop it dead in its tracks by cutting the pulse off. The biological mechanism that
underlay the inward sodium current was reversible (figure 3.18).

When Hodgkin and Huxley depolarized the membrane to the sodium equilibrium potential, and then repolarized it back to rest, a significant inward current appeared, which
exponentially decayed. This current disappeared in choline seawater (figure 3.19).

The discontinuities in current, such as those observed in figures 3.18 and 3.19, disappeared with the calculation of conductance, as shown in figure 3.20. Physically, this makes
sense. Change the potential discontinuously and the current changes discontinuously. But
conductance is continuous, physically as well as mathematically, and it takes a relatively
long time for a biological conductance to change in response to voltage changes. The ratios
of the two discontinuous processes tend to cancel each other out in the calculation of conductance. A continuous process would prove essential for them to be able to fit curves in
developing their model of these dynamics.

Now depolarize the membrane, and then quickly change to a different voltage. Recall
that conductance cannot change instantaneously. So the long time change in current with
the first depolarization, shown in the nonlinear curve with the negative resistance region
in figure 3.21, is very different from the I-V relationship measured just after the voltage

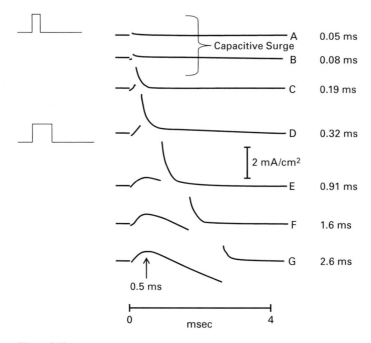

Figure 3.18
Membrane currents with depolarizing pulses of different lengths. Modified from [HH52b] with permission.

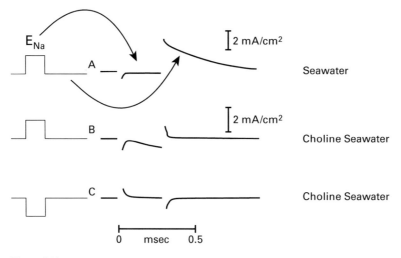

Figure 3.19
Inward sodium current *after* depolarizing pulse. Note the small outward current developing during depolarization in low-sodium external solution. Modified from [HH52b] with permission.

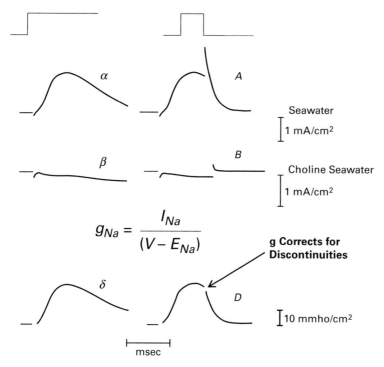

$$g_{Na} = \frac{I_{Na}}{(V - E_{Na})}$$

Figure 3.20
Current versus conductance traces for step and square pulse depolarizations. Modified from [HH52b] with permission.

is changed, when a linear relationship is observed. Note that when $V_1 = V_2$, the curves intersect—in this case both tests are the same, a depolarization to V_1.

The next observation is that the rate of decline in g_{Na} is proportional to the value of g_{Na} reached (figure 3.22).

In switching now to potassium currents, Huxley and Hodgkin use choline seawater, so that no sodium contributes to inward current, and after a depolarization, they rapidly repolarize to study the potassium currents.

Just as with sodium currents, they could contrast the nonlinear long-term versus the instantaneous chord conductance measurements for potassium from an instantaneous change in voltage. Since conductance cannot change instantaneously, the instantaneous I-V curve is linear in figure 3.23.

If one plots the turn-on and turn-off of g_K on a linear-linear scale, as in figure 3.24, these conductances are observed to turn on with a sigmoid-shaped function and to decay exponentially. As for g_{Na}, the rate of decay for g_K is proportional to the level of g_K reached (not shown).

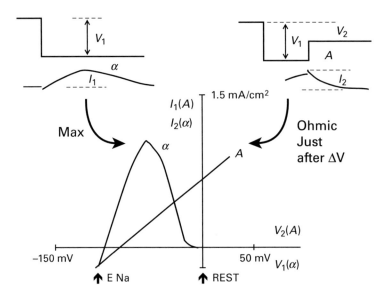

Figure 3.21
Contrast of slow nonlinear response to depolarization and rapid Ohmic response to second voltage change before the conductance can change. Modified from [HH52b] with permission.

Hodgkin-Huxley Paper 3

The third Hodgkin-Huxley paper concerns inactivation. Potassium turns on and stays on; sodium turns on and then off. Some process turns off sodium conductance when the membrane is depolarized. Sorting this out will take some more complex experiments.

Hodgkin and Huxley employed conditioning voltages—small changes in potential followed by large pulses. These small conditioning pulses were not associated with any appreciable inward currents, but they substantially altered the later response of the nerve. An example is shown in figure 3.25. Small conditioning depolarizations will progressively reduce the inward sodium current, but do not affect the subsequent outward potassium current.

A time course of this conditioning effect is evident in figure 3.25. Decreasing the ability of sodium to rush in was termed *inactivation*. It turns out that conditioning with small hyperpolarizing potentials increased sodium current, and this was termed *deinactivation*. By measuring the current produced and varying the conditioning pulse time, one could develop time courses of (de)inactivation and estimate a time constant for this process, as shown in figure 3.26. Deinactivation will be critical to our discussion of model-based control of Parkinson's disease in chapter 10.

If they perform a long enough conditioning pulse, they achieve the *infinite time* inactivation, described through variable h_∞ (or more precisely $1 - h_\infty$). Condition with a long

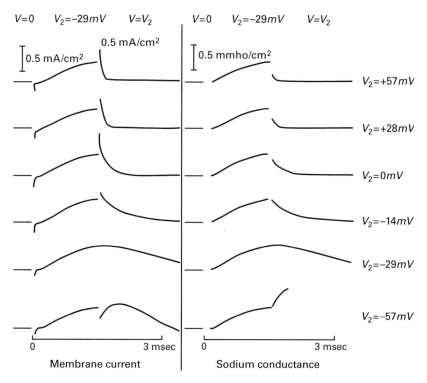

Figure 3.22
The rate of decay of conductance is proportional to the level of conductance reached. Reproduced from [HH52b] with permission.

enough hyperpolarizing pulse, and h_∞ goes to 1. It is as if the membrane is wide open. It fires with maximal sodium current strength. In figure 3.27 h_∞ traces a sigmoid curve. The normal rest potential (unconditioned) sits very near the inflection of the sigmoid (often assumed about $h = 0.6$). One positions systems near such inflection points (whether nerve membrane or transistor) in order to preserve maximum dynamic range and responsiveness of a system.

Last, they will study the recovery of inactivation by using paired pulses as in figure 3.28. If the paired pulses are too close together, there is no response. This property of inactivation is a major contributor to the phenomenon of the *refractory period*. The rate of recovery is very similar to the onset of inactivation—the time constant of these processes is about 12 msec. Both processes are dependent on the membrane potential.

Hodgkin-Huxley Paper 4

At present the thickness and composition of the excitable membrane are unknown. Our experiments are therefore unlikely to give any certain information about the nature of the molecular events underlying

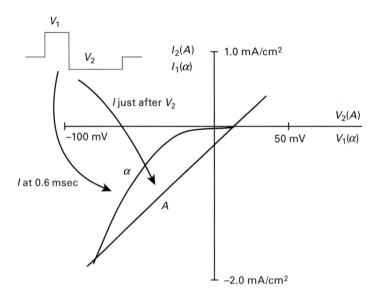

Figure 3.23
Contrast of delayed versus instantaneous I-V curves for potassium. Measured in choline seawater. From [HH52b] with permission.

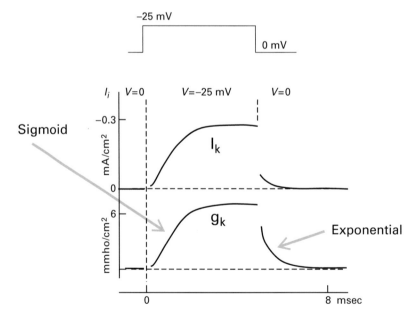

Figure 3.24
Sigmoid turn-on and exponential decay of turn-off of g_K. Modified from [HH52b] with permission.

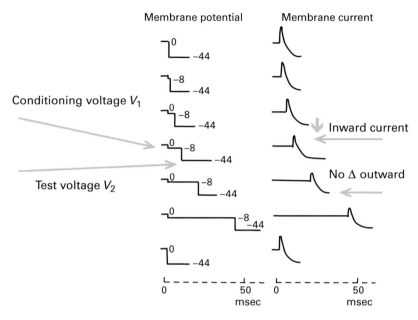

Figure 3.25
Demonstration of conditioning, V_1, and test, V_2, voltages on membrane current. Modified from [HH52c] with permission.

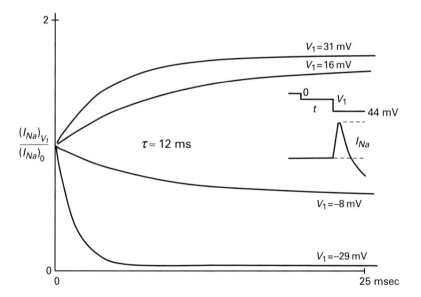

Figure 3.26
Time dependency of inactivation (curves going down) and deinactivation (curves going up). Modified from [HH52c] with permission.

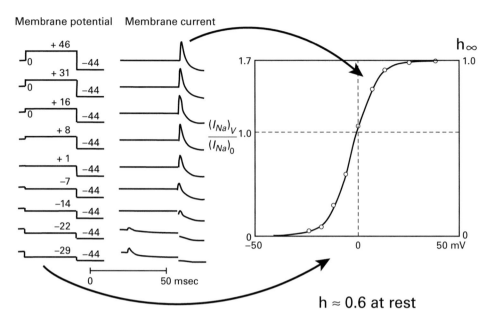

Figure 3.27
Plotting of h_∞. Modified from [HH52c] with permission.

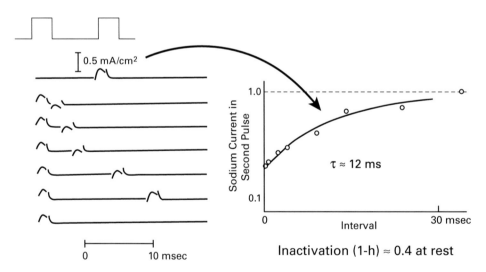

Figure 3.28
Paired pulses to measure recovery from inactivation. Modified from [HH52c] with permission.

changes in permeability ... for the sake of illustration we shall try to provide a physical basis for the equations, but must emphasize that the interpretation given is unlikely to provide a correct picture of the membrane. [HH52d]

It is important to understand what these equations represent. Hodgkin and Huxley construct a model that is capable of replicating their data, and will account for a range of phenomena that were known to take place in neurons. What this successful model fitting did not do was explain the underlying molecular events. There are no *channels*, *subunits*, *gates*, or other more modern terminology from molecular biology that influenced their theory. We are about to sketch out a set of deterministic equations that literally come from empirically fitting curves to their data. Therein lies their success and their limitations. Neuronal dynamics are stochastic. Fitting them with deterministic equations can be shown to dramatically fail [MS95]. Taking into account the underlying stochastic dynamics is at times crucial [Fit65] [SFS98], but remains insufficiently appreciated.

To summarize their evidence, Hodgkin and Huxley sketch out the model in circuit formalism that they think their data support (figure 3.29). There are two routes for current to pass—through capacitance or through ionic currents. The conductances for sodium and

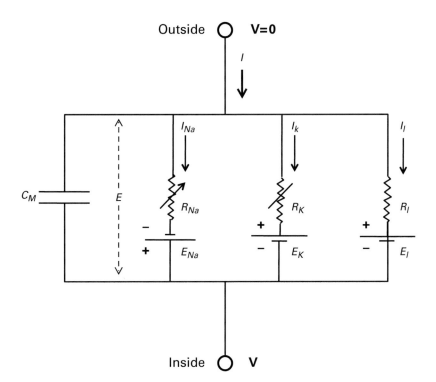

Figure 3.29
Model electrical circuit formalism for axon membrane that experimental data supported. Modified from [HH52d] with permission.

potassium are variable. The reversal potentials, E_{Na} and E_K, are represented as batteries; they are set by the concentration gradients and calculated through the Nernst potential. Last, there is a *leak* current, where all unaccounted for currents are lumped. Hodgkin and Huxley assume this leak has a conductance that is not voltage dependent, and a constant reversal potential. Later we will discuss *model inadequacy* in some detail (chapter 8). The leak was largely due to chloride, and in practice one could estimate it from the bias current on the voltage clamp at rest. But in a broader sense the leak was also their model inadequacy.

Hodgkin and Huxley assumed from their observations that changes in permeability depended on membrane potential rather than current. They suggested that these permeability changes had something to do with a dipole, a molecule in the membrane that was influenced by the electrical field. They could measure no movement (current) associating with such molecular effects (it would take another twenty years to demonstrate *gating currents* [AB73]). The absence of measurable gating current implied that the sites that controlled the current were sparse in relation to the charge carriers of the ionic current.

Assume, as before, that the total membrane current, I, is the sum of capacitive current, $C_M \frac{dV}{dt}$, and ionic currents, I_i. Furthermore, assume that the total ionic currents are the sum of sodium, I_{Na}, potassium, I_K, and lump the rest into leak current, I_l.

$$I = C_M \frac{dV}{dt} + I_i$$

$$I_i = I_{Na} + I_K + I_l$$

Further assume that the ionic currents are separable, based on the independence principle, into the separate currents

$$I_{Na} = g_{Na}(V - V_{Na})$$

$$I_K = g_K(V - V_K)$$

$$I_l = \bar{g}_l(V - V_l)$$

with variable conductances for sodium and potassium, g_{Na} and g_K, and a fixed conductance for leak, $\bar{g}_l$. The respective equilibrium potentials are written as V_{Na}, V_K, and V_l, where each $V = E - E_r$ is the difference in potential from resting membrane potential.

Recall the sigmoidal rise in g_K shown in figure 3.24, and the simple exponential fall. The simplest chemical system that can produce such dynamics assumes first-order kinetics—that a substance changes only in proportion to its own concentration to another form and back—with rates proportional to α and β[8]

---

8. Recognize that at the time, there was much prevailing literature that would have suggested a carrier molecule, or some other process interacting with their model of n, in order to transport ions across a membrane. It is one of the remarkable insights of Hodgkin and Huxley to use first-order kinetics as the absolutely simplest model capable of representing their findings. William of Occam would be pleased.

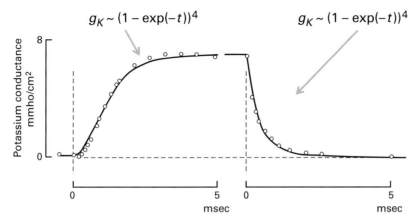

Figure 3.30
Fitting (solid lines) potassium conductance g_K rise and fall to data (circles) from figure 3.24. Modified from [HH52d] with permission.

$$n \underset{\alpha}{\overset{\beta}{\rightleftharpoons}} (1-n)$$

Such a process is accounted for by the differential equation

$$\frac{dn}{dt} = \alpha_n (1-n) - \beta_n n \qquad (3.4)$$

which literally says that the rate of change in concentration of n in time, dn/dt, is proportional to the *rates* α_n and β_n multiplied by their respective chemical species concentrations n and $1 - n$. The sigmoid of the g_K turn-on could be fit by a third or fourth power of such an equation, and Hodgkin and Huxley selected a fourth power fit. Using $g_K \sim (1 - \exp(-t))^4$ to fit the increase, and $g_K \sim \exp(-4t)$ for the decay, they achieved the fits shown in figure 3.30 using

$$g_K = \bar{g}_K n^4 \qquad (3.5)$$

where $\bar{g}_K$ is the maximum potassium conductance.

At rest, $dn/dt = 0$, so if

$$\frac{dn}{dt} = \alpha_n (1-n) - \beta_n n = 0$$

then

$$\alpha_{n_0} (1 - n_0) - \beta_{n_0} n_0 = 0$$

$$n_0 = \frac{\alpha_{n_0}}{\alpha_{n_0} + \beta_{n_0}}$$

provides a boundary condition. The solution is therefore

$$n = n_\infty - (n_\infty - n_0)\exp(-t/\tau_n)$$
(3.6)

where

$$n_\infty = \alpha_n/(\alpha_n + \beta_n)$$

$$\tau_n = 1/(\alpha_n + \beta_n)$$

and $1/(\alpha_n + \beta_n)$ defines the time constant τ_n. In fitting figure 3.30, they used a best τ_n, and realized that fitting the turn-on would have been better with a fifth or sixth power—not practical when computation was laborious with hand calculators. Applying this fit to a family of potassium conductance curves yielded the results in figure 3.31.

By using the value of τ_n that offered the best fits of the data, and estimating the maximal conductance g_K to be about 20% greater than the value at their most depolarized voltage

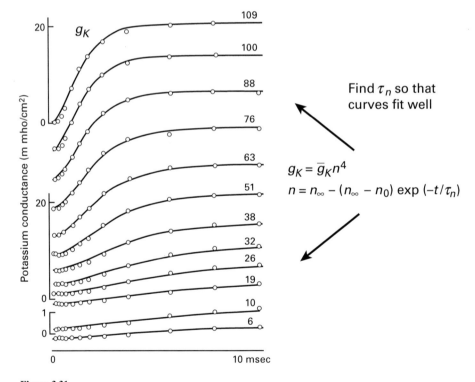

Figure 3.31
Fitting a family of potassium conductance curves with equations (3.5) and (3.6). Modified from [HH52d] with permission.

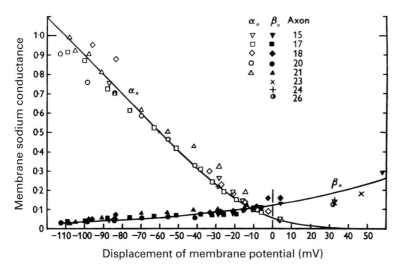

Figure 3.32
The reconstruction of curves for $\alpha_n(V)$ and $\beta_n(V)$. Modified from [HH52d] with permission.

clamp (-110 mV), they could estimate n_∞, and then estimate α_n and β_n from

$$\alpha_n = n_\infty/\tau_n$$

$$\beta_n = (1 - n_\infty)/\tau_n$$

Now all applicable measurement data are collected to piece together functions of $\alpha_n(V)$ and $\beta_n(V)$ as in figure 3.32. The black fit curves in figure 3.32 were empirically fit by

$$\alpha_n = 0.01(V + 10) \left/ \left[\exp \frac{V + 10}{10} - 1 \right] \right.$$

$$\beta_n = 0.125 \exp(V/80)$$

Hodgkin and Huxley struggled with whether to model g_{Na} by a second-order differential equation or assume two variables that obey a first-order equation. They chose the simpler, latter route, assuming that each variable, m and h, obeyed a first-order rate process

$$g_{Na} = m^3 h \bar{g}_{Na}$$

$$\frac{dm}{dt} = \alpha_m(1 - m) - \beta_m m \tag{3.7}$$

$$\frac{dh}{dt} = \alpha_h(1 - h) - \beta_h h$$

As above,

$$m = m_\infty - (m_\infty - m_0) \exp(-t/\tau_m)$$

$$h = h_\infty - (h_\infty - h_0) \exp(-t/\tau_h)$$

$$\tau_m = 1/(\alpha_m + \beta_m)$$

$$\tau_h = 1/(\alpha_h + \beta_h)$$

$$(3.8)$$

By experimenting with different ratios of τ_m and τ_h, they came up with the fits in figure 3.33. Using fit values of τ_m and τ_h, they could back out values of α_m, β_m, α_h, and β_h, as shown in figure 3.34. and again from curve fitting, they found

$$\alpha_m = 0.1(V + 25) \bigg/ \left[\exp \frac{V + 25}{10} - 1 \right]$$

$$\beta_m = 4 \exp(V/18)$$

and

$$\alpha_h = 0.07 \exp(V/20)$$

$$\beta_h = 1 \bigg/ \left(\exp \frac{V + 30}{10} + 1 \right)$$

The preceding is the process by which the Hodgkin-Huxley equations were generated. We summarize them here as follows:

$$I = C_M \frac{dV}{dt} + \bar{g}_K n^4 (V - V_K) + \bar{g}_{Na} m^3 h (V - V_{Na}) + \bar{g}_l (V - V_l)$$

$$dn/dt = \alpha_n (1 - n) - \beta_n n$$

$$dm/dt = \alpha_m (1 - m) - \beta_m m$$

$$dh/dt = \alpha_h (1 - h) - \beta_h h$$

$$\alpha_n = 0.01(V + 10) \bigg/ \left[\exp \frac{V + 10}{10} - 1 \right]$$

$$\beta_n = 0.125 \exp(V/80)$$

$$(3.9)$$

$$\alpha_m = 0.1(V + 25) \bigg/ \left[\exp \frac{V + 25}{10} - 1 \right]$$

$$\beta_m = 4 \exp(V/18)$$

$$\alpha_h = 0.07 \exp(V/20)$$

$$\beta_h = 1 \bigg/ \left(\exp \frac{V + 30}{10} + 1 \right)$$

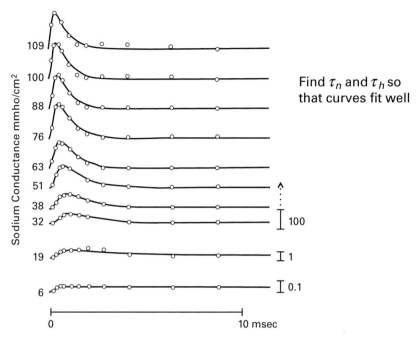

Find τ_n and τ_h so
that curves fit well

Figure 3.33
Fitting a family of sodium conductance curves with equations (3.7) and (3.8). Modified from [HH52d] with
permission.

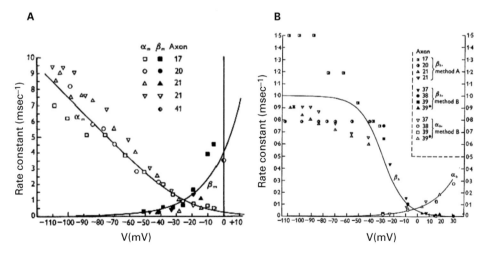

Figure 3.34
The reconstruction of curves (A) for $\alpha_m(V)$ and $\beta_m(V)$, and (B) for $\alpha_h(V)$ and $\beta_h(V)$. Modified from [HH52d]
with permission.

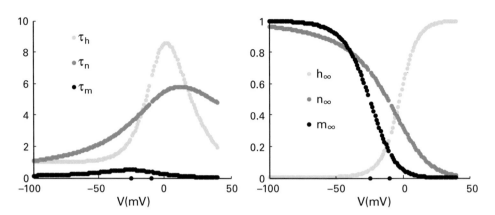

Figure 3.35
Relationship of the time constants and infinite time gate variables.

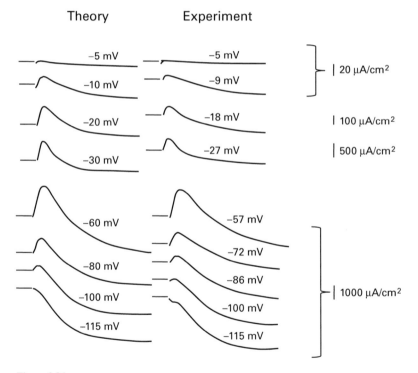

Figure 3.36
Predicting voltage clamp experiments from equations derived from voltage clamp data. Modified from [HH52d] with permission.

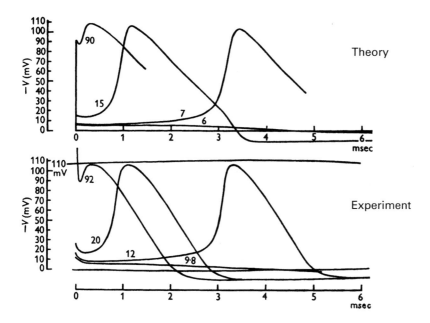

Figure 3.37
Predicting membrane action potential experiments from equations derived from voltage clamp data. Modified from [HHK52] with permission.

Let's start by gathering some intuition regarding the time constants and the long-time behavior of the rate constants.

It is useful to understand that the time constants and infinite time rate constants are functions of voltage, and the relative magnitudes of these time constants and infinite time gate variables are shown in figure 3.35. A small magnitude of τ_i indicates that it is fast. τ_m is fastest. You knew that. Note that τ_h is slowest near rest.

Now let's see if the equations will predict voltage clamp experiments that they were based on. Figure 3.36 shows that they do.

Next go back to the experiments of Hodgkin, Huxley, and Katz (figure 3.6), where they examined membrane action potentials that were *not* under voltage clamp. The replication in figure 3.37 is famous for having only one complete action potential.[9]

Calculating a propagating action potential was much more difficult than calculating the membrane action potential. In essence, one needs to solve a wave equation, by shrinking the membrane action potential (ordinary differential equation) to infinitesimally small elements, and linking these elements together to obtain the variations in space and time (partial

---

9. Numerical integration using a hand adding machine is laborious. Once they had one complete action potential, and the beginnings of the others, it was sufficient proof of replication.

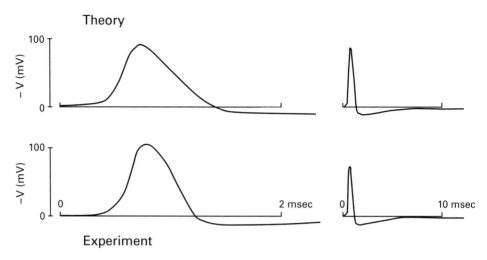

Figure 3.38
Propagated action potential. Modified from [HH52d] with permission.

differential equation). Their numerical integration of a propagated action potential is shown in figure 3.38.

Hodgkin and Huxley summarized their conductances versus voltage for their propagating action potential in figure 3.39, and compared this with Cole and Curtis's plot of the conductance change relative to the propagating action potential shown previously in figure 3.1. The qualitative relationship of total conductance, the sum of g_{Na} and g_K, were in excellent agreement with these previous experiments.

They explained the refractory period following a spike as a combination of inactivation and increased potassium conductance, as shown in figure 3.40. Comparing theory and experiment, the refractory period (relative and absolute) are both replicated in figure 3.41.

In examining the threshold region, note that the subthreshold regenerative response in the experimental traces at 11.8 to 12.0 mA are replicated in the theoretical curve 6 in figure 3.42. We will discuss this threshold behavior in more detail shortly.

One of the strange phenomena of neurons is that if you hold them at hyperpolarized potentials, and then release them quickly back to rest potential, they can exhibit a spike known as *anode break excitation*, which, indeed, the Hodgkin-Huxley model replicates in figure 3.43. They could now explain this phenomenon. Hyperpolarization decreases g_K and inactivation; when released, because the time constants shown in figure 3.35 show that neither n nor h can move quickly, g_{Na} is dominant, permitting an action potential to form. The essence of threshold is the region where inward current (sodium) is greater than outward (potassium), and the rate of change of g_{Na} increasing is greater than that of g_K increasing. This is what happens at the onset of anode break excitation. We will

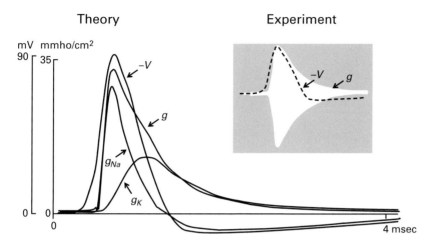

Figure 3.39
Conductance versus voltage for propagated action potential. Modified from [HH52d] with permission.

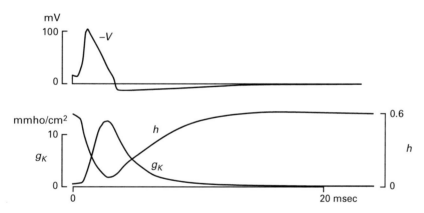

Figure 3.40
Origin of the refractory period is a combination of g_K and h. Modified from [HH52d] with permission.

return to this phenomenon of rebound activation of neurons in chapter 10 in our discussion of Parkinson's disease modeling and control.

Neurons *accommodate*; if you depolarize them too slowly, they will not fire. This is now explained. If the current depolarizes too slowly, then inactivation has a chance to override the increase in m, as does the increase in n.

Last, neural membranes oscillate. Indeed, Hodgkin-Huxley demonstrated that their equations could produce damped oscillations in response to a small short shock with high fidelity to the actual neuronal membranes.

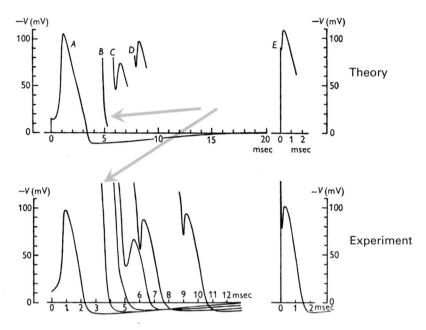

Figure 3.41
Absolute (arrows) and relative refractory period in theory and experiment. Modified from [HH52d] with
permission.

The agreement must not be taken as evidence that our equations are anything more than an empirical
description of the time-course of the changes in permeability to sodium and potassium ... certain
features of our equations were capable of a physical interpretation, but the success of the equations
is no evidence in favour of the mechanism of permeability change that we tentatively had in mind
when formulating them. [HH52d]

With the perspective of hindsight, their caution was prophetic. Many things were missing
from the Hodgkin-Huxley equations. The phenomena of membrane ion *pumps* were not
considered. Indeed, we will later discuss the dynamical implications of ion pumps for
tracking and controlling phenomena such as epileptic seizures in considerable detail (chapter
12). The *zoo* of ion channels that rain considerable complexity on mammalian neurons was
absent [GK02]. Also missing was the extraordinary geometrical complexity of neurons
beyond the giant squid axon [Asc02]. And the physics of macromolecular mechanisms
[MWC65] that lead to an understanding of stochastic excitability effects on the neuronal
membrane had yet to be considered [Fit65] [SFS98].[10]

10. A very nice alternative perspective on this history, with a more mathematical point of view, is John Rinzel's
[Rin90]

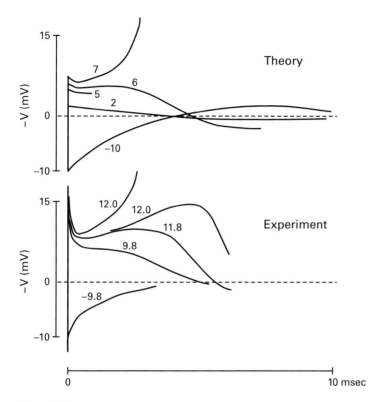

Figure 3.42
Subthreshold regenerative responses in theory and experiment. Modified from [HH52d] with permission.

All of this exploration of the nature of the excitable neuronal membrane is a prelude to reducing the complexity of the Hodgkin-Huxley equations (3.9), in the next chapter, to something more manageable: the Fitzhugh-Nagumo equations. The Fitzhugh-Nagumo equations provide the link to nonlinear control theory in chapter 5 [VTK04]. And all of this will lead us full circle, back to the Hodgkin-Huxley equations and their modifications in chapters 10 and 12.

Exercises

3.1. Prove the solution in equation (3.6) given (3.4).

3.2. These exercises will now guide you through writing your own integrator for the Hodgkin-Huxley equations.
    We begin with their original formulation. Use the following:

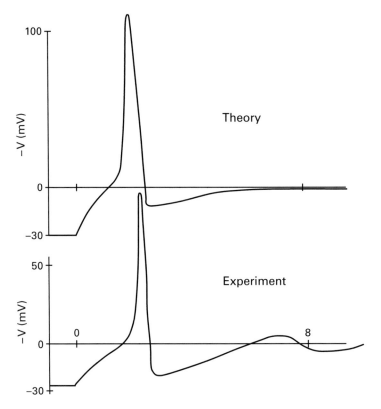

Figure 3.43
Anode break excitation in theory and experiment. Modified from [HH52d] with permission.

- Integration step time of $dT = 0.01$
- Initialize with t $= 1$, I(t) $= 0$, and V(t) $= 0$
- Pulse width of 15 out of 100 integration steps, starting at t $= 2$

$V_{Na} = 115$

$V_K = -12$

$V_L = 10.6$

$g_{Na} = 120$

$g_K = 36$

$g_L = 0.3$

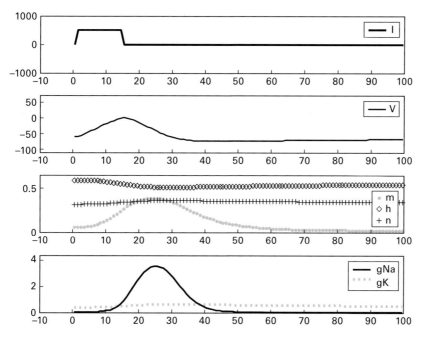

Figure 3.44
Sample output from integrating the Hodgkin-Huxley equations using the preceding suggestions. Pulse amplitude 500 for fifteen time steps.

- Initialize the rate constants, using the formulas from (3.9),[11] and then initialize the gating variables using three formulas as

$$q(1) = a_q(1)/(a_q(1) + B_q(1))$$

where

$$q = m, n, \text{ or } h.$$

- Initialize the conductances

$$g_{Na}(t) = g_{Na}m(t)^3h(t)$$

and

$$g_K(t) = g_K n(t)^4$$

11. The reader will question what might happen if the voltage approached the singularities of these rate constants, causing them to diverge to infinity. In practice, this rarely happens, and one can go most of a career ignoring this possibility. Perhaps a more cautious approach is to Taylor expand the rate constants near these indeterminate points [MSS82].

- You will need to make a loop that uses an integrator. Let's start with Euler integration.[12] We wish to evaluate $dy(t)/dt = f(t, y(t))$, using the first two terms of the Taylor series expansion of the function about a point t:

$$y(t+1) = y(t) + \frac{dy(t)}{dt}((t+1) - (t))$$

- We are not under voltage clamp, so the total membrane current is

$$I_m = C_m \frac{dV}{dt} + I_i$$

and if $I_m = 0$, then $C_m \frac{\partial V}{\partial t} = -\sum i_i$. The $\sum i_i$ includes your ionic currents, but also the added ionic current you impose from your current pulse. Write

$$dV = dT * (1/C_M) * \left(-I_{pulse} - \sum I_i\right)$$

and

$$V(t) = V(t-1) + dV$$

- Now update the rate constants and integrate the gating variables as

$$dq = dT\left(\alpha_q(1 - q(t-1)) - \beta_q q(t-1)\right)$$

$$q(t) = q(t-1) + dq$$

where $q = m, n,$ or h.

- If you want to plot with the modern convention, use $V = -(V + 60)$.
- Plot your results for current pulses of from -500 to $+500$. Where is the threshold?[13]
- If you have done this correctly, your results should look similar to figure 3.44:

3.3. Now that you have written an integrator, plot the relationships between the time constants and infinite time rate constants as in figure 3.35.

---

12. Using Euler integration like this is numerically abysmal [Pre07], but it is the place to start. There is actually a literature on the integration of the Hodgkin-Huxley equations, and the interested reader may wish to review this [MR74].

13. The threshold in the Hodgkin-Huxley equations is considered a *soft* one. We will discuss this in greater detail in the following chapter.

4 Simplified Neuronal Models

4.1 The Van der Pol Equations

... and perhaps also heart-beats.
—B. Van der Pol, 1926 [VdP26]

In the 1920s, Balthasar Van der Pol was working at the Philips Light Company in the Netherlands. He has an interesting background. Trained as a physicist, his mentors were J. J. Thompson, who discovered the electron, and H. A. Lorentz, whose theoretical work helped lay the mathematical foundation for special relativity.

At Phillips, Van der Pol was studying vacuum tubes—triodes. These had an internal filament surrounded by a grid, and a plate on the outside of this inner portion. One could charge the inner filament with a negative charge, and with a potential difference with respect to the plate, send electrons from one end to the other. One could either block or amplify such currents with the potential on the grid that lay in between. The arrangement is shown in figure 4.1.

In 1926, in the *London, Edinburgh, Dublin Philosophical Magazine and Journal of Science*, Van der Pol published a paper entitled *On Relaxation Oscillations* [VdP26]. The first thing he did was rather simple. He took the equation of a damped harmonic oscillator

$$\ddot{x} + k\dot{x} + x = 0$$

where $\dot{x}$ represents differential with respect to time (differentiating twice as $\ddot{x}$). He then changed the damping term k to give

$$\ddot{x} + \varepsilon(x^2 - 1)\dot{x} + x = 0 \tag{4.1}$$

This simple nonlinearity, x^2, introduced enormous complexity to the dynamics. Epsilon, ε, makes the damping term large or small.

Van der Pol, for small ε, describes limit cycles in this equation. In figure 4.2A he plots voltage versus the derivative of voltage. Starting at a point near the middle the trajectory spirals out and approaches a circle, and from a point further out on the diagram, spirals in

A   **B**

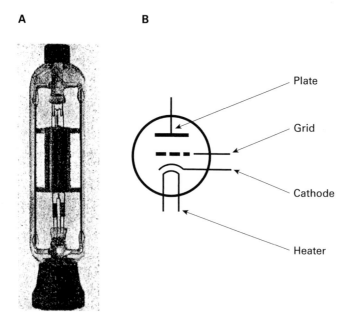

Figure 4.1
The triode. Panel A is a photograph [Kow21] of the first triode used in an electrosurgery circuit [GR03], and B shows its schematic. The grid in the middle, between the filament and the plate at the ends, controls the flow of electrons. Image in A reproduced from [Kow21] with permission.

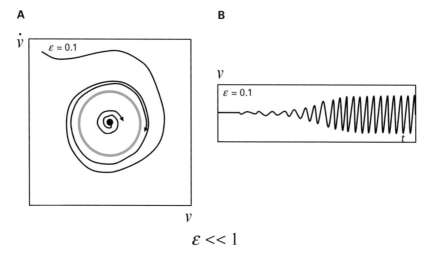

Figure 4.2
Limit cycle in the phase plane V versus $\dot{V}$ (A), and time series of voltage for $\varepsilon \ll 1$ (B). Modified from [VdP26] with permission.

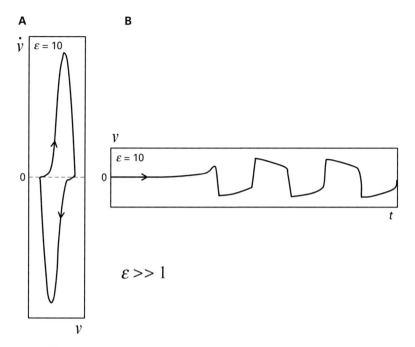

Figure 4.3
Phase plane V versus $\dot{V}$ (A), and time series of voltage for $\varepsilon \gg 1$ (B). Modified from [VdP26] with permission.

toward this same circular set. This is a stable limit cycle. Linear systems do not give stable limit cycles—one sees families of cycles parameterized by initial conditions that are not robust to perturbation. The limit cycle doesn't depend on the initial conditions; you settle down to it starting from an infinite number of potential initial conditions. Limit cycles are stable to perturbation. These are the properties that you want if you are building living systems that need to oscillate robustly. You want your heartbeat to settle down to something fairly similar to the setpoint frequency of heartbeat you want for your heart at rest, regardless of what you did to it this morning. Similarly, you would like your respiratory system to have a very deep limit cycle stable oscillator, so that you tend to breathe in a very cyclic manner.[1] *Life depends on limit cycles.*

If one plots voltage over time for small values of ε, one tends to see very stable sine wave oscillations, as in figure 4.2B—one frequency. Van der Pol, however, was interested in what happens when the damping coefficient is large, $\varepsilon \gg 1$. Multiple frequencies emerge. What was a limit cycle now looks strange and distorted in figure 4.3. Note in the time series plotted in figure 4.3 that the system appears to asymptotically approach one voltage, then

---

1. Why circadian rhythms from flies to humans do not cycle about a stable limit cycle is an open evolutionary question [PHK02].

takes a sudden rapid jump toward another voltage, and it flips and flops back and forth. It is bistable. Van der Pol termed these kinds of multistable systems *relaxation oscillators*. It is perhaps an unfortunate term.[2]

A key feature of the dynamics where ε is much different from 1 is shown in equation (4.1), where the dynamics have two different time scales (we will show this explicitly later in this chapter). In the time series plot of the relaxation oscillator in figure 4.3B, therefore, one could make the dwell time along the plateaus shorter (similar to neuronal action potentials), or longer (similar to heart cell action potentials), by tweaking the parameter ε. Van der Pol ends his 1926 paper speculating that the types of dynamics he has just described were similar to cardiac cycles—*and perhaps also heartbeats.*

4.2   Frequency Demultiplication

One might ask what would happen to the solutions of equations such as (4.1) if they were driven by, for instance, a sine wave. In 1927, Van der Pol and Van der Mark published a description of such driven dynamics in *Nature* [vdPVdM27]. What they found was that the frequencies of relaxation oscillators could produce a strange effect when driven by a given frequency, ω: They would *frequency demultiply*, that is, divide.

Using the simple circuit shown in figure 4.4, Van der Pol and Van der Mark drove it with $F sin(\omega t)$. By adjusting the capacitor, C, they found frequency division up to 1/200. As they increased the strength of the capacitor, they observed what sounded like noise before some of the transitions (figure 4.4). This driven nonlinear system was not generating random noise. They were observing chaotic dynamics—highly irregular deterministic dynamics with many different frequencies. We will examine chaotic behavior of nonlinear systems in more detail in the control frameworks discussed later. Such dynamics are a challenge for any nonlinear control system.[3]

4.3   Bonhoeffer and the Passivation of Iron

It simply is not enough to know that nerve fibers conduct . . . we need a much more basic and detailed understanding of these stimulus response systems. . . .
—M. Delbrück, 1969 Nobel Address

---

2. The *relaxation* part was because if you looked at RC circuits, the same time constant obtained from multiplying the effective resistance and capacitance of these circuits would give you the time constant. The reciprocal of the time constant would give you the period of these events, so it was like a relaxation circuit, or rather the relaxation of the RC circuit, and he called it a relaxation oscillator. There appears unfortunately to be some flexibility in the application of this phrase in the literature—but this is actually what it applies to.

3. At present, we leave out of this discussion whether neuronal systems generate chaotic dynamics [SJD+94]. Certainly they do not in an obvious sense. But if we are going to attempt to track nonlinear systems, we should be able to track them when they are at their most erratic. And the methods to be introduced later in this book can impressively track, and control, nonlinear systems even when they are chaotic.

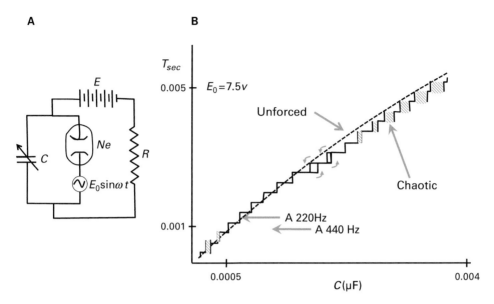

Figure 4.4
Driving a simple relaxation oscillator circuit would produce frequency demultiplication as well as chaotic dynamics. Modified from [vdPVdM27] with permission.

In 1947, Max Delbrück translated an article by Karl Bonhoeffer that discussed work performed during the war years in Germany [Bon48]. Delbrück had been studying lipid bilayers and was quite interested in nerve conduction.

Bonhoeffer would take a metal wire and have a reference electrode with a salt bridge and opposing electrode, as shown in figure 4.5A. Iron does some strange things. One of the reasons stainless steel does not rust is that chromium is added to it, forming an oxide on the surface of iron. Chromium oxide protects the iron (passivation) from corrosion. Nitric acid is used to remove iron atoms from the surface of the steel, and this quickly promotes formation of the chromium oxide. The oxide layer is visible when you watch this happen, and disappears during recovery.

The chemical activation of iron in nitric acid in response to an electrical pulse has a threshold and propagates as a wave; it seems to have an all-or-none response. A refractory period occurs, during which you have to wait for it to get ready to be activated again. It seems to accommodate, which means the reaction does not start if you gradually turn up the current pulse that you are using to activate passivation. And the reaction can oscillate.

Bonhoeffer likens this to ignition of an explosive compound. Figure 4.5B is heuristic. There is a stable point. Warm it a bit and it cools back down to the stable point. Warm it beyond a certain point, and it will actually begin to get hotter and hotter even if you take the

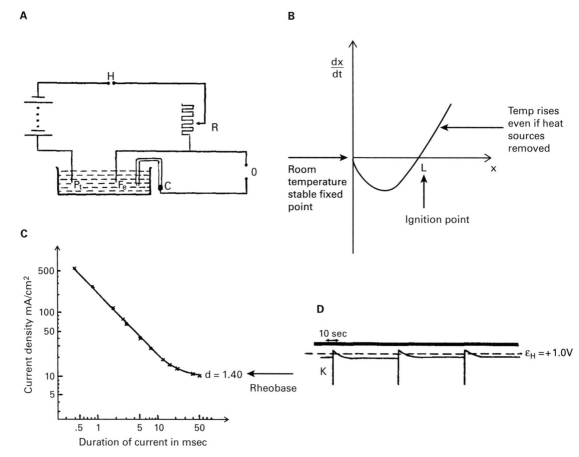

Figure 4.5
A Panel shows the chemical test apparatus for the electrical stimulation of passivation of iron. In B is a heuristic sketch for the nonlinear regenerative process for the ignition point of an explosion. The current-duration curve for activation of iron is shown in C, along with the equivalent of rheobase for analogous neuronal membrane properties. In D, the periodic relaxation oscillations of iron at a constant electrical current resemble features of the activation of neural tissue. Modified from [Bon48] with permission.

heat source away. Similar to this inflammation point of an explosive mixture, there appears to be an ignition point in the activation of iron.

There is also a point where, if you provide a low enough current, nothing happens at all (figure 4.5C). This is similar to a classic plot that you will see in the neuroscience literature, which is current versus duration of pulse required to activate a neuronal membrane. If you give a very short pulse, you need a big current to activate the system. As the pulse gets longer, you actually need less and less current until you reach a place where even an infinitely long pulse will not activate the process. This relationship holds in nerves

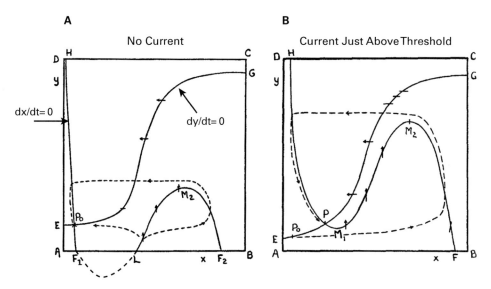

Figure 4.6
Chemical kinetics of the passivation of iron with nullclines plotted without current (A) and with current just above threshold (B).

and in iron bars. Bonhoeffer knew this and called this rheobase, after the neural pheno-menon.[4]

Figure 4.5D shows some of the oscillations that Bonhoffer recorded while electrically stimulating at a constant level. He recorded spike-like currents that spontaneously relaxed. It sort of looks like a heartbeat or a spiking reaction of a neural system.

Bonhoeffer knew of Van der Pol's work, and recognized these events as relaxation oscil-lations. He formulated the dynamics of iron passivation as chemical reactions, assuming that there would be an activation variable, x, and a refractory variable, y. Bonhoeffer rec-ognized that with analogy to electrical circuits, there needed to be a *negative resistance* to counter the tendency of such resistive circuits to dampen out oscillations (recall the dis-cussion of negative resistance in Hodgkin-Huxley dynamics in chapter 3). His intuition was sufficiently strong that he could qualitatively sketch the characteristic nullclines[5] from a generic set of differential equations, which would account for the characteristics of the passivation of iron (figure 4.6). His qualitative picture of such nullclines provide us with essential intuition regarding neuronal dynamics not only in this chapter, but in our later

---

4. The related term is chronaxie—the time duration corresponding to twice rheobase. It is a characteristic time for a pulse to activate a neural fiber.

5. Nullclines are plots where the values of the rate of change of a variable are equal to zero. The intersection between nullclines points to solutions of differential equations where nothing is changing. We will repeatedly examine nullcline plots throughout this book.

discussion of Parkinson's disease dynamics (chapter 10). Without current, the nullclines have only one intersection, as shown in figure 4.6A. Just above threshold current, one null-cline rises and a limit cycle emerges in figure 4.6B. The separation of time scales produces a relaxation oscillation similar to those in figure 4.3 and to be seen shortly when we modify these equations in the next section.

4.4 Fitzhugh and Neural Dynamics

The possibility of representing excitable systems by a generalization of the van der Pol equation was suggested to the author by Dr. K. S. Cole.
—R. Fitzhugh [Fit61]

Richard Fitzhugh was one of the first to pioneer the phase plane analysis of neuronal models with the view to understand their dynamical properties [Izh07]. He began working on a better understanding of the Hodgkin-Huxley equations, and was the first to recognize that the Hodgkin-Huxley model was *one member of a large class of nonlinear systems showing excitable and oscillatory behavior*—including both Van der Pol's triodes and Bonhoeffer's passivation of iron [Fit61]. Fitzhugh recognized that one of the values of *state space* diagrams is that each point on the diagram represented a particular state of the system. Different regions of the diagram corresponded to different physiological states of the nerve membrane. He further recognized that the study of the few *fixed points*, stable or unstable solutions that were in the phase plane, were the skeleton of the dynamics.[6] Understand the fixed points, and you understand a great deal about the qualitative nature of the system.[7]

Starting with Van der Pol's equation (4.1),

$$\ddot{x} + \varepsilon(x^2 - 1)\dot{x} + x = 0$$

Fitzhugh introduced a Liénard transformation,

$$\dot{x} = \varepsilon\left(y + x - \frac{x^3}{3}\right)$$

which lets you write (see exercise 4.1) the two first-order differential equations:

$$\dot{x} = \varepsilon\left(y + x - \frac{x^3}{3}\right)$$

$$\dot{y} = -x/\varepsilon$$

(4.2)

---

6. Large numbers of unstable fixed point solutions are characteristic of chaotic systems. The intuition that they form the *skeleton* of such dynamics was first recognized by Predrag Cvitanovic [Cvi88].

7. In a book about state estimation, these state space diagrams provide an invaluable way to visualize state geometrically, and also to construct geometrical versions of control laws. We are generally limited in their use, of course, to the cases where model reduction permits us to plot in the two or three dimensions that we can visualize.

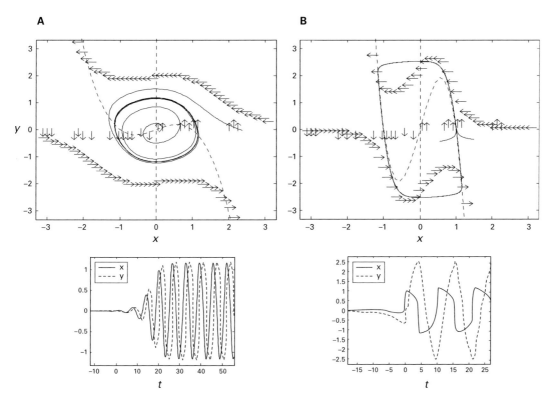

Figure 4.7
Nullclines (dashed lines) for the transformed Van der Pol equations (4.1) are $x = 0$ and $y = \varepsilon(x - x^3)$. In A are shown the results of two solutions with $\varepsilon = 0.5$, and below a set of time series for one of the solutions for x and y. In B are shown two solutions with $\varepsilon = 5.0$, and below a set of time series for one of the solutions. Prepared with pplane7 by John Polking, with permission.

The term $(x - \frac{x^3}{3})$ is a linear minus a cubic function, and this leads to an "N"-shaped function if plotted.

Setting the derivatives in equation (4.2) to zero gives us the nullclines plotted in figure 4.7. In panel A, the limit cycle is calculated for $\varepsilon = 0.5$, while in panel B, the relaxation oscillation is shown for $\varepsilon = 5.0$.

Notice that this particular transformation makes the relative time constants very clear; the rate of change in x, if ε is large, is fast; the rate of change of y, if ε is large, is slow. This shows the separation in time scales very clearly.

Fitzhugh next added a current, I, to the fast x variable to provide stimulation. He added two adjustable parameters, an a and b, putting y into the second equation so that y is represented in both. The slower variable y will become a *recovery* variable—it takes the role of inactivation and potassium in the Hodgkin-Huxley equations (chapter 12).

$$\dot{x} = \varepsilon \left(y + x - x^3/3 + I \right)$$
$$\dot{y} = -(x - a + by)/\varepsilon \tag{4.3}$$

When x goes up, y will decrease at a slower rate (an effective lag), and as y subsequently increases this causes x to decrease (see figure 4.7B time series).[8] Critical to adding by to the second equation in (4.3) is that he now had an adjustable slope on the $\dot{y} = 0$ nullcline, rather than the straight up and down line shown in figure 4.7.

He also assumed the following bounds on the constants:

$$1 - 2b/3 < a < 1$$
$$0 < b < 1 \tag{4.4}$$
$$b < c^2$$

Prescribe the current, I, and you had the analog of the *current clamp* of [HHK52]. The nullclines are now at

$$y = x^3/3 - x - I$$
$$y = -x/b + a/b$$

Let's examine the $\dot{x} = 0$ cubic nullcline first. If I were zero, then the $(x^3 - x)$ term would be "N"-shaped through $(0, 0)$. Conditions (4.4) require that at $I = 0$, the nullclines intersect at only one stable point—the resting state. Adding or subtracting I from the $\dot{x} = 0$ nullcline will raise or lower it. The $\dot{y} = 0$ nullcline is a straight line with an adjustable slope. Adding I will determine the stability of the intersection of these nullclines. Fitzhugh's stability diagram, labeling the different parts of the phase space with the nature of the system, is shown in figure 4.8.[9] If one applies an instantaneous pulse of current I to the resting point P, then the system solution instantly shifts along the dotted line to the left. There is labeled a *separatrix*—literally separating the phase space into two regions where the trajectories are different. If the pulse of current pushes the state across this separatrix, then a *regenerative* spike develops as a large excursion in the voltage x.

It is immediately apparent from figure 4.8 that there is no voltage x for which there is a single voltage threshold. Because the $\dot{y} = 0$ nullcline is tilted, unlike the special case of the Van der Pol equations in figure 4.6, there is a family of values of voltage that forms the threshold. Threshold for such systems is at least a line, or *manifold*. But thresholds for many neuronal models are often thicker than lines—they are *threshold sets*. Within the threshold

---

8. Fitzhugh, in a humble way, termed these equations the *Bonhoeffer–Van der Pol* model.

9. It is of interest that this diagram was produced by Fitzhugh using an analog computer. His term *no man's land* was a euphemistic reference to the difficulties of reproducing trajectories that reliably crossed into certain regions of the state space using the analog computation available.

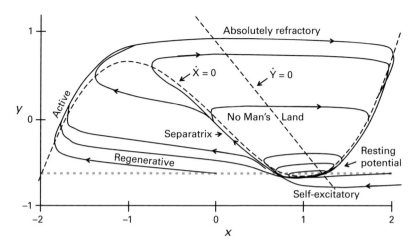

Figure 4.8
Phase plane diagram of the Fitzhugh modification of the Van der Pol equations. The nullclines are placed assuming
that the current $I = 0$. Reproduced from [Fit61] with permission.

sets, partial action potentials can be seen, as with the Hodgkin-Huxley equations (see
exercise 3.2). For both these equations and the Hodgkin-Huxley equations, *all intermediates
between all and none responses can be obtained by adjusting the stimulus with accuracy.*
A superb in-depth discussion of the different types of threshold can be found in [Izh07].

By applying a constant negative current I, the $\dot{x} = 0$ nullcline moves up, and this shifts
the relationship of the nullcline intersection and the threshold separatrix. In figure 4.9A, the
point at which constant current moves the intersection point P just to the position of the
new separatrix is the geometrical analog of *rheobase* (see figure 4.4C). If positive constant
current is applied, the $\dot{x} = 0$ nullcline moves downward, and on release of this current
the system state might be below the separatrix, resulting in anode break excitation (figure
4.9B).

And finally, Fitzhugh shows that a sufficiently large negative current raises the $\dot{x} = 0$
nullcline sufficiently high to make the intersection unstable, and a stable limit cycle appears
as in figure 4.10. An infinite series of periodic spikes ensues.

Fitzhugh noted in an earlier paper [Fit60] that the time courses of the Hodgkin-Huxley
variables n and h were nearly inverses of each other—projected on the $n - h$ plane the
trajectory followed close to a straight line. He experimented with reducing the four-
variable Hodgkin-Huxley equations to a two-dimensional form with the transformations
$w = (n - h)/2$, and $u = V - 36m$. Such reductions showed qualitative similarity with his
modified Van der Pol equations (4.3). But the two-dimensional phase space of (w, u) were
projections of four variables, and not a true phase space (where each point represents a
single state of the system). Improving this will need to await John Rinzel's later work
(discussed below).

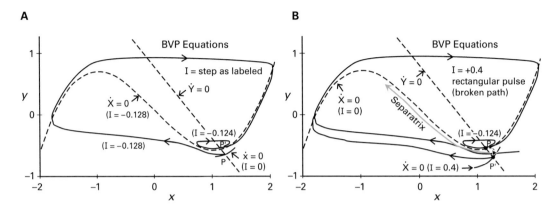

Figure 4.9
Effect of constant current moving the height of the $\dot{x} = 0$ nullcline. Rheobase is illustrated in A, and anode break excitation in B. Modified from [Fit61] with permission.

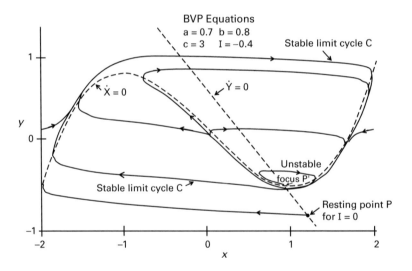

Figure 4.10
Effect of constant current moving the height of the $\dot{x} = 0$ nullcline sufficiently high so that the nullcline intersection becomes unstable and a stable limit cycle is produced. Reproduced from [Fit61] with permission.

Last, Fitzhugh noticed that neither the Hodgkin-Huxley equations nor his modified Van der Pol equations, replicated what real squid axons did in response to large constant current pulses. Real axons got tired and stopped firing—the models never showed such behavior. Fitzhugh recognized that work by Frankenhaeuser and Hodgkin [FH56] demonstrated that, with each action potential, there accumulated a small amount of potassium outside of the squid axon. This would serve to decrease the potassium equilibrium potential gradually over time, and would eventually lead to axonal unresponsiveness. Later, we will introduce a variable potassium into the Hodgkin-Huxley equations in order to be able to reconstruct and track such accumulating potassium. We will further show how the slow dynamics of potassium shifting out of and into cells is a critical piece of seizure dynamics, and how, in a control theoretic framework, not accounting for potassium leads to tracking failure in such networks (chapter 12).

The Bonhoeffer–Van der Pol model is not intended to be an accurate quantitative model of the axon, in the sense of reproducing the shape of experimental curves; it is meant rather to exhibit as clearly as possible those basic dynamic interrelationships between the variables of state which are responsible for the properties of threshold, refractoriness, and finite and infinite trains of impulses. (R. Fitzhugh, 1961 [Fit61])

4.5 Nagumo's Electrical Circuit

Jin-Ichi Nagumo and colleagues [NAY62] were interested in replicating the lack of attenuation and distortion characteristic of neural signal propagation in an electrical communication channel. Using Fitzhugh's formulation of equations (4.3), they used Kirchhoff's laws and the new tunnel diodes[10] to implement Fitzhugh's equations using the circuit shown in figure 4.11A. By chaining these elemental circuits, as in figure 4.11B and C, they could replicate the propagating action potentials shown in figure 4.11D and E. Note, in particular, that these propagating pulses were stable. Starting with a narrow pulse, it widens to an asymptotic width (figure 4.11D) that is the same as what a wide initial pulse contracts to (figure 4.11E).[11]

Fitzhugh's modification of the Van der Pol equation in (4.3) are now most commonly referred to as the *Fitzhugh-Nagumo* equations.

4.6 Rinzel's Reduction

John Rinzel overlapped for many years with Richard Fitzhugh at the Mathematical Research Branch at the U.S. National Institutes of Health. In 1985, Rinzel offered an alternative reduction of the Hodgkin-Huxley equations to the one Fitzhugh had suggested earlier.

---

10. The tunnel diode was invented in 1958 by Leo Esaki, who in 1973 received the Nobel Prize in Physics for discovering the electron tunneling effect used in these devices.

11. Competing with Fitzhugh for humility in describing their own results, they write that it *behaves similarly to the real axon but the representation is gross.*

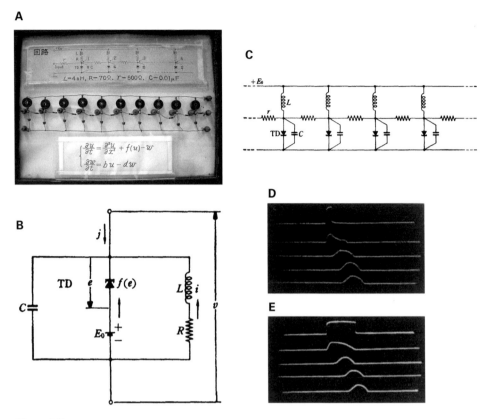

Figure 4.11
Nagumo's circuit is shown in A. Equivalent circuit to equations (4.3) are shown in B, and assembled to replicate a propagating action potential in C. The dynamics of the propagating pulse demonstrate asymptotic stability of the traveling wave as a narrow pulse widens, in D, and a wide pulse narrows, in E, each toward the same shaped action potential wave. B, C, D, and E reproduced from [NAY62] with permission. I am indebted to Professor Kazuyuki Aihara for kindly providing the photograph of Professor Nagumo's original circuit in A.

Taking advantage of the linearity of n versus h, he suggested fitting the line $h + Sn = 1$. The slope, S, was determined by the resting values of n and h as $S = (1 - h_0)/n_0$. This let him define a single recovery variable as $W = S[n + S(1 - h)]/(1 + S^2)$, from which two variables are replaced by one. He further notes that m is the fastest gating variable in Hodgkin and Huxley's equations, so one might as well just assume it is always at its infinite time value, or $m = m_\infty(V)$. Setting $h = (1 - W)$ and $n = W/S$, he develops a true two-variable reduction of the Hodgkin-Huxley equations:

$$I = C_M \frac{dV}{dt} + \bar{g}_K (W/S)^4 (V - V_K) + \bar{g}_{Na} m_\infty^3(V)(1 - W)(V - V_{Na}) + \bar{g}_l(V - V_l)$$

$$(4.5)$$

$$\dot{W} = \phi[W_\infty(V) - W]/\tau(V)$$

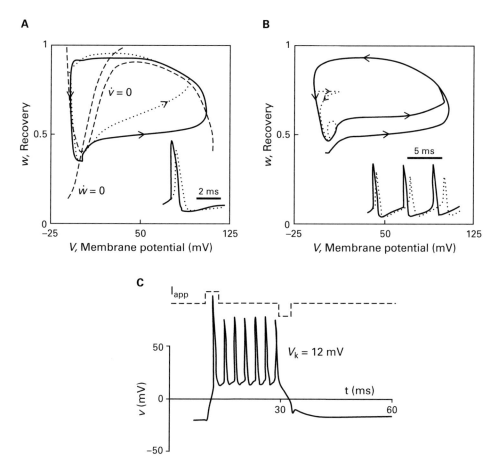

Figure 4.12
Similarities of full and reduced Hodgkin-Huxley equations to brief current pulse, A, and current step in B. Full model results are dotted, reduced Hodgkin-Huxley results are solid. C shows the bistability of reduced Hodgkin-Huxley equations under conditions of elevated potassium. No equivalent for Fitzhugh-Nagumo can replicate C. Reproduced from [Rin85] with permission.

where $W_\infty(V) = S[n_\infty(V) + S(1 - h_\infty(V))]/(1 + S^2)$, and $\tau(V)$ was a compromise between $\tau_h(V)$ and $\tau_n(V)$ as $\tau(V) = 5exp[-(V + 10)^2/55^2] + 1$. ϕ was a temperature correction factor. He further prescribed $h = 0$ if $W > 1$.

The first thing to notice was that Rinzel's reduction of the Hodgkin-Huxley equations gave a very good approximation to the Fitzhugh-Nagumo equations and the full Hodgkin-Huxley model (figure 4.12A and B). But now with explicit ions, such as potassium, one could perform simulations for which the Fitzhugh-Nagumo equations were not applicable. Figure 4.12C shows that, for elevated extracellular potassium, bistability could be shown in these reduced equations, similar to the bistability that John Moore had earlier demonstrated experimentally in squid axons [Moo59].

4.7   Simplified Models and Control

In 2004, Henning Voss, Jens Timmer, and Jürgen Kurths [VTK04] applied an unscented Kalman filter to estimate variables and parameters of the Fitzhugh-Nagumo equations. This will be discussed at length in chapter 5. Nevertheless, it is important to recognize that there are many other ways to boil down the dynamics of a neuronal membrane, or the Hodgkin-Huxley equations, to a smaller number of dimensions. To the extent that such models preserve relevant dynamics of a given neuronal system, the computational efficiency gained may introduce substantial benefit. The interested reader is urged to consult [GK02] or [Izh07] for a variety of other useful formulations. But from a philosophical control perspective, we will in a later chapter analyze in some detail the advantages and disadvantages of model fidelity to the biophysics, versus fidelity to the dynamics. The entire issue of *model inadequacy* (chapter 8) lays at the cutting edge of control theory. But we are getting ahead of ourselves. First, we need to apply the unscented Kalman filter to the Fitzhugh-Nagumo equations in the next chapter.

Exercises

4.1. Derive equations (4.2) using (4.1) and the Liénard transformation $\dot{x} = \varepsilon \left(y + x - \frac{x^3}{3} \right)$.

4.2. Write a program to integrate the Fitzhugh-Nagumo equations (4.3). Use an Euler integration scheme. Set values of $a = 0.7$, $b = 0.8$, and $\varepsilon = 3$. I would suggest starting with an integration time of 0.01, and integrating over perhaps 100 units of time. Try applying pulses for the middle half of this time. Use both positive and negative pulses between -1 and $+1$.

• What range of current gives subthreshold responses?

• What range of current produces limit cycles?

• Can you find a range of current that produces anode break excitation?

4.3. Euler integration, as we have been using, calculates the derivative at the endpoint of an interval, and assumes that it is a good approximation to a function throughout that interval. Euler assumes that the derivative of a function, dx/dt, evaluated at x_n and t_n, can be approximated as

$$\frac{x_{n+1} - x_n}{t_{n+1} - t_n} \approx \frac{dx}{dt}(x_n, t_n)$$

so that the next value being approximated in the integration, x_{n+1}, is equal to

$$x_{n+1} \approx x_n + \Delta t \frac{dx}{dt}(x_n, t_n)$$

where $\Delta t = t_{n+1} - t_n$ as illustrated in figure 4.13. A more accurate way to numerically integrate is to reestimate the derivative along the way from time t_1 to t_2. The second-order Runge-Kutta method (RK2) first evaluates the same product of derivative and time interval as in Euler integration

$$k_1 = \Delta t \frac{dx}{dt}(x_n, t_n)$$

which gives an increment in x. But we then take half of this increment, at half of the time interval, and calculate

$$k_2 = \Delta t \frac{dx}{dt}\left(x_n + \frac{k_1}{2}, \ t_n + \frac{\Delta t}{2}\right)$$

such that

$$x_{n+1} \approx x_n + k_2$$

This intermediate derivative is more accurate than Euler, and the geometry is illustrated in figure 4.13. A detailed authoritative source on numerical integration is Press et al. [Pre07].

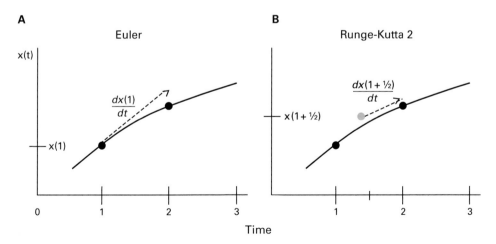

Figure 4.13
Schematic of Euler integration (A) versus Runge-Kutta of second order (B).

• Write a function that will let you pass parameters and variable values to and just calculate the Fitzhugh-Nagumo equations (derivatives). Such a function will be useful in writing RK2 and higher-order routines. Let the function return a vector of the two Fitzhugh-Nagumo variables, $\dot{x}$ and $\dot{y}$. This vector of derivatives can then be multiplied by the time interval, to generate k_1. By adding $k_1/2$ to the vector of variables, you can reuse this same function to calculate k_2.

• Now modify your program from exercise 4.2 so that it integrates the Fitzhugh-Nagumo equations using RK2 integration.

4.4. Last, the standard for numerical integration in this book is the fourth-order Runge-Kutta (RK4) integrator.[12] This is an extension of the RK2 routine introduced in exercise 4.3.

• You will need an algorithmic routine that calculates the RK4 intermediate quantities, $k_1 \ldots k_4$. One way of structuring this is to use the Fitzhugh-Nagumo derivatives function introduced in exercise 4.3, and call it with the following routine:

```
k1=dT*Fitzhugh-Nagumo_dot(Variables,Parameters);
k2=dT*Fitzhugh-Nagumo_dot(Variables+k1/2,Parameters);
k3=dT*Fitzhugh-Nagumo_dot(Variables+k2/2,Parameters);
k4=dT*Fitzhugh-Nagumo_dot(Variables+k3,Parameters);
xy(:,t)=Variables+k1/6+k2/3+k3/3+k4/6;
```

where dT is the time interval, Fitzhugh-Nagumo_dot is the function that calculates $\dot{x}$ and $\dot{y}$, Variables and Parameters are column vectors of the variables and parameters in the model, and xy is the full time series of the variables values (each column are the values at one point in time).

---

12. Press et al. [Pre07] suggest that use of RK4 is for *ploughing fields*. But they also recognize that, properly used, such a workhorse algorithm can serve us well.

5 Bridging from Kalman to Neuron

This linearity assumption is a serious one. Its justification lies in the fact that the patterns produced in the early stages when it is valid may be expected to have strong qualitative similarity to those prevailing in the alter stages when it is not.
—Alan M. Turing, 1952 [Ala52]

5.1 Introduction

Richard Feynmann suggested that the goal of physics was to create mathematical models that explain observations [Fey65]. Observations are data. Our goal in this book is to use models of neurons and brains to fit to brain data. We will do this in order to observe and track brain activity. And we seek to use this observability to develop control algorithms for neural activity.

So our first mission is to determine: Can we fit models to brain data?

The problem is not whether we can insert data into a model. Mindless computation is an occupational hazard of computational neuroscience.

The problem is that neurons, and brains, are not linear systems. And the problem with nonlinear systems is that they can do things that linear systems never do, such as not having any truly periodic behaviors. If your brain is generating sine waves, you are in trouble.

By the mid-twentieth century, it had become apparent that even simple nonlinear systems might generate deterministic yet aperiodic behavior [Lor63]. Such nonlinear systems might have no stable states. Instead, they were characterized by a plethora of unstable states, each with a different frequency. Another hallmark of such systems was sensitivity to initial conditions. If the system were in one of two nearby states, the future evolution might exponentially diverge. This extreme form along the spectrum of the dynamics of nonlinear systems would be termed *chaos* [LY75]—dissipative, bounded, uniformly unstable with sensitivity to perturbation, and aperiodic behavior that is anything but random.

Why bring up chaos in a book about the nervous system? It is not at all clear that brains, or even simple nervous system networks, generate behaviors that we can classify as chaotic.

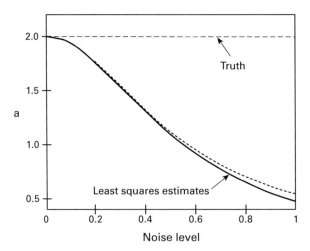

Figure 5.1
Demonstration of least squares failure to fit a simple nonlinear model. The true parameter $a = 2$ is severely underestimated whether added noise is uniform (dotted line) or Gaussian (solid line). Reproduced with permission from [MS04].

I say this not because there does not exist a huge literature arguing that brains do.[1] It is because we have almost universally failed to demonstrate that our analysis of brain states is of true states—where the recurrence of similar states led to similar subsequent events. The same can be true of the atmosphere and oceans, where the lack of a true *analog* [Lor63], a state repeating a previous state, leaves the ultimate issue of whether such complex physical systems are chaotic or not an infinitely open question.

But finding a brain state analog is not the point. The point is that brains are made up of floridly nonlinear elements. Neurons have membranes that generate action potentials that can be modeled with the Hodgkin-Huxley or Fitzhugh-Nagumo equations. Their synaptic connections are nonlinear. Their ion concentration dynamics are nonlinear, as will be discussed later (chapter 12). In general, we are pretty safe in assuming that any method to fit models to brain data using linear models might not be the optimal choice.

And the problem with fitting models to nonlinear systems is that it is notoriously hard to do. If there is noise (When is there not noise in a system and in its measurements?), a system that is sensitive to small perturbations causes trouble. A beautiful example is offered by McSharry and Smith [MS99a]. They started with a *logistic map*. This is a simple equation that has long had a role in modeling in population biology.

$$x_{i+1} = ax_i(1 - x_i) \tag{5.1}$$

1. The author has, for better or worse, contributed to some of this controversy.

For suitable values of the parameter a, this almost trivially nonlinear map generates chaotic trajectories. And using Gauss's least squares method from chapter 1, and knowing the model that generated the data, gives the *wrong* value of a—dramatically wrong, as figure 5.1 shows. One can find that fitting a cost function to determine such parameters might well reject the very model that generated the data [MS04].

Our goal is a bit less severe than this. *We are not out to reengineer the brain.* That is, it would be splendid to have a model that truly represented all of the brain's components, or even a single neuron. But no such model exists. And if it did, it would be so complex that, were you to have a computer large enough to run the model fitting for some fraction of eternity, you still would not get the correct parameters. There would be too many parameters to fit any of them with much certainty. Our goal is to fit dynamics. Observe the dynamics of the nervous system, and you can potentially control it based on that tracking.

So let's track nonlinear dynamics.

5.2   Variables and Parameters

We are interested in this book in tracking natural systems using autonomous systems of differential equations. Autonomous systems have time only present in the derivatives, as follows:

$$\dot{x}_1(t) = f_1(x_1(t), x_2(t), \ldots x_n(t), \lambda_1, \ldots, \lambda_m)$$

$$\dot{x}_2(t) = f_2(x_1(t), x_2(t), \ldots x_n(t), \lambda_1, \ldots, \lambda_m)$$

$$\vdots$$

$$\dot{x}_n(t) = f_n(x_1(t), x_2(t), \ldots x_n(t), \lambda_1, \ldots, \lambda_m)$$

where we allow for n variables x, m parameters λ, and $\dot{x}$ indicates derivative with respect to time.

Assuming vectors in the notation, one would write

$$\dot{x} = F(x, \lambda) \tag{5.2}$$

for the model, leaving off noise for now. We assume, as before, a measurement function

$$y = A(x) + r \tag{5.3}$$

where r is random noise.

If the measurements are a vector of values y, then our goal is to find the highest probability, P, of observing these data y, given the model x and parameters λ such that $P(y|x, \lambda)$ is maximized. This probability peak yields the parameters that are most likely to have accounted for the data observed—the *maximum likelihood*.

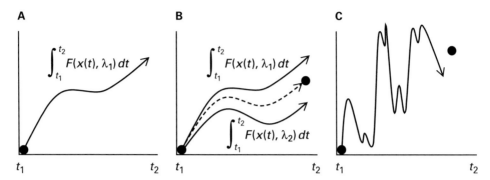

Figure 5.2
Direct integration of a differential equation using an initial value boundary condition shown as a filled circle in
A. In B, a second boundary condition is known, a true value that the underlying system produced, and one can
iterate the integration of $F(x(t), \lambda_i)$ to hit the target and optimize the parameters λ_i. C shows a chaotic trajectory
where sensitivity to initial conditions renders a single shooting effort almost impossible to optimize.

The situation is complicated if there is uncertainty in the variables as well as in the
parameters. The *errors-in-variables* problem has been well studied (see reviews in [VTK04,
Sod07]). Throughout this book, we will need to estimate both the variables and parameters
in our model fitting to data—the *dual-estimation* problem. It will turn out, later, that an
ensemble Kalman filter is ideal for such dual-estimation problems through the use of an
augmented state, which includes both variables and parameters to be estimated. But we are
getting ahead of ourselves.

For now let's first assume that you actually know the model that generated the dynamics
to be tracked, and that there is no uncertainty in the independent variables. Even with this
advantage, for noisy nonlinear systems, fitting the model to the data is not easy.

One way to do this is to start with an initial condition, x_0, and integrate the model
forward from time t_1 to a final time t_2. This is a single boundary value problem. If we know
the model (differential equation), and there is no noise, then the solution is exact. The
situation is illustrated schematically in figure 5.2A. We did this, for instance, when we
integrated the Hodgkin-Huxley or Fitzhugh-Nagumo equations in previous chapters. Are
the trajectories of these integrations valid? Well, valid for what? They are trajectories
from numerical integration. Do they reflect anything from nature? Doubtfully. Do they
reflect anything from even those models, if we did not know the parameters of the models
(e.g., the maximal conductances for sodium, potassium, and leak in the Hodgkin-Huxley
equations)? No. And if there was noise in the models, but you did not know the noise at each
point in time, would your trajectory happen to be the the same as the model's? Of course
not. This is the solution of the *direct problem* (as opposed to the *inverse problem*, where
you are given data and want to calculate the most likely model trajectory underlying the
measurements).

So the next more complex thing to do is to introduce a two-point boundary value problem. You have initial conditions, but you also know the final state. You want to fit the parameters to the model so that the solution is consistent with the endpoint. The first boundary condition is initial—you aim the equations. The trajectory then shoots, and your final value is compared with the value of the system. So you adjust the *aim* of the equations by resetting the parameters and try again. This *shooting method* is schematized in figure 5.2B. For fairly well-behaved equations, shooting methods are very good means of picking the best set of parameters λ_i. An excellent discussion of shooting methods can be found in [Pre07].

But what if nonlinearities in the dynamics increase as schematized in figure 5.2C? Chaotic dynamical systems shred information as they evolve, so that any sets of initial conditions and parameters close to the true solution will wildly diverge after a relatively short time. Even if you had the exact initial condition and parameter values, if you introduce noise you are again on the path to exponential divergence from the truth. In fact, using computers introduces trouble—the roundoff error of digital precision will create errors of the size of the system being simulated after a certain number of iterations. This is the *shadowing problem*: You want to know if any of your calculated trajectories remain close to the true trajectory [SGY97]. The bottom line is that you can shoot for a solution, but not for very long (not long in comparison with the instability in the equation).

So if you cannot shoot for a relatively long time in a chaotic system and come close to the true solution, a logical choice is to shorten the shooting range. *Multiple shooting* improves tracking by shooting over shorter distances, and repeating the process in small saltatory steps. These methods are well developed, and one can measure data and use multiple shooting to solve the inverse problem for nonlinear systems. Nevertheless, multiple shooting methods for chaotic systems are well known to be computationally expensive to implement, difficult to tune, and restricted to models that are differentiable [SSKV02].

5.3   Tracking the Lorenz System

Since Edward Lorenz wrote his paper entitled *Deterministic Nonperiodic Flow* in 1963 [Lor63], his model has been a paradigmatic system of nonlinear dynamics. Why introduce a highly simplified model of atmospheric convection at this point? Much of the literature that feeds the developments to follow in this and subsequent chapters come from physics and meteorological literature where the cutting-edge research in system tracking, observation, and prediction often takes place. It often appears as if no geophysicist or numerical meteorologist can write a paper without thinking about using the 1963 Lorenz model.[2] So we will take some time to describe it here, as we will use it later.

---

2. There is a more complex model that Lorenz published in 1998 [LA98], which we will not use in this book.

A

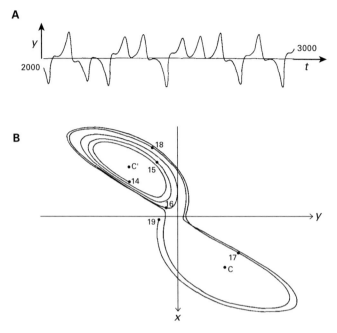

Figure 5.3
Solutions of Lorenz equations shown in A for y versus time, after skipping initial transient, and B shows plot of y versus x. Reproduced from [Lor63] with permission.

The differential equations of the Lorenz model are

$$\dot{x} = -\sigma x + \sigma y$$

$$\dot{y} = -xz + rx - y \qquad\qquad (5.4)$$

$$\dot{z} = xy - bz$$

The variables in this set of equations refer to convection of fluid in an experiment where fluid is heated from below and convection rolls develop. The variable x refers to the velocity of convective flow, y the temperature difference between ascending and descending fluid, and z the distortion of the vertical temperature profile.[3] The parameter σ refers to the Prandtl number, r the Raleigh (Reynolds) number, and b is an adjustable parameter. For values of these parameters $\sigma = 10$ and b $= 8/3$, Lorenz found that values of $r > 24.74$ gave chaotic solutions (figure 5.3).

3. Although Lorenz's original paper [Lor63] is readable on the meaning of these variables, it is a very technical paper. An excellent overall description of these dynamics can be found in [ASY97].

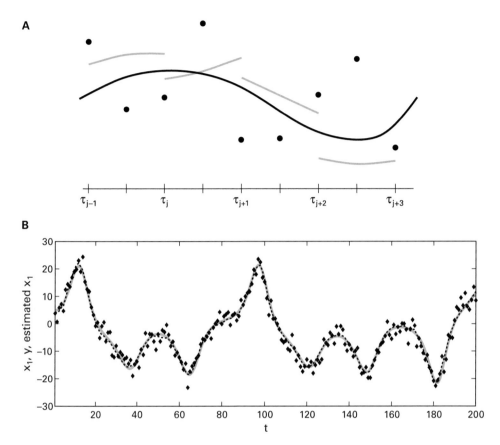

Figure 5.4 (plate 1)
Multiple shooting schematic for solving inverse parameter estimation shown in A, and an actual example of noisy Lorenz system is shown in B. Reproduced from [VTK04] with permission.

Let's assume that you knew you were observing data from a Lorenz system, and you had the equations. You are given noise-contaminated measurements of the x variable only, and you do not know the parameters. Voss and colleagues have examined the problem in detail [VTK04]. First, they start with an initial condition and known parameters and generate a true trajectory. It is important here to remember that these parameters are fixed. Then they measure x with noise mixed in, and ask whether they can solve for the parameters—an inverse problem. Figure 5.4A (plate 1) shows a schematic of multiple shooting to solve this inverse problem. Figure 5.4B shows their actual multiple shooting fit to these data after three passes over the data. Following twenty iterations over the full data set, the fit will be excellent, and their parameter estimates are close to the actual parameters, within 2.5 to 3.4%. This example has nothing about real-time use.

Now repeat this measurement and estimation scenario using the unscented Kalman filter. We use the same noisy x measurements and let the parameters be unknown, first forming an *augmented state vector* by adding the parameters to the state variables. This forms a six-dimensional vector for the Lorenz system of equations (5.4). We will use augmented state vectors throughout this book in dealing with neuronal systems.

Voss and colleagues [VTK04] make a first pass over the data with a known bad choice of parameters, and use the final parameter values to make another pass at fitting the data, as illustrated in figure 5.5 (plate 2). Their parameter estimates are now superb, within 0.3 to 1% error. Although they did not use the unscented Kalman filter in a real-time setting, it could have been used as such.

5.4 Parameter Tracking

So far we have discussed parameter estimation. There are no dynamics involved, and we (and the citations we have referred to) have focused on parameters that must be found. Once found, they become system constants.

Nothing is constant in the nervous system. Since Sherrington's classic text analyzing the dynamics of reflex variability [She06], spinal reflexes, nerve propagation through branch points, synaptic probabilistic release, basal ganglia rhythmicity, cortical rhythms—nothing ever seems to stay constant.[4] So it makes a great deal of sense to let the parameters change in our efforts to computationally track neural dynamics.

In *parameter tracking*, one assumes that the parameters, like the variables, change their values as a function of time. What is the difference? Implicit in the parameter tracking formulations is that the time scale of the rate of change of parameters is much slower than the changes in variables. One can then apply trivial dynamics to parameters in the sense that

$$\lambda(t+1) \approx \lambda(t)$$

In the recursive formulation of the Kalman filter, however, fitting the augmented state vector with such trivial dynamics on the propagation of the parameters still permits the resetting of the parameters to adapt as the data is fit with each step of the recursive algorithm. Recall that we employed static models in our examples in the exercises in chapter 2.[5]

In implementing this strategy, one needs to add noise to the parameters in order to ensure that the algorithm will *search* about for better parameters with each step [VTK04]

$$\lambda(t+1) = \lambda(t) + r(t)$$

where $r(t)$ is drawn from a random distribution or estimated from the expected variance.

---

4. An interesting facet of such variability has been seen in the use (and misuse) of reflex measurements in the treatment of spasticity [WS93].

5. A nice discussion of the use of trivial dynamics for static models in Kalman filtering is also found in [Str86].

A

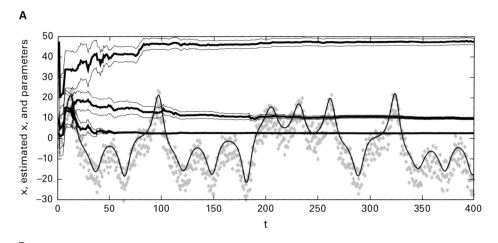

B

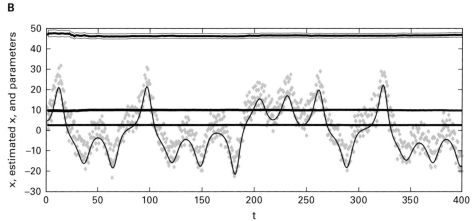

Figure 5.5 (plate 2)
A shows the first pass of the unscented Kalman filter through the noisy x output (dots) of the Lorenz system. The three parameters being estimated are shown as solid lines, with thin line confidence limits gradually converging from left to right. The values at the right are then used to initiate another pass at parameter estimation, shown in the lower panel, B. Although parameter estimation improves a bit more, the upper panel demonstrates the potential power of this technique to work with one pass in real-time applications. Modified from [VTK04] with permission.

5.5   The Fitzhugh-Nagumo Equations

The bridge to neuroscience from this world of nonlinear control theory appears to occur in five paragraphs toward the end of the tutorial paper by Voss and colleagues [VTK04]. In prior work, they had explored unscented Kalman filtering of the stochastic Van der Pol equations [SSKV02]. But no neuroscientist (certainly not the author) would have recognized the nervous possibilities at a reading of that earlier article. They chose the Fitzhugh-Nagumo equations (4.3), and set up the following scenario. They assumed that all one could do was to measure voltage, x. They added considerable amounts of noise to the measurement and set parameters for this equation at the same levels we used in exercise 4.2: $a = 0.7$, $b = 0.8$, $c = 3$. But they allowed the applied current, I, in equations (4.3) to float and become a function of time, $I(t)$ with the trivial dynamics of a parameter.

In generating their model, Voss and colleagues [VTK04] applied a nontrivial dynamics to $I(t)$—a sequence of connected cosine half waves as shown in figure 5.6 (plate 3). In the figure, a constant noise variance was applied to the measured parameter $I(t)$ with each time step. The initial values of $I(t)$, and the unobserved variable, y, were just set to zero, as shown in panel B of the figure. The initial value for the variable x, the voltage, was set to the actual value of x used in the model simulation.[6]

The results in figure 5.6 are remarkable. There is no analog to accomplish this using previous techniques such as multiple shooting (although in all fairness, the author is unaware of comparable alternative efforts).

The importance of these results is that it demonstrates in a simplified neuronal model that the estimation of inaccessible parameters, and tracking of unknown parameters, is possible in the case of neuronal dynamics. It is the starting point for our explorations of using ensemble Kalman filtering techniques on data from single cells, reconstructing cellular networks, tracking cortical dynamics, Parkinson's disease control, and the other scenarios to be explored in later chapters. Nevertheless, we have so far discussed no model capable of accounting for dynamics more extensive than a piece of neuronal membrane. To work with the nervous system, we need to be able to account for spatially extended systems. At the end of [VTK04], the authors suggested that *it should also be possible to perform some kind of parameter estimation for spatiotemporal systems.* In the next chapter, we explore how to begin to accomplish this for cortical dynamics.

---

6. The authors [VTK04] graciously made available their insightful code structure with publication of the article. Without this generous offer, many of the developments to be derived later in this book would have been much delayed and not performed as well. It is a strong argument for the publication of code as supplementary material with all computational papers. Otherwise, one of the key aims of scientific publication, a sufficient description of methods such that others can replicate the work, can be significantly impaired. It's easy to write papers that no one can replicate.

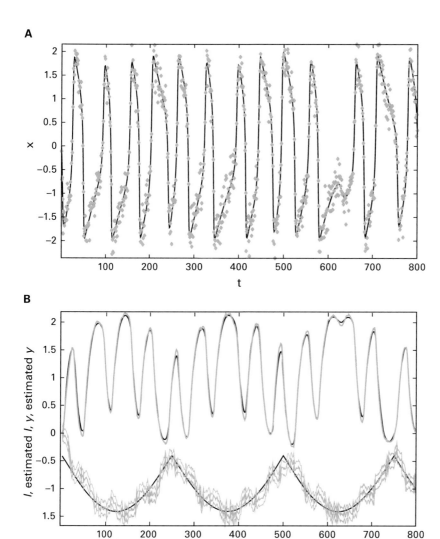

Figure 5.6 (plate 3)
Variable and parameter tracking in the Fitzhugh-Nagumo equations. A shows the underlying trajectory (solid line), the noisy measurements (dots), and B shows the reconstruction (red) of the unmeasured variable y (black, nearly overlapping) in the upper half of the panel, and the applied current $I(t)$ and its estimation as a tracked parameter (magenta) with 1 standard deviation confidence limits (thin magenta lines). Reproduced from [VTK04] with permission.

Exercises

Here we will explore the Voss et al. [VTK04] algorithmic structure in detail. Although we will make many modifications to their original algorithm throughout this book, we have not found a better overall subroutine structure than the one presented in their original calculations. What we present below is both Matlab and Octave compatible.

There are five scripts in this algorithm, which we denote as voss.m, vossFNfct.m, vossFNint.m, vossobsfct.m, and vossut.m. Four are callable subroutines. The main program, voss.m, is:

```
% Algorighm originally made available by Henning U. Voss, 2002
% "Nonlinear dynamical system identification
%   from uncertain and indirect measurements"
% HU Voss, J Timmer, J Kurths - International Journal
%   of Bifurcation and Chaos, Vol. 14, No. 6 (2004) 1905-1933
% Reproduces Figure 9 from paper
% Modifications by S. Schiff 2009 - Matlab and Octave Compatible
clear all; close all;
   global dT dt nn % Sampling time step as global variable
   dq=1; dx=dq+2; dy=1;
        % Dimensions: (dq for param. vector, dx augmented state, dy observation)
   fct='vossFNfct'; % this is the model function F(x) used in filtering
   obsfct='vossFNobsfct'; % this is the observation function G(x)
   N=800; % number of data samples
   dT=0.2; % sampling time step (global variable)
   dt=dT; nn=fix(dT/dt); % the integration time step can be smaller than dT
% Preallocate arrays
   x0=zeros(2,N);       % Preallocates x0, the underlying true trajectory
   xhat=zeros(dx,N);    % Preallocate estimated x
   Pxx=zeros(dx,dx,N);  % Prallocate Covariance in x
   errors=zeros(dx,N);  % Preallocate errors
   Ks=zeros(dx,dy,N);   % Preallocate Kalman gains
% Initial Conditions
   x0(:,1)=[0; 0]; % initial value for x0
% External input current, estimated as parameter p later on:
   z=[1:N]/250*2*pi; z=-.4-1.01*abs(sin(z/2));
% RuKu integrator of 4th order:
for n=1:N-1;
  xx=x0(:,n);
  for i=1:nn
    k1=dt*vossFNint(xx,z(n));
    k2=dt*vossFNint(xx+k1/2,z(n));
    k3=dt*vossFNint(xx+k2/2,z(n));
    k4=dt*vossFNint(xx+k3,z(n));
    xx=xx+k1/6+k2/3+k3/3+k4/6;
  end;
```

```
  x0(:,n+1)=xx;
end;
x=[z; x0]; % augmented state vector (notation a bit different to paper)
xhat(:,1)=x(:,1); % first guess of x_1 set to observation
% Covariances
  Q=.015; % process noise covariance matrix
  R=.2^2*var(vossFNobsfct(x))*eye(dy,dy);
        % observation noise covariance matrix
  randn('state',0);
  y=feval(obsfct,x)+sqrtm(R)*randn(dy,N); % noisy data
  Pxx(:,:,1)=blkdiag(Q,R,R);% Initial Condition for Pxx
% Main loop for recursive estimation
for k=2:N
  [xhat(:,k),Pxx(:,:,k),Ks(:,:,k)]=...
        vossut(xhat(:,k-1),Pxx(:,:,k-1),y(:,k),fct,obsfct,dq,dx,dy,R);
  Pxx(1,1,k)=Q;
  errors(:,k)=sqrt(diag(Pxx(:,:,k)));
end; % k
% Results
  chisq=...
    mean((x(1,:)-xhat(1,:)).^2+(x(2,:)-xhat(2,:)).^2+(x(3,:)-xhat(3,:)).^2)
  est=xhat(1:dq,N)'; % last estimate
  error=errors(1:dq,N)'; % last error
  meanest=mean(xhat(1:dq,:)')
  meanerror=mean(errors(1:dq,:)')
% Plot Results
subplot(2,1,1)
  plot(y,'bd','MarkerEdgeColor','blue', 'MarkerFaceColor','blue',...
        'MarkerSize',3);
  hold on;
  plot(x(dq+1,:),'k','LineWidth',2);
  xlabel('t');
  ylabel('x_1, y');
  hold off;
  axis tight
  title('(a)')
subplot(2,1,2)
  plot(x(dq+2,:),'k','LineWidth',2);
  hold on
  plot(xhat(dq+2,:),'r','LineWidth',2);
  plot(x(1,:),'k','LineWidth',2);
  for i=1:dq; plot(xhat(i,:),'m','LineWidth',2); end;
  for i=1:dq; plot(xhat(i,:)+errors(i,:),'m'); end;
  for i=1:dq; plot(xhat(i,:)-errors(i,:),'m'); end;
  xlabel('t');
  ylabel('z, estimated z, x_2, estimated x_2');
  hold off
```

```
   axis tight
   title('(b)')
```

The script that performs the unscented transform calculations is vossut.m:

```
% Unscented transformation from Voss et al 2004
% This Function has been modified by S. Schiff and T. Sauer 2008
function [xhat,Pxx,K]=vossut(xhat,Pxx,y,fct,obsfct,dq,dx,dy,R);
N=2*dx;        %Number of Sigma Points

Pxx=(Pxx+Pxx')/2; %Symmetrize Pxx - good numerical safety
xsigma=chol( dx*Pxx )'; % Cholesky decomposition - note that Pxx=chol'*chol

Xa=xhat*ones(1,N)+[xsigma, -xsigma]; %Generate Sigma Points
X=feval(fct,dq,Xa); %Calculate all of the X's at once

xtilde=sum(X')'/N; %Mean of X's
X1=X-xtilde*ones(1,size(X,2)); % subtract mean from X columns
Pxx=X1*X1'/N;
Pxx=(Pxx+Pxx')/2; %Pxx covariance calculation

Y=feval(obsfct,X);
ytilde=sum(Y')'/N;
Y1=Y-ytilde*ones(1,size(Y,2)); % subtract mean from Y columns
Pyy=Y1*Y1'/N + R; %Pyy covariance calculation

Pxy=X1*Y1'/N; %cross-covariance calculation

K=Pxy*inv(Pyy);
xhat=xtilde+K*(y-ytilde);
Pxx=Pxx-K*Pxy'; Pxx=(Pxx+Pxx')/2;
```

And three small callable functions round out this algorithm in vossFNint.m:

```
%This function calculates the Fitzhugh-Nagumo equations
function r=vossFNint(x,z)
a=.7; b=.8; c=3.;
r=[c*(x(2)+x(1)-x(1)^3/3+z); -(x(1)-a+b*x(2))/c];
```

   vossFNobsfct.m:

```
%This function strips out and returns just the observation variable, voltage
function r=vossFNobsfct(x)
r=x(2,:);
```

   and vossFNfct.m:

```
% RuKu integrator for FitzHugh system with parameters
% Modified from Voss et al 2004.
function r=vossFNfct(dq,x)
```

```
global dT dt nn
p=x(1:dq,:);
xnl=x(dq+1:size(x(:,1)),:);
for n=1:nn
  k1=dt*fc(xnl,p);
  k2=dt*fc(xnl+k1/2,p);
  k3=dt*fc(xnl+k2/2,p);
  k4=dt*fc(xnl+k3,p);
  xnl=xnl+k1/6+k2/3+k3/3+k4/6;
end
r=[x(1:dq,:); xnl];

function r=fc(x,p);
a=.7; b=.8; c=3.;
r=[c*(x(2,:)+x(1,:)-x(1,:).^3/3+p); -(x(1,:)-a+b*x(2,:))/c];
```

5.1. For the first exercise, let's modify this program so that it tracks all parameters, a, b, and c, in addition to the input current.

• *Hint*: The original integration to get x0 is unchanged. You need to increase the value of dq as you add parameters. I suggest augmenting the state vector and estimating the additional parameters one at a time. Start with a. Once this is working, add b and c. You need to modify the setting of the covariance Pxx prior to the recursive estimation. I might suggest starting with the same process noise Q used for the original tracking parameter. You then would change

```
Pxx(:,:,1)=blkdiag(Q,Q,R,R);
```

and after the recursive call to vossut, add

```
Pxx(1,1,k)=Q;
Pxx(2,2,k)=Q;
```

Don't forget to pass dq to the observation function, and also to the integrator (vossFNfct) within the vossut function where the sigma points are integrated in a vectorized fashion (a brilliant programming step by Voss). Within vossFNfct, strip out the parameters as a matrix p, pass them all to fc, and set up vectors within fc now. Don't forget to use .* and ./ for the final integration step in vossFNfct.

• Are the results resonable? Probably not! One of the unknown issues with such tracking is setting the relative process noises Q. We will find this again dramatically shown later in the book when dealing with Parkinson's disease models (chapter 10).

• Now try setting the relative Q's differently after each vossut call:

```
Pxx(1,1,k)=Q;
Pxx(2,2,k)=Q*0.1;
```

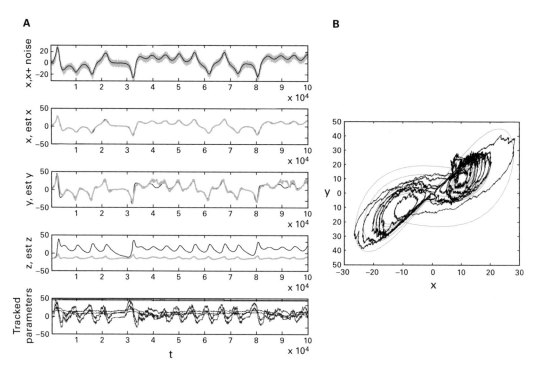

Figure 5.7 (plate 4)
Results of reconstructing the Lorenz equations using the unscented Kalman filter from measuring only noisy x variable and estimating variables and parameters. In this simulation, the integration time step is 0.0001, $\sigma = 10$, $r = 46$, and $b = 8/3$. A shows plots of reconstructed Lorenz variables. Upper panel is true (black) and noisy x variable (blue), second panel is true (black) and reconstructed x variable (red), third panel is true (black) and reconstructed y variable (red), fourth panel is true (black) and reconstructed z variable (red), and lower panel shows three reconstructed parameters with confidence limits. B shows true (black) and reconstructed (red) Lorenz x versus y trajectories. Note the difficulties with the z variable—similar troubles are seen in synchronizing Lorenz equations through this variable [PC90].

```
Pxx(3,3,k)=Q*0.1;
Pxx(4,4,k)=Q*0.1;
```

What happens to the tracking?

5.2. Now let's modify the algorithm from exercise 5.1 to track the Lorenz equations. There are three parameters and three variables to track.

• *Hint*: The Lorenz equations are very sensitive to initial conditions and thus round off error. So use a small integration time step—say, 0.001. Run the program for about 10,000 steps. Use the following to help get started:

```
Parameters [10; 46; 8/3]% sigma, r, and b
```

• Try using Q's of 0.015 like for the Fitzhugh-Nagumo equations. Examine at least the tracking of x and y, as well as the parameters. Can you track well?

• Now try using Q's proportional to the magnitude of the parameters, such as

```
Q1=10
Q2=46
Q3=8/3
```

Is this better?

• If you plot the true x variable against the estimated $\hat{x}$, you might get something that looks like figure 5.7 (plate 4), if you have done this well. Note in the figure that I have included the difficulties with the Lorenz z variable—not unexpected [PC90].

6 Spatiotemporal Cortical Dynamics—The Wilson Cowan Equations

It is hoped that the relative simplicity of the model may serve as a basis for a better understanding of the functional significance of cortical complexity.
—Hugh Wilson and Jack Cowan, 1973 [WC73]

6.1 Before Wilson and Cowan

So far, we have focused on models that represent the excitability of single cells (or components of single cells). Brain dynamics emerge from the interactions of single cells—in the case of vertebrates, the interactions between millions and billions of those cells. So at first glance, modeling the complexity of the patterns of activity from this complex network seems daunting. Nevertheless, it was not long after Hodgkin and Huxley's work with the squid axon that such larger-scale brain dynamics were seriously addressed:

Cells having some properties similar to those of neurons are considered. A mass of such cells, randomly placed together with a uniform volume density, appears capable of supporting various simple forms of activity, including plane waves, spherical and circular waves and vortex effects. The propagation of a plane wave of activity has been considered in some detail. (R. L. Beurle, 1956 [Beu56])

If it were not for the fact that the date on the above quotation were 1956, these sentences would fit well within any number of papers published recently on the theory and experimental findings of cortical activity.

Beurle was interested in describing the properties of a mass of cells in a brain. He understood clearly that through their interactions, the properties of the mass would be distinct from the activities of cells in isolation. Perhaps they would bear similarities to the activities that drive organismal behavior.

There is an implicit recognition of *emergence* throughout Beurle's writing. This rather attractive word, with a rather faddish emergence in scientific circles over the past decade, probably has no better description than Ablowitz's 1939 paper [Abl39]. Incredibly, one can find this concept, as *emergents*, used in the modern sense in Lewes's 1875 book, as properties that cannot be deduced by the sum of its parts:

No one supposes that the properties of each element combined in water could be deduced from the observed properties of the combination; no one supposes that [from] the observed properties of oxygen and hydrogen, separately considered, the properties of water could be deduced. (G. H. Lewes, 1875 [Lew75])[1]

An ensemble of linear dynamical systems, coupled linearly, will produce macroscopic dynamics readily accounted for by the (linear) summation of the individual dynamics. Nonlinear systems, on the other hand, can have emergent properties whereby the whole behaves differently than the sum of its parts would have suggested. Neurons, with thresholds, are paradigmatic examples of nonlinear networks where we expect emergent dynamics to emerge.

Beurle assumed that the behavior of the mass of cells (idealized in his formulation) would have a threshold.

He assumed that these model neurons would have very long refractory periods, so that once they fired, in some kind of pattern of activity, they did not fire again during that pattern. Figure 6.1A shows a traveling wave consuming sensitive cells and turning them refractory in its wake.

In figure 6.1B Beurle shows waves colliding and reflecting from this encounter to go back along the tracks of the incident waves. In figure 6.1C, he predicts a wave encountering a cortical boundary and refracting as its propagation speed at the boundary slows. Fifty years later, Xu and colleagues would find this phenomenon in the compression and reflection of waves of activity in visual cortex as they transit from visual cortical regions V1 to V2 [XHTW07]. Such waves compress, as would light crossing an analogous boundary from an area of low to one of higher refractive index. And along with the compression, reflected brain waves are observed [XHTW07]. Many of these wave properties in visual cortex were envisioned in Beurle's early insight.

6.2   Wilson and Cowan before 1973

Hugh Wilson was a postdoc with Jack Cowan in the early 1970s. Jack had an unusual legacy for a neuroscientist. His mentors included Claude Shannon and Dennis Gabor—giants in the mathematics and theory of communication (Shannon) and signal processing and holography (Gabor). His thinking about the brain was influenced by his association with Norbert Wiener, Warren McCulloch, and Walter Pitts, a pivotal group of scientists in the mid-twentieth century who initiated efforts at fusing engineering and mathematics with our

---

1. Lewes was the lover of Marian Evans, who wrote under pen name George Eliot in Victorian England. Upon her death, Evans established the George Henry Lewes Trust at Cambridge to support students studying physiology. Among the list of students so supported who became central figures in twentieth-century science included Nobel laureates H. H. Dale, C. S. Sherrington, and A. V. Hill. Importantly for our purposes here is that W. A. H. Rushton was supported by the Lewes Trust when he performed the seminal work upon which our chapter 11 is based.

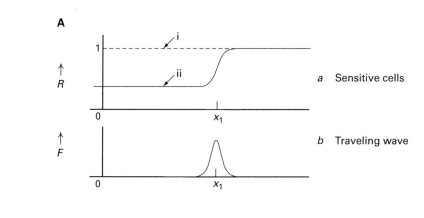

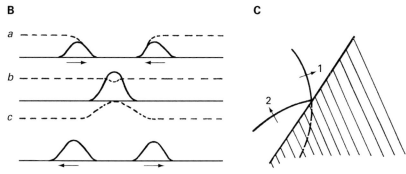

Figure 6.1
Traveling wave (A), reflected waves (B), and interaction with boundary (C). Modified from [Beu56] with permission.

understanding of the brain (exemplified in [MP43, Wie48, LMMP59]. The collaboration of Wilson and Cowan produced two seminal papers in computational neuroscience from their collaboration. In the first [WC72], they sought a deterministic model of cortex.[2]

Their quest was to develop a deterministic model of cortex that used mean values from the statistical processes of the underlying cellular spiking. There are no spikes in their analysis. They discuss the local redundancy in cortex. For the physicists, this is essentially a local mean field approach to looking at cortical dynamics, and if they were doing this a decade or two later, they might have talked about renormalizing the individual cells into larger, more equivalent cells in cortex. *It is just such local redundancy which must be invoked to justify characterizing spatially localized neural populations by a single variable* [WC72].

Wilson and Cowan made a major break from Beurle by assuming that there were excitatory and inhibitory cells that were ubiquitous in the brain. I will take that assumption

---

2. A detailed personal history and discussion of this work can be found in [Cow10].

a little further: Composites of interacting excitatory and inhibitory cells are ubiquitous in spinal cord, brainstem, cortex—in every component of the vertebrate central nervous system. *All nervous processes of any complexity are dependent on the interaction of excitatory and inhibitory cells* [WC72]. Wilson and Cowan realized that such excitatory-inhibitory interplay would be essential to establish dynamical stability of the brain. Indeed, J. C. Eccles would suggest that the entire purpose of inhibitory neurons might be to prevent seizures (personal communication, 1985). This principle of the balance of excitation and inhibition has taken on increasing importance following the experimental work on *up* states [SHM03, HDHM06]. And such excitatory-inhibitory interplay will play a crucial role later in our efforts to reconstruct epileptic network dynamics (chapter 12).

Dale's principle says that one neuron secretes only one type of transmitter—that "the chemical function, whether cholinergic or adrenergic, is characteristic for each particular neurone, and unchangeable" [Dal35]. This required Wilson and Cowan to separate independently their populations of cells, because some populations are only excitatory and some are only inhibitory. This is an *independence principle*, which is in part analogous to the independence of the ion channels that Hodgkin and Huxley applied to nerve membrane dynamics (chapter 3). In a qualitative sense, sodium is excitatory and potassium is inhibitory in a patch of the neuron membrane, but one cannot carry that analogy too far. Nevertheless, we are about to discuss nonlinear wave propagation and nonlinear pattern formation in cortex, and those processes do share dynamical properties with the mechanisms of excitability and propagation of single-action potentials.

Wilson and Cowan also assumed that every kind of interconnection that is possible between these two types of cells would be present. This lets their model be general enough to encompass all possible ways that excitatory and inhibitory populations can interact. By setting the connection strengths and parameters of the model, it could represent all possible topologies that cortex might contain.

They assumed that the excitatory-to-excitatory connections were tighter and more localized than the reach of inhibitory connections. The key variable they use is Beurle's—the proportion of cells that act per unit time, and that will be included in the following integral formulations they propose.

Let $E(t)$ be the proportion of excitatory cells active per unit of time. Similarly, $I(t)$ will be the proportion of inhibitory cells active per unit of time. $E(t) = 0$ and $I(t) = 0$ will be the rest state of the subpopulations when their background activity is low. Note that this gives us the prospect of *negative* values of E or I that are meaningful—the activity is depressed relative to baseline.

If the absolute refractory period is r, the proportion of excitatory cells E that are refractory could be obtained by integrating from $t - r$ to the present time t

$$\text{Proportion refractory} = \int_{t-r}^{t} E(t')dt'$$

If those are the proportion of cells that are refractory, then the ones that are sensitive, the ones Beurle had talked about, are 1 minus this fraction.

$$\text{Proportion sensitive} = 1 - \int_{t-r}^{t} E(t')dt' \tag{6.1}$$

Wilson and Cowan sought a *subpopulation response function*, S_e for excitatory (or S_i for inhibitory), that would represent the proportion of the subpopulation receiving at least threshold activation per unit time. This is the proportion of cells in a subpopulation that *would* respond to a given level of excitation if none of them were refractory at that time.

They next assumed that the population has a distribution D of *thresholds* θ, $D(\theta)$, which might be represented by a Gaussian distribution. It did not really matter what the distribution was, but it had to have a hump, and it had to be focused around a mean distribution of thresholds. If all these cells have the same average excitation $x(t)$, then the fraction of these cells S that were over the threshold at time t would be

$$S(x) = \int_{0}^{x(t)} D(\theta)d\theta \tag{6.2}$$

Such a formulation led to sigmoid activation curves. An alternative way to formulate this would be to assume that the thresholds of all cells were the same, but that they received a different number of synapses distributed according to a similar one-humped distribution function.

By the time of Wilson and Cowan's work, it was well known that a neuronal population might have a definite threshold for excitation, and that the input-output relation often appeared to be of sigmoidal shape (figure 6.2).

Wilfrid Rall's studies on the quantification of the recruitment within the motoneuron pool sought to place such sigmoidal activations on a sound quantitative footing.[3] Rall was interested in addressing how the spatial summation of synaptic excitation interacted with the distribution of thresholds within the motoneuron pool to produce the amplitude of the monosynaptic reflex output [Ral55a]. His statistical theory to account for these findings explained how simple Gaussian activation functions, shown in figure 6.3A, combined with thresholds for neuronal activation within the population, would generate sigmoidal population activation curves, shown in figure 6.3B.[4]

---

3. Rall would later spend the bulk of his career at the Mathematical Research Branch at NIH for several decades, performing seminal work in the physics of neurons. But his previous doctoral thesis was done in Dunedin with J. C. Eccles, and his subsequent work in Dunedin is reflected in the results shown here.

4. The nature of correlated and uncorrelated fluctuations seen in these input-output curves became of renewed interest when the surgical treatment of children with spasticity involved measurement of reflex input-output strengths in selective dorsal rhizotomy [WS93]. A more detailed analysis of fluctuations in single cell and population activity can be found in [GFK+94].

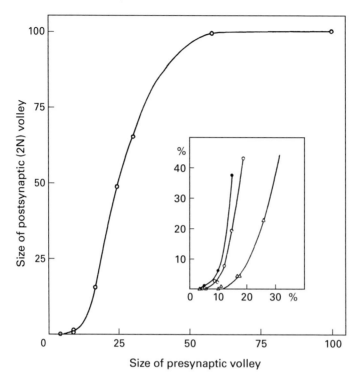

Figure 6.2
Input-output curve representing monosynaptic reflex amplitude in response to the size of the afferent volley. The inset shows that in multiple experiments there was a clear threshold of afferent input, between 5 and 10%, that formed a threshold for the population response. Reproduced from Lloyd [Llo43] with permission.

Single populations give sigmoidal activation curves (figure 6.4A). Multiple populations destroy the simple sigmoid shape, but reflect the multiple inflection points generated by the different population contributions (figure 6.4B).

If one assumes that the level of excitation of a cell decays as $\alpha(t)$, then the average level of exciation generated in an E cell at time t is

$$x(t) = \int_{-\infty}^{t} \alpha(t - t')[w_1 E(t') - w_2 I(t') + P(t')]dt' \tag{6.3}$$

where w_i is the average number of synapses on each cell, and $P(t)$ is the external input to the E population ($Q(t)$ will be the input to the I population). The activity in a subpopulation at time $t + \tau$ will then be the product of the cells at time t that are both sensitive and above threshold. Given the proportion of cells not refractory in (6.1), the value of $S(x)$ in (6.2), and the formula for $x(t)$ in (6.3), Wilson and Cowan wrote

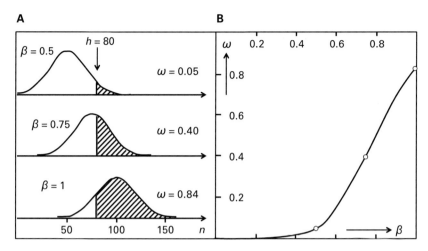

Figure 6.3
Demonstration of how the fractional activation, β, of Gaussian activation curves of afferent neurons, and threshold, h, shown in A, would integrate the cumulative activation to produce sigmoidal output functions shown in B. Modified from [Ral55b] with permission.

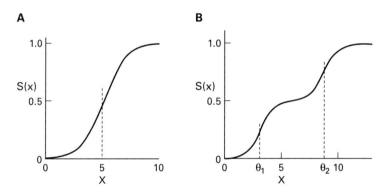

Figure 6.4
Population activation curves from unimodal, (A) and multimodal population input (B). Multiple modes can be produced by multiple distributions of thresholds, or afferent synapses. Reproduced from [WC72] with permission.

$$E(t+\tau) = \left[1 - \int_{t-r}^{t} E(t')dt'\right] \cdot S_e \left\{\int_{-\infty}^{t} \alpha(t-t')[w_1 E(t') - w_2 I(t') + P(t')]dt'\right\}$$

$$(6.4)$$

$$I(t+\tau) = \left[1 - \int_{t-r}^{t} I(t')dt'\right] \cdot S_i \left\{\int_{-\infty}^{t} \alpha(t-t')[w_3 E(t') - w_4 I(t') + Q(t')]dt'\right\}$$

Temporally coarse graining these functions E and I so that they are replaced by a moving time average for a small interval will remove rapid temporal fluctuations that are not essential for the overall population dynamics that they sought

$$\int_{t-r}^{t} E(t')dt' \rightarrow \bar{E}(t)t'\big|_{t-r}^{t} = r\bar{E}(t)$$

$$\int_{-\infty}^{t} \alpha(t-t')E(t')dt' \rightarrow \bar{E}(t)t'\big|_{t-k}^{t} = k\bar{E}(t)$$

where the r and k reflect the times that the refractory period and synaptic activation lingered, respectively. Taylor expanding about $\tau = 0$ for $E(t+\tau)$ and $I(t+\tau)$

$$E(t+\tau) = E(t) + \frac{dE}{dt}(t+\tau-t) + \ldots = E(t) + \tau\frac{dE}{dt}$$

$$I(t+\tau) = I(t) + \frac{dI}{dt}(t+\tau-t) + \ldots = I(t) + \tau\frac{dI}{dt}$$

leads to

$$\tau\frac{d\bar{E}}{dt} = -\bar{E}(t) + (1 - r\bar{E})S_e \left\{k[w_1\bar{E}(t) - w_2\bar{I}(t) + P(t)]\right\}$$

$$(6.5)$$

$$\tau'\frac{d\bar{I}}{dt} = -\bar{I}(t) + (1 - r\bar{I})S_e \left\{k'[w_3\bar{E}(t) - w_4\bar{I}(t) + Q(t)]\right\}$$

Has all of this slogging through this original derivation been worthwhile to the reader? I know of no other way to, for instance, have any idea why in the Wilson-Cowan equations as they are now used (see, e.g., equations (6.7)) there appear equivalent terms to $-\bar{E}(t)$ and $-\bar{I}(t)$ as shown in (6.5). Some things cannot be hand-waived away.

Is the coarse graining used in (6.5) reasonable? Wilson and Cowan compared the results in figure 6.5. They argued that the fast time scale oscillations from (6.4) are not important.

Finally, they proposed that the rest state was stable to small perturbations. So they shifted their sigmoid activation functions so that $E = 0$ and $I = 0$ when $S_e(0) = 0$ and $S_i(0)$, respectively. We now know that the so-called rest state of cortex is highly active [Len03, DRFS01]. Nevertheless, we also suspect that such states still have some structural

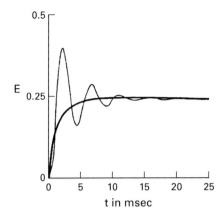

Figure 6.5
Comparison of (6.4) (light line) with (6.5) (heavy line). Reproduced from [WC72] with permission.

stability to perturbations (a more detailed consideration of such perturbations can be found in [UCJBS09]).

These equations could have nullclines plotted on a phase plane, illustrating the case that the excitatory (P) and inhibitory (Q) inputs are zero. Two stable and one unstable equilibrium are shown in figure 6.6.

The effect of positive P or Q is to shift these nullclines and alter the stability of these fixed points. This generates changes in the qualitative nature of the cortical dynamics modeled. When Wilson and Cowan varied the excitatory population excitation P up and down, they noted hysteresis about two stable states in their solutions, as shown in figure 6.7A. A considerable amount of relative excitation P was required to trigger the high-activity state—a *population threshold* that was an emergent property of these equations.

The results, in figure 6.7A, were not dependent on inhibition being present. The more complex multiple hysteresis loops shown in panel B depend on the presence of inhibition. Hysteresis is one way to envision modeling short-term memory in such models. Wilson and Cowan were aware of hysteresis effects seen in visual cortex experiments when they modeled this result.

The next phenomenon that they observed in this model was that the duration of inputs affected the population response, as shown in figure 6.8A. In fact, such a strength-duration curve is very similar to that of accommodating single neurons, including an effective rheobase for the population.

Other important phenomena that the model demonstrated of physiological significance are damped oscillations in response to brief stimulation, similar to those of evoked potentials recorded from cortex (figure 6.8B).

Wilson and Cowan found a range of excitatory input P for which the nullclines would produce an unstable fixed point, about which would emerge a stable limit cycle, as shown in figure 6.9.

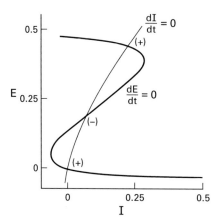

Figure 6.6
Phase plane representation of Wilson-Cowan equations showing nullclines with two stable ($+$) and one unstable ($-$) fixed points for the case that the excitatory (P) and inhibitory (Q) inputs are zero. Reproduced from [WC72] with permission.

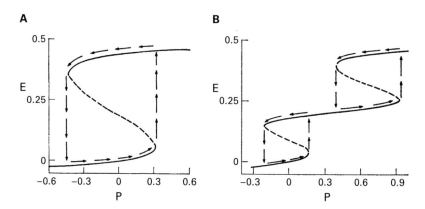

Figure 6.7
Hysteresis observed in excitatory population active, E, in relation to excitatory activation, P, shown in A. Multiple hysteresis loops in B require inhibition. Solid lines are stable states, whereas dashed lines are unstable states. Reproduced from [WC72] with permission.

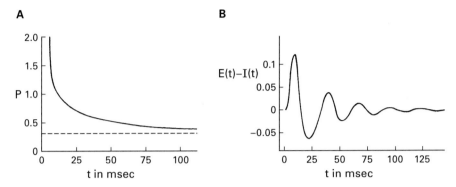

Figure 6.8
Excitatory current strength, P, versus duration required to activate cortex (A), so as to excite it to a high level of activity as in figure 6.7. B shows the ability of the model to demonstrate damped oscillations. Reproduced from [WC72] with permission.

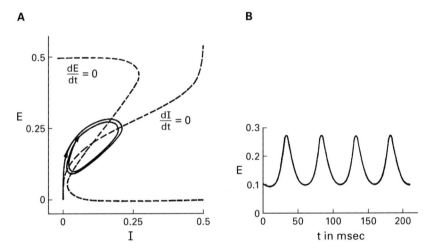

Figure 6.9
Emergence of stable limit cycle about unstable fixed point in A, and time series of E shown in B. Reproduced from [WC72] with permission.

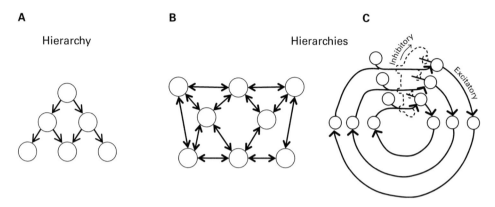

Figure 6.10
Contrast between hierarchy (A) and two different types of heterarchy, the fishnet network (B) and McCulloch's circular heterarchy, shown in C, modified from [McC45] with permission.

The implication of limit cycles for cortex was that their amplitude and frequency could encode stimulus intensity. Furthermore, limit cycle stability implied noise insensitivity to such encoding. All of their dynamical findings, Wilson and Cowan stress, are consequences of the nature of the nullclines in their phase space representation. The precise form of the model and value of parameters are of secondary importance.

6.3   Wilson and Cowan during 1973

In the previous section, we studied a general model of cortical subpopulation interactions that had no spatial extent. The analogy is somewhat like the Hodgkin-Huxley equations of the space-clamped membrane patch (chapter 3). The Hodgkin-Huxley equations needed to be extended into the one-dimensional propagating action potential. For the dimensionless Wilson-Cowan equations, we need to extend them to the two-dimensional cortex.

An important concept that Wilson and Cowan introduced in their 1973 paper [WC73] was that cortex was more likely organized as a *heterarchy*, rather than a *hierarchy*. The concept of heterarchy is generally attributed to Warren McCulloch in his 1945 paper, and the difference between hierarchies and heterarchies is schematized in figure 6.10.[5]

In their 1973 paper, Wilson and Cowan developed a mathematical model that represented cortex as a complex, interconnected heterarchy of quasi-two-dimensional sheets. Neurons were assumed to be uniformly distributed within the sheet. They assumed that lateral connectivity was a function of distance only. Furthermore, they assumed that the sheet was

---

5. McCulloch's papers are famously difficult to read. An interesting philosophical discussion can be found in the Web-published paper by von Goldammer and colleagues [VGPN03].

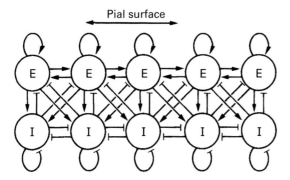

Figure 6.11
Schematic of general two-dimensional model of cortex. The aggregates of E and I cells are shown with all possible interconnectivities. Reproduced from [WC73] with permission.

isotropic (there was no direction preference). Again as in 1972, they assumed that there were two types of neurons—excitatory and inhibitory—and that there were equal numbers of each.[6] Importantly, they presumed that reliability from unreliable components would be achieved by redundancy.

The schematic of their 1973 model is shown in figure 6.11. In the figure, the excitatory, E, and inhibitory, I, nodes are not representative of single cells, but rather spatially localized aggregates of such cells. All possible interconnectivities are represented, but it is assumed that the probability of connection falls off exponentially with distance. Some rather severe assumptions enter into this model. They assume that connectivity is similar between all neurons. They assume that the conduction velocities were effectively infinite—in other words much faster than the dynamics of cortical patterns that emerge from this model.

The evolution of the equations from the 1972 paper,

$$\tau \frac{d\bar{E}}{dt} = -\bar{E}(t) + (1 - r\bar{E})S_e \left\{ k[w_1 \bar{E}(t) - w_2 \bar{I}(t) + P(t)] \right\}$$

$$\tau' \frac{d\bar{I}}{dt} = -\bar{I}(t) + (1 - r\bar{I})S_e \left\{ k'[w_3 \bar{E}(t) - w_4 \bar{I}(t) + Q(t)] \right\}$$

to the 1973 paper,

$$\tau \frac{\partial \bar{E}(x, t)}{\partial t} = -\bar{E}(x, t) + (1 - r_e \bar{E}(x, t))S_e \left\{ k[w_{ee} \otimes \bar{E}(t) - w_{ie} \otimes \bar{I}(t) + P(t)] \right\}$$

$$\tau' \frac{\partial \bar{I}(x, t)}{\partial t} = -\bar{I}(x, t) + (1 - r_i \bar{I}(x, t))S_i \left\{ k'[w_{ei} \otimes \bar{E}(t) - w_{ii} \otimes \bar{I}(t) + Q(t)] \right\}$$

$$(6.6)$$

6. Equal numbers of excitatory and inhibitory neurons is not a very accurate assumption, as we know today, but probably not a major problem if the parameters of the model are properly adjusted.

are shown here, where $\otimes$ indicates convolution. In regions of densely packed neurons, such as within cortical columns, the convolutions can be replaced by products of appropriate averages.

Wilson and Cowan make three important assumptions on their parameters: (1) the rest state is stable to small perturbations, (2) there is no uniformly excited state in the absence of a maintained stimulus (otherwise it would be in a seizure-like state), and 3) the inhibitory connections are longer range than the excitatory connections.

Assumption 3 is critical to form inhomogenous patterns in excitable media. The underlying physics were laid down by Alan Turing in 1952 for the chemical basis of morphogenesis. Ordinarily diffusion is stabilizing—it smooths things out. But for multiple chemical species, this requires the speed of diffusion of the species to be similar. In chemical reaction-diffusion systems with fast- and slow-diffusing reactants, an instability can be created whereby inhomogenous patterns may form. Neural networks formally share some of the same physics as these chemical systems, in that the fast-diffusing reactant may be likened to the further connectivity reach of the inhibitory neurons.[7]

These Wilson-Cowan equations produced a wide variety of dynamical modes, which included active transients, spatially localized limit cycles, and spatially inhomogenous stable steady states.[8] In the spirit of Hodgkin and Huxley, Wilson and Cowan extensively show that their model can account for a wide variety of neuronal phenomena that had been previously experimentally observed. In contrast to Hodgkin and Huxley, they started from general principles, and then sought what phenomena they could simulate. Hodgkin and Huxley started from experimentally derived equations.

Both temporal summation (figure 6.12A) and spatial summation (6.12B) could be seen in the threshold to fire self-regenerative active transients. Such bumps of activity are analogous to single-action potential spikes in Hodgkin-Huxley or Fitzhugh-Nagumo equations.

Edge enhancement could be observed. This phenomena was increased as the strength of stimuli were increased, as shown in figure 6.13.

Spatially localized limit cycles, in response to constant but spatially localized stimuli, are shown in figure 6.14. The frequency of the limit cycles encoded stimulus strength.

They noted frequency demultiplication (recall Van der Pol 1926 [VdP26], chapter 4), which had by then been observed in driving the thalamus, and illustrated in figure 6.15.

For sufficiently *large* values of inhibitory activation, Q, they observed traveling waves, as shown in figure 6.16. Such traveling waves will be essential in our modeling of cortical waves later in this chapter. Intriguingly, Wilson and Cowan predicted propagation velocities

---

7. It is hard to improve on the reaction-diffusion analogy of grasshoppers and forest fires in James Murray's fine book on *Mathematical Biology* [Mur03].

8. A superb and authoritative description of the dynamics of these equations, can be found in Hugh Wilson's book *Spikes, Decisions, and Actions*, accompanied by computer code carrying out many of the difficult computations using the original formulation of the equations [Wil99]. A concise and deep mathematical perspective on these equations can be found in Ermentrout and Terman's recent book [ET10].

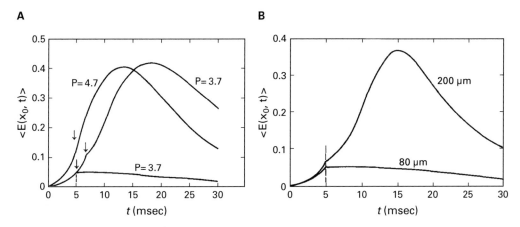

Figure 6.12
A shows temporal summation. At the same stimulus intensity, the longer stimulation time leads to an active transient. In B, the larger spatial stimulus application with the same current intensity per node leads to an active transient. Modified from [WC73] with permission.

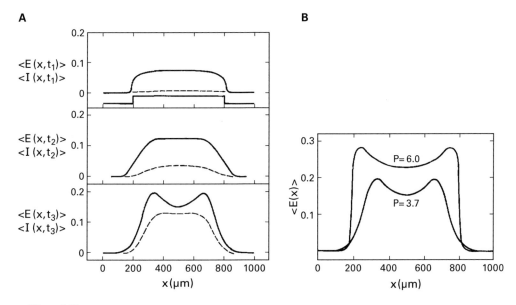

Figure 6.13
Example of edge enhancement shown in the temporal profiles of excitation (solid) and inhibitory (dashed) lines in A, and increasing spatial disparity in the maximal edge enhancement is brought out by increasing stimulus strength in B. Modified from [WC73] with permission.

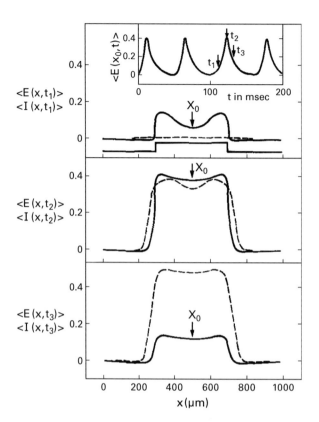

Figure 6.14
Spatially localized limit cycles. Reproduced from [WC73] with permission.

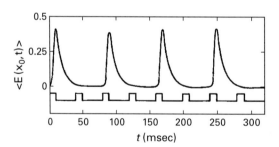

Figure 6.15
Example of frequency demultiplication. Reproduced from [WC73] with permission.

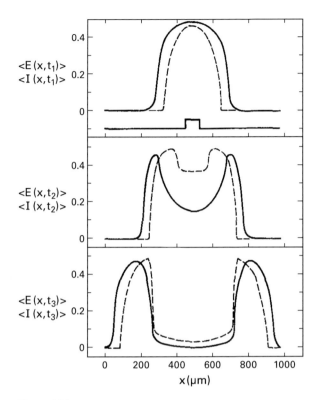

Figure 6.16
Traveling waves in Wilson-Cowan equations. Reproduced from [WC73] with permission.

of such traveling waves in the neighborhood of 4 cm/sec, exactly what we observe now with cortical traveling waves [RSG05].

Finally, spatially inhomogeneous localized activity could be observed as shown in figure 6.17. Such stationary bumps are what would be, if the individual neuron activities were observed, reverberant firing activity. Wilson and Cowan likened this to the neural correlate of working memory observed by Fuster and Alexander [FA71].

6.4   Wilson and Cowan after 1973

In 2001, David Pinto and G. Bard Ermentrout modified the Wilson-Cowan equations [PE01a, PE01b]. First, in the region of sparse activity, that is when $r_{e/i}$ is small, $(1 - r_e \bar{E}(x, t)) \to 1$ in equations (6.6), and similarly for the $\bar{I}$ equation.[9] For potentially disinhibited cortex,

---

9. This is a reasonable assumption from what we are learning about the actual sparseness of neuronal activity [Len03].

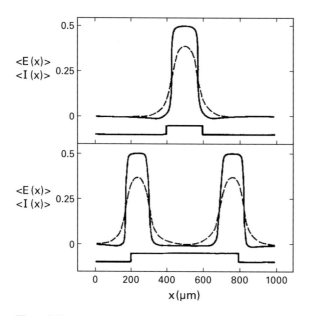

Figure 6.17
Spatially inhomogeneous localized activity. Reproduced from [WC73] with permission.

as they were seeking to model, they added a slow recovery variable to just the excitation dynamics as (switching notation a bit)

$$\dot{u}(x,t) = -u(x,t) + \int_{-\infty}^{\infty} w(x-x')P(u(x',t)-\theta)dx' - a(x,t)$$

$$(6.7)$$

$$\frac{1}{\varepsilon}\dot{a}(x,t) = -\beta a(x,t) + u(x,t)$$

Here, u is activity (equivalent to Wilson and Cowan's $\bar{E}$), a is the slow recovery variable, θ is threshold, P is a synaptic firing rate relative to the threshold, and $\epsilon < 1$ slowing down $\dot{a}$ relative to $\dot{u}$. Note the $-u(x,t)$ term is the same as the $-\bar{E}$ term in equation (6.5). Variable a is equivalent to the action of both potassium, as well as inactivation dynamics and spike frequency adaptation. But keep in mind that we are in the Wilson-Cowan world—there are no individual neurons, only neuronal aggregates and local averages of dynamics. This formulation will be particularly useful if one were to perform experiments where traveling waves would be incited in cortex under conditions where inhibition was pharmacologically blocked [PE01a, HTY+04]. For the same reason that it was useful to Pinto and Ermentrout, we will shortly apply this version of the following Wilson-Cowan equations to cortical wave experiments.

In a second paper [PE01b], Pinto and Ermentrout extended this model to include inhibition. The most general way of doing this would be to have two sets of two-dimensional lattices, formulated as

$$\dot{u}(x,t) = -u(x,t) + \int_{-\infty}^{\infty} w_{ee}(x-x')P(u(x',t)-\theta)dx' - a(x,t)\cdots$$

$$- \int_{-\infty}^{\infty} w_{ie}(x-x')P(v(x',t)-\theta)dx'$$

$$\frac{1}{\varepsilon}\dot{a}(x,t) = -\beta a(x,t) + u(x,t)$$

$$(6.8)$$

$$\dot{v}(x,t) = -v(x,t) + \int_{-\infty}^{\infty} w_{ei}(x-x')P(u(x',t)-\theta)dx' - b(x,t)\cdots$$

$$- \int_{-\infty}^{\infty} w_{ii}(x-x')P(v(x',t)-\theta)dx'$$

$$\frac{1}{\delta}\dot{b}(x,t) = -\lambda b(x,t) + v(x,t)$$

where u represents excitatory and v inhibitory neurons, and w_{ee}, w_{ie}, w_{ei}, w_{ii}, the connectivity footprints from $e \to e$, $i \to e$, $e \to i$, and $i \to i$, respectively. I have written two recovery variables, a and b, with potentially separate recovery time constants $1/\varepsilon$ and $1/\delta$.

If you are modeling full brain dynamics, such as EEG from epilepsy patients, then you need to start from this more general formulation for excitatory and inhibitory interactions in equations (6.8). Not all of the terms may be necessary ([PE01b] often eliminated w_{ii}, for instance). And the symmetry considerations, to be discussed later, are also critical to take into account. If, however, you are going to start with cortex creating oscillatory waves when inhibition is blocked, then the formulation in equations (6.7) will serve us well for the remainder of this chapter.

6.5 Spirals, Rings, and Chaotic Waves in Brain

My colleague William Troy has long been interested in oscilatory and wave phenomena. When he began to be interested in models of neural cortex [LTGE02, LT03], the complex bumps and spiral waves were intriguing. After the work of Beurle mentioned earlier, there were demonstrations in homogenous integrate and fire networks that target waves evolving toward spiral waves could be observed, often following periodic driving of the network (see [MC93] and note references within to previous work in inhomogenous excitable networks). But it was not at all apparent that these phenomena were seen in the brain to the degree that they were seen in mathematics. It is not that it was hard to produce waves of activity in neural tissue, but the optical techniques capable of replacing traditional electrode measurements to image complex waves were relatively recent [BW03]. The problem was that most brain slice preparations were done coronally or axially [TC98], and because of the layered strucure, waves were not produced in a homogenous two-dimensional media as in Troy's mathematics. From a math meeting we placed a phone call to an experimentalist that we knew was doing experiments in a different type of brain slice and interested in wave physiology—Jian-young Wu.

Wu was using special tangential slices that were not commonly studied physiologically. Several years before, Michael Gutnick and colleagues [FBG98] had used such tangential slices in their studies of layer 4 in barrel cortex physiology (figure 6.18A), and had noted that in the presence of pharmacological disinhibition with bicuculline, all-or-none traveling waves could be initiated (figure 6.18B). In distinction to [FBG98], Wu removed the superficial layers of cortex, and then cut a tangential slice from visual cortex (figure 6.18C) that contained layers 3, 4, and 5 (of the six-layer cortex). These layers contained some of the largest neurons, and wavelike phenomena were known to occur preferentially in these layers and likely were instigated in layer 5 in in vitro studies [TC98]. We asked him if he could try to find spiral waves. It was not many weeks before he, and his postdoc Xiaoying Huang, had found many.

Spiral waves are generic in physical, chemical, and mathematical studies of two-dimensional excitable media. Such structures emerge in the presence of isotropy, where there are no preferred directions in the two-dimensional plane. It is the anisotropy that makes coronal or axial slice cuts problematic for correlating brain activity waves with the abundance of physical and mathematical theory available. The rodent visual cortex has no ocular dominance columns [GSL99], and is a particularly attractive cortex if a homogenous and isotropic one is sought.[10]

Rotational waves had been observed before in turtle brain experiments [PCP+97]. But they were difficult to see in the intact cortex, and the signal processing required to extract them from optical imaging data were cumbersome.

These new waves in mammalian tangential cortical slices were robust [HTY+04]. They were observed using voltage-sensitive dye. Such measurement technique gives a local mean field average of transmembrane potential through a full-thickness, small cylindrical core (about 100×100 μm) of the piece of brain under each photodiode (figure 6.18D). The full photodiode array imaged 128 or, later, 472 pixels. The data were exactly what the Wilson-Cowan equations, or the Pinto-Ermentout variant, was designed to model.

---

10. Whether the barrel cortex gives the same wave properties as visual cortex, or might impose more inhomogeneity to wave propagation much like bumper pool from the barrels, remains an open question.

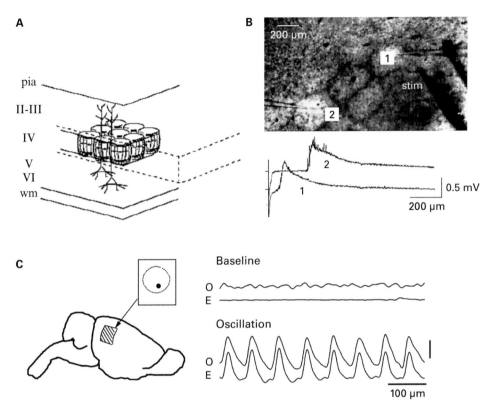

Figure 6.18
A shows layer-specific tangential slice from barrel cortex, and B shows all-or-none traveling waves. A and B reproduced from [FBG98] with permission. C shows the tangential slices from visual cortex, with oscillations that are similar whether electrically (E) or optically (O) recorded. Reproduced from [HTY⁺04] with permission.

Brain slices have most of their connections to the rest of the brain cut, and one always seems to need to add something to activate them in order to restore a level of network activity. Wu and colleagues used carbachol, which activates acetylcholine receptors, and bicuculline, which blocks inhibitory gamma-aminobutyric acid type-a (GABAa) channels that are fast inhibitory synaptic receptors. Exposed to these compounds, robust oscillations with organized wave activity could be readily seen in figure 6.19 (plate 5). These oscillatory episodes could be seen spontaneously, but they were typically triggered with small stimuli for aligning them with the optical recordings and to minimize photobleaching effects. Just like action potentials, they had a refractory period afterwards, within which new episodes could not be triggered with stimuli. A variety of events are seen in vertebrate nervous systems such as up states [SHM03], spinal cord burst firing [CMO06], animal [ZCBS06] and human [SSKW05] epileptic seizures, and these cortical wavelike events [HTY⁺04]

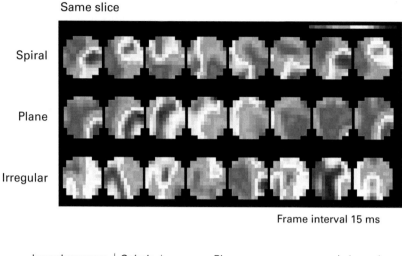

Figure 6.19 (plate 5)
From a single oscillatory episode, a progression is seen of chaotic irregular, to spiral, to plane, and back to chaotic waves. Reproduced from [HTY$^+$04] with permission.

share the properties of being episodic, oscillatory, and having apparent refractory periods, following which small stimuli can both start and stop such events.

Interestingly, we saw no clear anatomic interference with the waves that formed. Although we understood that some degree of irregularity or symmetry breaking was likely to nucleate a wave, the spiral centers wandered about the two-dimensional field when we observed them. Sometimes a spiral would rotate clockwise, and the next time one formed, it would rotate counterclockwise. There appeared to be a true singularity of phase—one pixel could be found that was surrounded by the greatest spatial phase gradient. From a mathematical point of view, our observations were consistent with a degree of translation invariance that would permit such waves to be analyzed with many of the tools used to study other excitable physical systems.

We collaborated with Carlo Laing and William Troy to attempt to simulate such waves using the Wilson-Cowan equations.[11] They employed a partial differential equation solution to solve these equations, and produced qualitatively similar ring, plane, and spiral wave

---

11. I suppose it would be appropriate were I to designate the form of these equations as the Pinto-Ermentrout variant of the Wilson-Cowan equations. But for the rest of this text, I will lump and designate all variants and derivatives of the Wilson-Cowan equations as Wilson-Cowan.

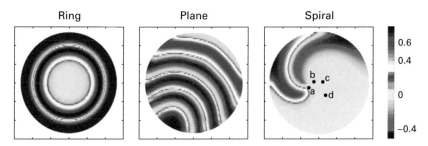

Figure 6.20 (plate 6)
Simulation of ring, plane, and spiral waves with the Wilson-Cowan equations. Reproduced from [HTY+04] with permission.

patterns as were observed experimentally (figure 6.20, plate 6). Simulation is not proof of mechanism, but these calculations show that the rather simple generic principles of cortical organization within the Wilson-Cowan equations are capable of emulating a wide range of the dynamics we observed in these experiments. What we did not simulate at this time were the complexities of the irregular and chaotic spatiotemporal patterns that often initiated and resolved such oscillatory episodes. One can, indeed, see much more irregular patterns within the Wilson-Cowan system, but it requires some careful setting of the initial conditions prior to integration. A full exploration of the basin of attraction of these equations has not, to the author's knowledge, been done. But the other feature prominent in the experiments that is left out of the simulation is that these wave episodes *evolve* with time. The Wilson-Cowan equations are deterministic, and with a given set of initial conditions, will (after transients resolve) yield one dynamical pattern. Real slices, and real brains, never settle down to just one stable pattern. We will cover this subject later in considerably more detail (chapter 7).

Intriguingly, the difficulties in earlier experiments in observing waves of activity in brains [PCP+97] could be overcome in a variety of ways. One is to employ optical imaging such as just discussed, using slices of brain to focus on relevant regions in the tissue. In recent years, waves of activity in intact brains in the visual cortex have been observed optically [XHTW07]. An alternative is to employ arrays of sharp electrodes in intact cortex, using the electrodes to penetrate down to the layers exposed here in slices. A recent study of waves in the motor cortex employed such penetrating electrodes to considerable success [RRH06], as did a recent study of hippocampal traveling waves during navigation [LS09].

In terms of modeling and tracking, it is important to perhaps reiterate that our goal in this book is *not* to reengineer the brain. We don't particularly care if the Wilson-Cowan equations actually reflect the detailed dynamical mechanism going on within our observed cortex. We want to find efficient models that have the dynamical repertoire of the brain. We want them to be only as complex as needed to perform such emulation. This latter need is so that the parameters can be most accurately determined, so that the tracking is most accurate. Employing an accurate but hugely complex model, even though it might be a true reflection of the brain dynamics, is not going to serve us well if the parameter and

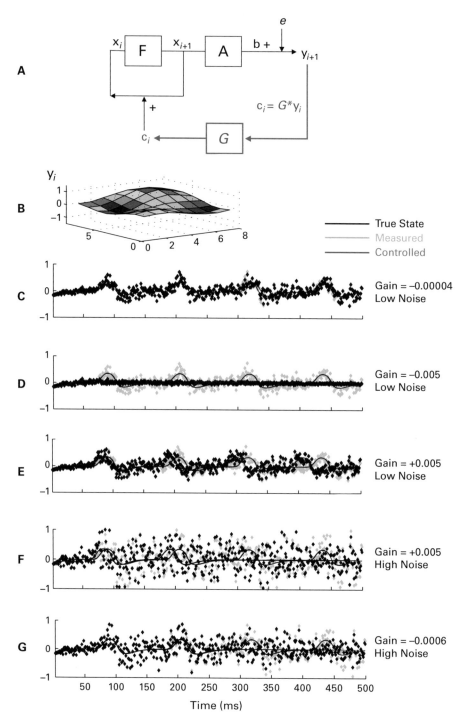

A

B

y_i

True State
Measured
Controlled

C Gain = −0.00004
Low Noise

D Gain = −0.005
Low Noise

E Gain = +0.005
Low Noise

F Gain = +0.005
High Noise

G Gain = −0.0006
High Noise

Time (ms)

Figure 6.21 (plate 7)
(Caption on facing page)

variable fits are so poor that the model cannot track the observed dynamics well. Last, we need prediction—short-term prediction of the brain dynamics. We assume that a model that is fundamental and based on our growing knowledge of neuroscience will, at some point, perform better than an empirical statistical model. And given all of the above, the Wilson-Cowan model is perfect for our next efforts at using them to track and control brain dynamics.

6.6   Wilson-Cowan in a Control Framework

We now take the Wilson-Cowan formalism and wrap it within a control formalism.

We begin with an 8×8 grid of discrete Wilson-Cowan equations as in [SS08]

$$\dot{u}(x, y, t) = -Cu(x, y, t) - a(x, y, t) + K \sum_{\substack{i,j \\ i \neq j}} w(x, y, p_i, q_j) H(u(p_i, q_j, t) - \theta)$$

(6.9)

$$\tau \dot{a}(x, y, t) = Bu(x, y, t) - a(x, y, t)$$

where

$$w(x, y, p, q) = e^{-k\left((x-q)^2 + (y-p)^2\right)}$$

In reading this notation, (x, y) represents the location that the function is evaluated at, and (p, q) are dummy variables that represent all other positions in the two-dimensional plane that are examined for whether the activity, u, is at least as large as threshold θ. The footprint function $w(x, y, p, q)$ falls off as a symmetrically decreasing Gaussian function with distance. H is the Heaviside function, equal to 1 when the activity at $u_{i,j} \geq \theta$, and equal to 0 when $u_{i,j} < \theta$, where θ is the threshold. τ is the differential time constant between the two equations—the rate of change of a is slower than the rate of change of u. C, K, B, and k are adjustable parameters.

In figure 6.21A (plate 7), we show the schematic for the Wilson-Cowan equations represented as the dynamical process F. We observe the dynamics through the process A, which

Figure 6.21 (plate 7)
(A) Schematic of dynamical system F, which acts on state variable vector x, and observation function A, which produces output vector b, to which measurement noise e is added, generating output observable y. When direct proportional control is used, control function G is employed, which generates the control vector c. This control vector is directly "injected" into the neuronal elements, adding vector c to the activity variable u (nothing is added to the unobserved variable a or to the parameters of the system). (B) An instantaneous snapshot of the 8×8 output y during a rotational spiral. (C) The output of the observable y under conditions of small negative gain and low noise, from a single element of the grid in B, shown as a time series without (blue) and with (red) control applied. The underlying state vector activity element u_i is plotted as a thin black line. In contrast with C, D shows outputs under conditions of higher negative gain and low noise, E shows positive gain and low noise, and F and G show high noise conditions with positive and negative gains, respectively. Low noise corresponds to 1 to 5, and high noise corresponds to 10 to 14 on an arbitrary scale. Note that under high noise conditions, the application of direct proportional control as illustrated in the schematic renders the system dynamics unstable and destroys the rotational spiral regardless of positive or negative feedback. Reproduced from [SS08] with permission.

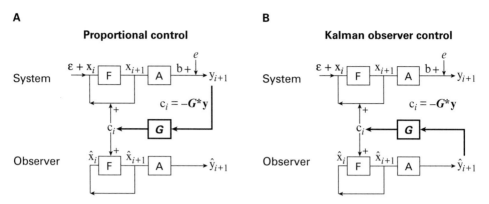

Figure 6.22
Schematic of the use of a Kalman filter observer in control. A illustrates the use of an observer to estimate state variables and parameters while control is derived from the actual noisy output y. The control vector c is applied to both the system and the observer. B shows the use of the observer system to calculate proportional control through an estimation of the observable $\hat{y}$. Again, the control vector is fed back into both the experimental system and the Kalman observer.

includes adding random noise e. We first assume that there is no model that observes this system, so we take measurement y and multiply it by G to generate control signal c, which is added to the dynamics F. This is the typical scenario where a noisy measurement is used in a *proportional control* feedback scheme. We examine the output of figure 6.21A and, for suitable initial conditions, we observe spiral waves, a snapshot of which is shown in figure 6.21B. For relatively low noise, and negative feedback ($G < 0$), we can observe slight slowing of the oscillations recorded from one node in figure 6.21C. If the strength of the negative feedback is increased, the oscillations are stopped completely, as shown in figure 6.21D. There is actually little safety margin between slowing the system and crushing the waves. In figure 6.21E, we use positive feedback ($G > 0$) to speed up the waves in low noise. It is always easier to speed waves up than to slow them down, whether in such models or in real experiments (e.g., [RSG05]). In high noise, the situation looks bleak. Whether one applies positive (figure 6.21F) or negative feedback (figure 6.21G), the system loses stability in the presence of feedback and the oscillations are destroyed.

Now let's introduce an *observer* system, as shown in figure 6.22. There are two general control schemes. One can keep the proportional control scheme as shown earlier, but observe the dynamics using an observer system. The control signal c is pumped into both the original system and the observer system. *Kalman filters really are filters.* They can extract from corrupted signals a signal reflective of the dynamics that the Kalman filter seeks to emulate. So a better way to reconstruct the control signal might be to calculate the control signal from the observer system, taking advantage of its filtering properties, and pump this signal into both the real and the observer system, as shown in figure 6.22B.

We should be very clear about several things. If a simple system has low noise, then one has a better chance to effectively use the measured output in proportional control feedback scenarios to modulate the system (figure 6.21C and 6.21E). If all you want to do is to stop the activity of an excitable system, you can deliver a large current and stop the activity, as shown in figure 6.21D. In the brain, seizures can be stopped, and fibrillation in the heart, by delivering a large enough current. *You don't need a Kalman filter for mindless wave crushing.* But if you are in a regime where the amount of noise (or uncertainty in your model) is sufficiently large that adding feedback control destabilizes the system, you can improve the situation with model-based observation. In particular, the best scenario is when control is calculated from a Kalman observer when the observer system is built from a reasonable model. In figure 6.23 (plate 8), the left side shows a region where there is sufficient noise so that proportional control with positive gain calculated from the noisy measurement destabilizes the system (left column), whereas control calculated from the Kalman observer is capable of speeding up the system without destabilization.

An analogous situation is shown in figure 6.24 (plate 9). Here again, use of a proportional control signal with negative gain calculated from the noisy measurements destabilizes the system (left), whereas use of the Kalman observer calculated control leads to controlled slowing of the system (on the right).

In both of these numerical examples, we calculate the amount of control energy expended to achieve a given level of control. As one might imagine, when the control law $c = Gy$ is buffeted constantly by the noisy output y, this leads to wild excursions in the modulated activity, and the energy pumped into the system is large. When the control law is calculated from the Kalman observer, an amount of control energy more pertinent to the control law is expended, and the energy used is considerably reduced (we will show that it can be exponentially reduced in chapter 12). Although the control energy is not as critical when dealing with numerical simulations, it is crucial when dealing with brains [FSBG+07, FGR+07].

Critical in demonstrating the results in figures 6.23 and 6.24 was the use of *covariance inflation* to tune the measurement uncertainty used in the covariance matrix of the estimated state. Recall that Julier and Uhlmann [JU97a, JU97b] employed an arbitrary parameter on the spread of their sigma points (chapter 2). Here we follow a similar strategy as proposed by Hunt and colleagues [HKS07]. Without employing such adjustable covariance inflation, our results would have been unstable.

The philosophy here is that, in a perfect world, you would create a narrow uncertainly ball of sigma points that would enclose within the true state, and track this true state accurately as the model iterated. But the danger is that if you lose track of the system, you have no idea how to recover the trajectory. The alternative is to expand the uncertainty ball. The extreme of this would be that in a navigation scenario, were the uncertainty ball as large as the planet, you would never lose track of your position, but your uncertainly over where you are would encompass the entire planet! In the results here, we arbitrarily tuned a covariation parameter (to be discussed in this chapter's exercises). Improving the way we

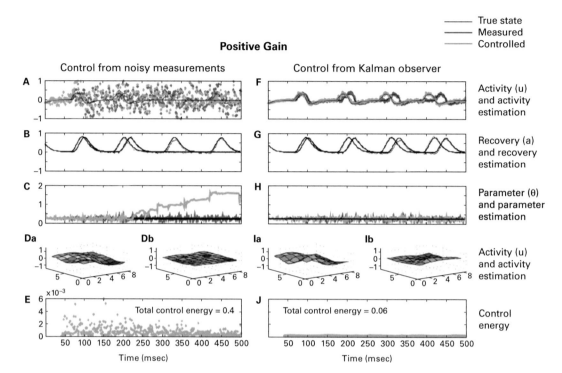

Figure 6.23 (plate 8)
Comparison of the use of positive gain proportional control calculated directly from high noise observations, left column A–E, and from the Kalman observer, right column F–J. (A) Output variable y_i from single element of grid, is the same as in figure 6.21F. The application of positive feedback gain at this high level of noise results in instability of the rotational oscillation, with loss of both the underlying oscillation u_i (black line) and measured observation y_i (red). (B) Without control being applied, the Kalman observer can accurately estimate (blue) the unobserved variable a (black). With this level of noise, when control is applied, not only is the observable a oscillation lost (red), but the true system state loses stability and the wave is destroyed after two oscillations (black). (C) Without control, despite the high noise level, the parameter θ (black) is accurately tracked (blue). When the stability of the dynamics is lost during control, the parameter θ is no longer accurately estimated (red). Da shows the variable y, and Db the estimated underlying activity $\hat{u}$. The snapshots in Da and Db are taken from the last frame in the 500 msec simulation, and the full simulation can be viewed as Supplementary Movie 1 from [SS08]. In E is shown the energy (sum of squared voltage) of the control vector c_i, and the total sum of all control vector energy for the 500 msec is 0.4 (arbitrary scale). F shows the estimated observable $\hat{y}$ without (blue) and with (red) control derived from the Kalman observer. (G) With Kalman observer control, the unobserved recovery variable a is now well tracked and the rotational spiral is stably increased in frequency, and in H, the parameter θ is now well tracked. Ia and Ib show the experimental variable y and estimated underlying variable $\hat{u}$. In J is shown that the control energy required is markedly reduced, with total control energy 0.06 (arbitrary scale).

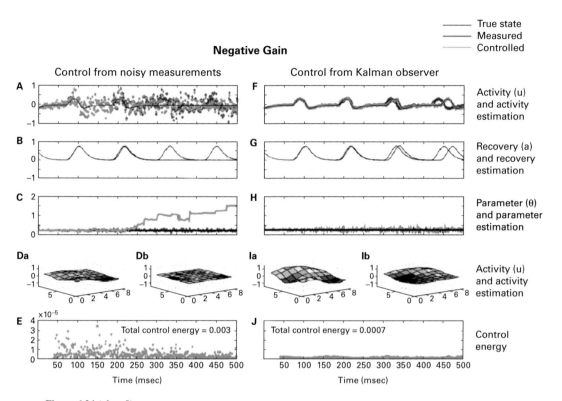

Figure 6.24 (plate 9)
Comparison of the use of negative gain proportional control calculated directly from high noise observations. In left column A–E, and from the Kalman observer, in right column F–J. (A) Output variable y_i from single element of grid, same as in figure 6.21G. The variables in this figure are similar to the description for figure 6.23. Note that with control derived from the Kalman observer (right column), negative gain in high noise conditions stably slows the frequency of the spiral, and requires considerably less control energy than the unstable case where direct proportional control is calculated from the noisy observations (left column).

handle covariation inflation adaptively is at the cutting edge of such algorithm design. We will return to this issue in chapter 8 on model inadequacy.

Note, also, that our Wilson-Cowan model had homogenous parameters—a rather ridiculous assumption if the underlying system did not (it did in this case). We will address this issue more realistically in chapter 8.

Our use of a fundamental model, the Wilson-Cowan equations, trusted that we knew a lot about the underlying system. The alternative is to use an empirical model based only on the data itself. This is a classic tussle between arrogance and ignorance. I am not sure what is the best way to handle such system tracking at present, and we will cover empirical modeling of these same dynamics in chapter 7. If you were designing a control system, hedging your bets is probably the safest route. Since each model, fundamental or empirical, can have prediction errors compared at each time step, one can run both in parallel if you have the computational overhead. If you don't this year, you might next year. And such parallel algorithms are natural for parallel computational implementation. Nevertheless, as we continue to learn more about the nervous system, our fundamental models will eventually overtake our empirical ones in control accuracy.

The preceding calculations implicitly employed direct injection of current to the elements of the discretized Wilson-Cowan equations. Subtleties of true electric field polarization of extended neurons, with the counter-fields generated within each neuron, and the ephaptic electrical crosstalk present between real neurons ([PBG+05]) will complicate the picture during experiments with real neuronal networks. Such complications may dictate that a more detailed model, incorporating the polarization characteristics of neurons as well as the electrically resistive properties of brain tissue within which the neurons are embedded, will be required to provide more accurate performance.

Exercises

6.1. Let's take the Wilson-Cowan equations as shown in (6.9) and work out some ways of calculating them.

- First let's set the parameters as

$k = 0.05$; $B = 1000$; $K = 10$; $C = 15$; tau $= 0.01$; Threshold $= 0.1$

with an integration time of delt $= 0.001$.
Use a 12×12 grid, and perhaps 200 iterations.

- Now let's do the full brute force version of these equations. You can set the initial conditions for u and a as

u0 = rand(n, p);

a0 = zeros(n, p);

where $n = p = 12$.

- Here is a code fragment that will get you started:

```
for step=1:numsteps
  for i=1:n
    for j=1:p
      integ=0;
      for a=1:n
        for b=1:p
          if (a==1|a==n)|(b==1|b==p);
          u0(a,b)=0; end %Neumann boundary conditions
```

```
            if u0(a,b)>Threshold
              r=(i-a)^2+(j-b)^2;
              if Gaussian == FALSE%exponential
                integ=integ+K*exp(-k*sqrt(r));
              else%is Gaussian
                integ=integ+K*exp(-k*r);
              end
            end
        end
      end
    a1(i,j)=a0(i,j)+delt*(B*u0(i,j)-a0(i,j))/tau;
    u1(i,j)=u0(i,j)-delt*C*u0(i,j)-delt*a1(i,j)+delt*integ;
    end
    u0=u1;a0=a1;
    [YOUR FAVORITE PLOT ROUTINE HERE]
end
```

- Try replacing the initial conditions with a small bump, such as

```
u0=zeros(n,p);% this gives initial condition bump
  for i = 2
    for j = 3
      u0(i,j)=1;
    end
  end
```

- The above algorithm was accurate to the equations but a bit miserable in terms of efficiency. Here, courtesy of my colleague and collaborator Tim Sauer, is a much more compact and efficient version:

```
q=2; %or anything up to 12
for step=1:numsteps
  ue=[zeros(p,q) u0 zeros(p,q)]; %pad the grid with zeros
  ue=[zeros(q,n+2*q);ue;zeros(q,n+2*q)];
  integ=zeros(n,p);
  for i=-q:q %check firing (i,j) steps away on grid
    for j=-q:q
      integ=integ+K*exp(-k*(i*i+j*j))* ...
      (ue(i+1+q:i+n+q,j+1+q:j+p+q)>Threshold);
    end
  end
  integ=integ-K*(u0>Threshold); %subtract self-input i = j = 0
  a1=a0+delt*(B*u0-a0)/tau;
  u1=u0-delt*C*u0-delt*a0+delt*integ;
  u0=u1;a0=a1;
  [YOUR FAVORITE PLOT ROUTINE HERE]
end%numsteps
```

• Did you see any spirals? Probably not. Try these initial conditions and set the grid size to n = 8; p = 8;

```
u0=[-0.1637  -0.2244  -0.1982  -0.1410  -0.1029  -0.0811  -0.0656  -0.0396
    -0.2593  -0.3580  -0.3021  -0.2008  -0.1321  -0.1028  -0.0840  -0.0516
    -0.2444  -0.3386  -0.2746  -0.1726  -0.0950  -0.0742  -0.0637  -0.0417
    -0.0055  -0.0383   0.0012  -0.0657  -0.0395  -0.0388  -0.0387  -0.0270
     0.2361   0.3685   0.2616   0.2237   0.0279  -0.0174  -0.0239  -0.0183
     0.4698   0.7002   0.5627   0.2726   0.0939  -0.0071  -0.0171  -0.0135
     0.3618   0.5613   0.4323   0.2754   0.0673  -0.0056  -0.0130  -0.0104
     0.2442   0.2669   0.2571   0.1067   0.0152  -0.0049  -0.0077  -0.0061
    ];
```

```
a0=[ 0.4104   0.5339   0.4345   0.2786   0.1904   0.1454   0.1157   0.0694
     0.7058   0.9183   0.6974   0.4049   0.2427   0.1821   0.1465   0.0893
     0.9263   1.1846   0.8515   0.4104   0.1757   0.1283   0.1085   0.0707
     1.0750   1.4204   1.0146   0.4695   0.0921   0.0646   0.0636   0.0444
     1.0367   1.4612   0.9507   0.4993   0.0532   0.0309   0.0380   0.0293
     0.7878   1.1079   0.7914   0.3707   0.0468   0.0205   0.0267   0.0213
     0.4156   0.5573   0.4218   0.1886   0.0255   0.0150   0.0202   0.0162
     0.1386   0.1823   0.1406   0.0437   0.0083   0.0090   0.0119   0.0095
    ];
```

Setting different initial conditions, or sequences of initial conditions, produces a wide range of dynamics from these equations.

6.2. Now let's explore how to modify the Voss algorithm, presented in the exercises in chapter 5, so that we can apply it spatiotemporally.

• We add covariance inflation to the ut subroutine.

```
ip=.0001;% ep is the inflation parameter
Pxx=(Pxx+Pxx')/2+ip*eye(size(Pxx));
```

The rest of the ut subroutine is remarkably unchanged.

• The observation function is easy:

```
function r=kalmanwc_obs(x,dq)
global K C B tau f g dt k nn
r=x(dq+1:f*g+dq,:);%just the u's
```

• The integrator function is a bit more complex.

```
function r=kalmanwc_int(x,z)
global K C B tau f g dt k nn

u = reshape(x(1:f*g),f,g);
a = reshape(x(f*g+1:2*f*g),f,g);
```

```
q=2;
p=[K,C,B,tau,z];
ue=[zeros(g,q) u zeros(g,q)];
ue=[zeros(q,f+2*q);ue;zeros(q,f+2*q)];
integ=zeros(f,g);
for i=-q:q                          %check firing (i,j) steps away on grid
 for j=-q:q                         %add to input
   integ=integ+p(1)*exp(-k*(i*i+j*j))*(ue(i+1+q:i+f+q,j+1+q:j+g+q)>p(5));
 end
end
integ=integ-p(1)*(u>p(5));      %subtract self-input i = j = 0
udot=(-p(2)*u-a+integ);         %ODE right hand side
adot=(p(3)*u-a)/p(4);           %for Wilson-Cowan
r=[udot(:);adot(:)];%a column vector with both variables
```

- The vectorized integrator called from the ut subroutine is now

```
function r=kalmanwc_fct(dq,x) %x here is [z ; u ; a]
%original z will be returned!
global K C B tau f g dt k nn

dT=nn*dt;

p=x(1:dq,:); %strips out p from first row
xn=x(dq+1:size(x(:,1)),:);
for n=1:nn  %Runge-Kutta order 4
  k1=dt*fc(xn,p);
  k2=dt*fc(xn+k1/2,p);
  k3=dt*fc(xn+k2/2,p);
  k4=dt*fc(xn+k3,p);
  xn=xn+k1/6+k2/3+k3/3+k4/6;
end
r=[p; xn]; %returns original p

function w=fc(x,p1);
global K C B tau f g dt k
[Rows,Columns]=size(x);
w=zeros(Rows,Columns);
for NumberOfColumns = 1:Columns
  u = reshape(x(1:f*g,NumberOfColumns),f,g);
  a = reshape(x(f*g+1:2*f*g,NumberOfColumns),f,g);
  q=2;p=p1(:,NumberOfColumns);
  p=[1.38;3;10;4.85;p];
  ue=[zeros(g,q) u zeros(g,q)];
  ue=[zeros(q,f+2*q);ue;zeros(q,f+2*q)];
  integ=zeros(f,g);
  for i=-q:q                          %check firing (i,j) steps away on grid
    for j=-q:q                        %add to input
```

```
        integ=integ+p(1)*exp(-k*(i*i+j*j))*(ue(i+1+q:i+f+q,j+1+q:j+g+q)>p(5));
    end
end
integ=integ-p(1)*(u>p(5));       %subtract self-input i = j = 0
udot=(-p(2)*u-a+integ);          %ODE right hand side
adot=(p(3)*u-a)/p(4);            %for Wilson-Cowan
w(:,NumberOfColumns)=[udot(:);adot(:)];%column vector with both variables
end
```

- Finally, here is a simplified version of the main script to control the Wilson-Cowan equations using an observer system. The full version of the program is archived with [SS08]. For initial conditions, use the matrices at the end of exercise 6.1.

```
%Kalman filter applied to Wilson-Cowan dynamics
%Steven Schiff and Tim Sauer, 2007-2009
%Simplified from Original Archived with Journal of Neural Engineering Paper
%"Kalman Filter Control of Spatiotemporal Cortical Dynamics"
% We have followed structure from Voss et al 2004
% Octave and Matlab compatible
clear all;close all
TRUE = 1; FALSE = 0;
global K C B tau f g dt k nn
%***SELECT the Gain, NoiseFactor, to reproduce calculations in paper
   NoiseFactor = 10
   EnergyPlotYAxis=0.00004
%%%%%%%%%%%%%%%%%%%%%%%%%%%%%%%%%%%%%%%%%%%%%%
%n.b.: increasing negative gain will quench the waves
%For instance, try Gain=-0.05 to see this clearly.
%%%%%%%%%%%%%%%%%%%%%%%%%%%%%%%%%%%%%%%%%%%%%%
% p=[K,C,B,tau,Threshold]
   K=1.38;C=3;B=10;tau=4.85;k=0.91;%5 parameters plus 2 variables each cell
%Dimension of Grid
   f=8;g=f;
% Dimensions: (dq for param. vector, dx augmented state, dy observation)
   dq=1; dx=dq+2*f*g; dy=f*g;
fct='kalmanwc_fct'; % this is the model function F(x) used in filtering
obsfct='kalmanwc_obs';  % this is the observation function G(x)
N=500% number of data samples
dt=0.06;nn=1;dT=nn*dt;  %Step size dt
MaxPlotEnergy = eps;
ControlTime = 40
EnergyThreshold = 4
%%%%%%%%%%%%%%%%%%%%%%%%%%%%%%%%%%%%%%%%%%%%%%
for WithControl = 1:2
   if WithControl == 1
      Gain = 0%get baseline %use with noise 10 - shows advantage of Kalman Observer
```

```
  else
    Gain = -0.0006 %can set this + or -
  end%otherwise use Gain as set
% Initial conditions for estimation
  MSEsum = 0; %initialize
  uEnergy_y_sum = 0; uEnergy_yhat_sum = 0;
  u = load('-ascii', 'u_initial_8x8_spiral.txt');
  a = load('-ascii', 'a_initial_8x8_spiral.txt');
  clear x0
  x0(:,1)=[u(:); a(:)]; %1 big column vector of all variables for initial time
% External input, estimated as parameter p(5) later on:
  z=ones(1,N)*0.24; %z=[1:N]/50*2*pi; %z=[0.24+0.00*sin(z)];
  x(:,1)=[z(1,1); x0]; % augmented state vector (notation different from paper)
%%%%%%%%%%%%%%%%%%%%%%%%%%%%%%%%%%%%%%%%%%
t=1; %initialize
  xhat(:,1)=x(:,1); % first guess of x_1 set to observation
  xhat(1,1)=0.55;  % set first guess of first parameter arbitrarily
  yhat= xhat(dq+1:dq+f*g,1);
  Q=.0001 % process noise covariance matrix
  errors=zeros(dx,N);
  Energy=zeros(1,N);
  tempx=x(:,1)*ones(1,N);
  R=.2^2*cov(kalmanwc_obs(tempx(:,1:N),dq)'); % observation noise cov matrix
  R=R+eps*eye(size(R));
  Pxx(:,:)=blkdiag(Q,R,R); % need one copy of Q for each tracked parameter
  randn('state',3);
  y=feval(obsfct,x(:,1),dq)+sqrtm(R)*randn(dy,t); % noisy data
  randn('state',0); %Not ysing this - set state above
  Energy_y(t)   = sum( y(:, t) .* y(:, t) ); % y
  Energy_yhat(t) = sum( yhat(:, t) .* yhat(:, t) ); % y
%%%%%%%%%%%%%%%%%%%%%%%%%%%%%%%%%%%%%%%%%%
  for t = 1:N-1
    if t>1
      xx=x0(:,t-1); %Pick column (at start, only 1 column of initial conditions)
      if t>=ControlTime
        uVector_yhat = Gain * yhat(:,t-1);
        uEnergy_yhat(t) = uVector_yhat' * uVector_yhat;
        ControlVector_yhat = [uVector_yhat ; zeros(f*g,1)];
        xx = xx + ControlVector_yhat;
        xhat(:,t-1) = xhat(:,t-1) + [0 ; ControlVector_yhat];
      else
        uEnergy_yhat(t) = 0; %no control yet
      end
      for i=1:nn %nn determines how many times to loop the RK - can use 1
        k1=dt*kalmanwc_int(xx,z(:,t-1));%feed a col of x and a value of z
        k2=dt*kalmanwc_int(xx+k1/2,z(:,t-1));
```

```
      k3=dt*kalmanwc_int(xx+k2/2,z(:,t-1));
      k4=dt*kalmanwc_int(xx+k3,z(:,t-1));
      xx=xx+k1/6+k2/3+k3/3+k4/6;
   end;
   x0(:,t)=xx;
   x=[z(1:t); x0(:,1:t)]; % augmented state vector
   R=.2^2*cov(kalmanwc_obs(x(:,1:t),dq)'); % observation noise cov matrix
   R=R+eps*eye(size(R));
   y=feval(obsfct,x,dq)+NoiseFactor*sqrtm(R)*randn(dy,t); % noisy data
   [xhat(:,t),Pxx]= ...
               kalmanwc_ut(xhat(:,t-1),Pxx,y(:,t),fct,obsfct,dq,dx,dy,R);
   Pxx(1,1)=max(Pxx(1,1),Q); % keep the parameter cov from fading to 0
   yhat(:,t)= xhat(dq+1:dq+f*g,t);
   Energy_y(t) = sum( y(:, t) .* y(:, t) ); %the rms at time t
   Energy_yhat(t) = sum( yhat(:, t) .* yhat(:, t) ); % y
   errors(:,t)=sqrt(diag(Pxx(:,:)));
 end
%%%%%%%%%%%%%%%%%%%%%%%%%%%%%%%%%%%%%%%%%%%%%%%
   subplot(5,4,1:4)
   hold on;
   plot(x(dq+1,1:t)','k','LineWidth',.5);
   if Gain == 0
      plot(t,y(1,t)','bd','MarkerEdgeColor','blue', ...
         'MarkerFaceColor','blue','MarkerSize',5);
   else
      plot(t,yhat(1,t)','rd','MarkerEdgeColor','red', ...
         'MarkerFaceColor','red','MarkerSize',5);
   end
   xlabel('t','fontsize',14);
   axis([1 N -1 1])
   title('Control from est observations yhat','fontsize',14)
   ylabel('x state, y obs','fontsize',14);
   drawnow
 subplot(5,4,5:8)
   plot(x(dy+2,1:t)','k','LineWidth',.5);%unobserved variable
   hold on
   plot(xhat(dy+2,1:t)','r','LineWidth',.1);%est unobserved variable
   axis tight
   axis([1 N -1 1])
   ylabel('a, est a','fontsize',14)
 subplot(5,4,9:12)
   hold on
   plot(xhat(1,1:t),'r','LineWidth',2);
   plot(x(1,1:t),'k','LineWidth',2); %parameter
   axis([1 N 0 2.0])
   ylabel(['thr',', ',' est ','thr'],'fontsize',14);
```

```
    subplot(5,4,13:14)
     u = reshape(y(:,t),f,g);%Display noisy observations
     surf(u);title('u','fontsize',12);axis([0 f 0 g -1.5 1.5])
     drawnow
    subplot(5,4,15:16)
     uhat = reshape(xhat(dq+1:f*g+dq,t),f,g);
     surf(uhat);title('uhat','fontsize',12);axis([0 f 0 g -1.5 1.5])
     drawnow
    subplot(5,4,17:20)
     if t>ControlTime
       if uEnergy_yhat(t)>MaxPlotEnergy; ...
           MaxPlotEnergy = uEnergy_yhat(t); end %store max graph value
       uEnergy_yhat_sum = uEnergy_yhat_sum + uEnergy_yhat(t);
       if Gain == 0
         plot(t,uEnergy_yhat(t),'bd','MarkerEdgeColor','blue', ...
           'MarkerFaceColor','blue','MarkerSize',5);
       else
         plot(t,uEnergy_yhat(t),'rd','MarkerEdgeColor','red', ...
           'MarkerFaceColor','red','MarkerSize',5);
       end
       hold on
       xlabel(['total control engy =...
                          ',num2str(uEnergy_yhat_sum)],'fontsize',14)
       ylabel('u Energy','fontsize',14);
       axis([1 N 0 MaxPlotEnergy])
     end
  end%t
end%WithControl
```

6.3. So can you substitute Fitzhugh-Nagumo dynamics for Wilson-Cowan in the preceding example? It is really a replacement of the dynamics at each node in the network—all the rest stays the same. The parameters are different.

• *Hint:* Try setting global variables to

global A B C f g dt K k nn

and use parameter settings as

K = 1.38; A = 0.7; B = 0.8; C = 3; k = 0.91;

Later, we will explore using the one model to track a different (wrong) model. It is not that unreasonable a suggestion. In real life, all of these models are *wrong* in that they do not represent nature correctly, but that is a subject for a later chapter.

7 Empirical Models

Empirical. *In matters of art or practice: That is guided by mere experience, without scientific knowledge.*
—Oxford English Dictionary, 2002 [Dic02]

7.1 Overview

Much of the book so far has focused on fundamental models of nervous system dynamics. But the control framework we have discussed is applicable regardless of the derivation of the model.

Using fundamental models assumes that you know what you're doing. It assumes that you know so much about the system that you really can write down from first principles models of neurons, models of connections for those neurons and the functions of their synapses, and models of the ions around them, with a better end result than you would get from guessing and, in a sense, better than what you could do from analyzing the data itself without the model. Only in the past few decades can we feel justified in the hubris that the previous sentence might be true.

In this chapter, on the other hand, we discuss models that assume your prior knowledge is really inadequate for observing and controlling neuronal systems. I focus on a class of decompositional techniques that often flexibly described in the literature as principal component analysis, singular value decomposition, Karhunen-Loève decomposition, and principal orthogonal decomposition. The underlying fundamentals of these techniques are similar, and I focus on those fundamentals here.

We consider a matrix, A, acting on a vector, x, as Ax. The matrix A is a set of scalar coefficients that will transform the vector by changing the coordinates that the vector components represent. These changed coordinates take the vector from one vector space, the domain, to another vector space, the range. The transformation can remain in the original space, but often (the focus of this chapter) the transformation is to a different and smaller space.

Singular value decomposition is a matrix decompositional technique. One starts with any matrix A.[1] A square matrix is symmetric if

$$A^T = A$$

where T indicates transpose. If i and j are rows and columns, then $A^T_{ij} = A_{ji}$. For any *symmetric* $m \times m$ matrix A with real entries, there is an orthonormal eigenbasis, a set of real-valued eigenvectors[2] $w_1, \ldots, w_m$. The orthonormality says that $w^T_i w_j = 0$ if $i \neq j$, and otherwise $w^T_i w_i = 1$. For any A, not necessarily symmetric, $A^T A$ is symmetric.

In previous chapters, we dealt with covariance matrices in Kalman filtering. If we start with a matrix of data values A, where each row represents a different sample in time, and each column represents a different sensor or electrode, then forming $A^T A$, or AA^T, will turn this matrix into a square symmetric covariance matrix.[3] Recalling that $(AB)^T = B^T A^T$, then to show that $A^T A$ is symmetric (even if A is not symmetric) we can write

$$(A^T A)^T = A^T (A^T)^T = A^T A$$

Furthermore, the eigenvalues of $A^T A$ must be nonnegative. In fact, if v is a unit eigenvector, and $Av = \lambda v$, then

$$0 \leq |Av|^2 = v^T A^T Av = \lambda v^T v = \lambda$$

There are then two ways to make square symmetric correlation matrices from typically rectangular $n \times m$ data matrices A: $A^T A$ and AA^T. Each has an orthonormal set of eigenvectors. One set of eigenvectors comes from the rows of A, and one comes from the columns. These rows and columns span different spaces, in general, so we will distinguish the set of m unit eigenvectors of $A^T A$ as $v_1, \ldots, v_m$, and the n unit eigenvectors of AA^T as $u_1, \ldots, u_n$. We are going to use the left and right multiplied matrices throughout this chapter to create these different covariance matrices.

Singular value decomposition computes these two orthonormal bases from a matrix A. In essence, it tells us how to transform from one set of these coordinates to the other. If the columns represent points in space (voltages from electrodes), and the rows points in time, then we will transform from one to the other. If the rows and columns were Cartesian x and y coordinates, it makes no difference to the mathematics.

---

1. There is a marvelous discussion of singular value decomposition geometry in the small chapter entitled "Matrix Times Circle Equals Ellipse" in [ASY97]. We will follow their logic here, but the original is well worth the reader's time.

2. An eigenvector v of A is a vector such that $Av = \lambda v$; that is, eigenvectors are not rotated when operated on by a matrix, but rather are stretched or contracted by the scalar eigenvalue λ.

3. Recall in the chapter 5 exercises that we enforced symmetry numerically in a covariance matrix P by writing in Matlab or Octave Pxx = (Pxx + Pxx')/2.

If $v_1, \ldots, v_m$ are the eigenvectors of $A^T A$, we define

$$s_i u_i \equiv A v_i$$

where u_i are unit vectors and s_i are scalars. Multiplying on the left by AA^T,

$$s_i AA^T u_i = AA^T A v_i = \lambda_i A v_i = \lambda_i s_i u_i$$

and that the u_i are eigenvectors of AA^T.

In [ASY97] the geometry of Ax is described as *matrix times circle equals ellipse*. The eigenbasis, the vectors that span the circle x, are the v's. The orthonormal eigenbasis that spans the ellipse and is oriented along the semimajor and semiminor axes is the singular value basis. The magnitude of the axes is defined by the eigenvalue s's—the singular values. Figure 7.1 illustrates the geometry of $Av_i = s_i u_i$. In the figure, each $s_i u_i$ represents a scalar eigenvalue s_i times a column u_i. If we combine all of the columns, we have a diagonal matrix of the scalar values times a matrix of the orthonormal column vectors, as $AV = US$. Right multiplying by V^T, we get the form of the singular value decomposition that we will use

$$A = USV^T \tag{7.1}$$

where

$$U^T U = I$$

$$V^T V = I$$

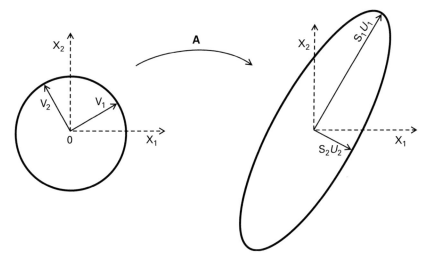

Figure 7.1
The geometry of singular value decomposition $Av_i = s_i u_i$. Reproduced from [ASY97] with permission.

It is interesting to note that this form of the singular value decomposition is not new. By 1874, Camille Jordan had described this well [Ste93]. By 1907, Erhard Schmidt had described the approximation formulas that we will employ later in this chapter. As with Cholesky's decomposition, described in chapter 2, the mathematics that make the algorithms in this book possible often preceded by decades, or the better part of centuries, our technical ability to incorporate them into digital computers.

One always encounters confusion with comparable techniques in the literature. In *principal components analysis*, the mean of data is often subtracted off, the v_i's are called *components*, and the $s_i u_i$ the *scores*. In practice, principal components are typically calculated from square symmetric matrices, often calculated by forming $A^T A$ or $A A^T$ from rectangular data matrices A. Of course, a strong method of calculating principal components is to use singular value decomposition. Similarly, physicists have a predilection to use Karhunen-Loève decomposition in dealing with discrete data such as we are describing here, although historically this nomenclature began with a focus on infinite dimensional stochastic systems. Psychologists often use *factor analysis*,[4] which is a variation of the principal components (and can be equivalent under certain conditions). Factor analysis arose from the analysis of intelligence, seeking to uncover the fundamental factors responsible for the multifactorial dimensions of intellect. Both mathematicians and physicists have a love of the term *principal orthogonal decomposition*, often used in a deep literature on dynamical systems and coherent structures [HLB96]; it will of tremendous value later in this chapter.

One thing that we need to get out of the way early on is the somewhat touchy subject of the mean—to remove it or not. If you have a cloud of data points that looks like an ellipse, then it might or might not be at the origin, as shown in figure 7.2. If you leave the mean in, the first singular vector or principal component will be along the axis from the origin to the mean. If you are a financial analyst, the market average might be a critical component to track. If you are a neuroscientist, the small changes in activity, riding on top of the ongoing baseline level of cortical activity in an optical imaging experiment, might be the relevant dynamics that are of concern. In biological image analysis, the average level of illumination might have no relevance. On the other hand, if you were studying cortical *steady potentials*, the mean might be a critical component to retain. The message is simple—there is no simple answer. But it is critical to decide whether or not the mean is meaningful in the data to be decomposed.

Let's take a break from single neurons and cortical networks and now examine the dynamics of a network of nine brains.

7.2 The Second Rehnquist Court

The U.S. Supreme Court is typically denoted by the name of the Chief Justice. In what is known as the Second Rehnquist Court, over the eight years from 1994 through 2002, the

---

4. "Factor analysis is, to put it bluntly, a bitch" (Stephen J. Gould, *The Mismeasure of Man*, 1981 [Gou81]).

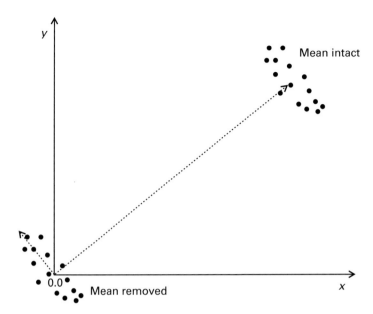

Figure 7.2
Example of the differences in the first singular vector or principal component (gray arrows) based on whether the data mean is left in or taken out.

composition of the nine judges was unchanged (figure 7.3). The court heard almost 500 cases in which all nine judges voted.

The mathematics of singular value decomposition do not care whether we are dealing with images, voltages, or justices. In 2003, Lawrence Sirovich, a colleague who was a household name in the applied mathematics of turbulence and now works on the dynamics of biological systems, published a pattern analysis of the second Rehnquist court using singular value decomposition. Most classroom audiences stop texting (briefly) at this point in a lecture on singular value decomposition mathematics.

The first thing that Sirovich did is to label the justices. The mathematics of singular value decomposition, whether justices or images, do not care about the order in which the data are arranged. So Sirovich used alphabetical order as, well, *justice space*:

R = [Breyer, Ginsburg, Kennedy, O'Connor, Rehnquist, Scalia, Souter, Stevens, Thomas]

Next he creates a *decision vector*, assigning a +1 if the justice voted with the majority, or a −1 if the justice voted against the majority. Note that this is a remapping of the true $2^9 = 512$ possible decisions. With a display of mathematical political correctness, Sirovich considers + and − votes to be equivalent, mapping both to +1 when they agree with the majority (dissension will be enforced as minority, a −1 vote). This removes a degree of freedom, with $2^{(9-1)} = 256$ possible decisions remaining. This extra degree of

Figure 7.3
Official group photo of Supreme Court Justices taken in 1998. Photo in public domain.

freedom needs to be accounted for shortly when we calculate the effective dimension of the court.

$$n = [n_B, n_G, n_K, n_O, n_R, n_{Sc}, n_S, n_{St}, n_T]$$

So if all justices agree, then the decision is *u*nanimous

$$u = [1, 1, 1, 1, 1, 1, 1, 1, 1]$$

Sirovich gives the example of the vote that helped decide the U.S. *p*residential election in 2000

$$p = [-1, -1, 1, 1, 1, 1, -1, -1, 1]$$

So there are two ideal models. The *omniscient court*, where they always make unanimous correct decisions, and the justices are clones

$$u = [1, 1, 1, 1, 1, 1, 1, 1, 1]$$

or the *platonic court*, where each justice sees equally compelling arguments, each justice's decision is as independent as the toss of a fair coin, and there are $2^{(9-1)} = 256$ possible

Table 7.1
The most frequent patterns of decision of the second Rehnquist court

Br	Gi	Ke	O'C	Re	Sc	So	St	Th	Margin	Frequency % (n)
1	1	1	1	1	1	1	1	1	9	47 (220)
−1	−1	1	1	1	1	−1	−1	1	1	9.6 (45)
1	1	1	1	1	1	1	−1	1	7	4.5 (21)
1	1	−1	1	−1	−1	1	1	−1	1	3.8 (18)
1	1	1	1	1	−1	1	1	−1	5	3.0 (14)
1	1	1	1	−1	−1	1	1	−1	3	2.6 (12)
1	−1	1	1	1	1	1	−1	1	5	2.4 (11)
−1	1	1	1	1	1	1	−1	1	5	1.9 (9)
1	1	1	−1	−1	−1	1	1	−1	1	1.9 (9)
1	1	1	−1	1	−1	1	1	−1	3	1.5 (7)
1	−1	1	1	1	1	−1	−1	1	3	1.3 (6)
1	1	−1	1	1	−1	1	1	−1	3	1.1 (5)

Br, Breyer; Gi, Ginsburg; Ke, Kennedy; O'C, O'Connor; Re, Rehnquist; Sc, Scalia; So, Souter; St, Stevens; Th, Thomas. Data from [Sir03].

decisions. One would get the same dynamics were nine monkeys to be randomly casting votes instead of our idealized Platos.

Using the *information* metric, I, described by Claude Shannon [SW49], one can characterize the novelty of n outcomes as

$$I = -\sum_n p_n \log_2 p_n \qquad (7.2)$$

For the omnicient court, there is one outcome, so the $log_2 1 = 0$ and $I_o = 0$ bits of information.[5] For the platonic court, there are 2^8 outcomes, so each $p \log_2 p = 1/256 \cdot (-8)$, and the sum yields $I_p = 8$ bits of information. Just as in Shannon's communications information, increasing randomness in a communications channel yields *maximal* information [SW49].

Sirovich estimates a *dimension* $= (I + 1)$ by adding 1 to the information. He notes that out of the 256 possible decisions, 12 account for more than 99% of the decisions, as shown in table 7.1.

Now we ask within what space do the patterns of the justices' decisions live? Using formula (7.2), we add the calculated probabilities of each decision occurring out of the total n, and sum the products of $p_n \log_2 p_n$. There are 3.68 bits of information, suggesting that the court acts as if there were only 4.68 instead of 9 justices.

Now let's perform a singular value decomposition to further investigate which justices vote with other justices. We seek a description of whose votes are correlated with whose.

5. A *bit* is the amount of information that you gain from deciding between 0 and 1.

We stipulate that we will assign these groups in a manner that is independent, and orthogonal to each other. We seek a set of *coherent structures* among the voting patterns. Later in the chapter we will seek coherent structures among patterns of firing neurons rather than among whole voting brains.

First, we set up a data matrix that is structured as shown in figure 7.4. The justices are arranged in columns—they are the *sensors* in our spatiotemporal system (the decisions are, after all, sensing of neural activity within each justice's brain). The rows are the 468 decisions.

Next, Sirovich calculates the first several singular vectors and their singular values as shown in table 7.2. The relative weightings, w_i, are the singular values associated with each vector divided by the sum of all of the singular values, $\lambda_i / \sum_n \lambda_i$. The first thing to notice is that the first two weightings dominate the decomposition. They account for 79% of the variance in the decisions. This means that most of the decision patterns can be expressed by a linear combination of these first two singular vectors. Especially interesting is that we have seen similar vectors before: The first vector is very close to the omniscient decision vector u above, and the second vector is very similar to the presidential decision vector p above. *A linear combination of these two patterns accounted for most of the dynamics of this court.*

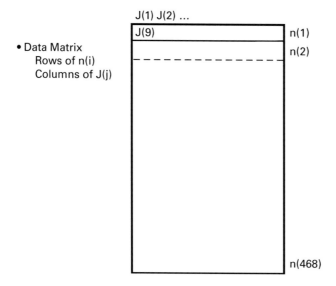

Figure 7.4
Schematic of the data matrix for singular value decomposition. Each decision is represented by a row of nine columns, one for each justice. There are 468 rows for the 468 decisions. Singular value decomposition does not care about the ordering of the justices, but later we will wish to keep the spatial order of neural data ordered so that we can map such data back to their original geometry. For justices, the geometry of their yearly group photos is meaningless.

Table 7.2
Relative singular values, w_i, and first three singular vectors, v_i, from the singular value decomposition of the justices' decisions

w_1	w_2	w_3
0.57	0.22	0.05

V_1	V_2	V_3
0.34	−0.33	−0.12
0.33	−0.37	0.11
0.36	0.17	−0.35
0.36	0.10	−0.53
0.35	0.30	−0.22
0.31	0.40	0.46
0.35	−0.31	0.08
0.26	−0.44	0.35
0.32	0.41	0.43

Weights $w_i = \lambda_i / \sum_i \lambda_i$. Data from [Sir03].

Last, Sirovich illustrates the decisions plotted graphically, and shows in figure 7.5 how the first two singular vectors span this space. I have sketched into this plot the u and p decisions, which lie very close to these first two vectors. Note that subtracting the mean would give very different results. One would need to consider not the votes, but the deviation of the voting patterns from the average voting patterns. Leave the mean in for a supreme court analysis.

7.3   The Geometry of Singular Value Decomposition

We now need to switch to analyzing spatiotemporal patterns from individual brains. Unlike with the ensemble of justices, we will now need to be kept track of the geometry of the sensor positions. And understanding the geometry of singular value decomposition will be critical.

We go carefully through the geometry of the matrix form of singular value decomposition from equation (7.2) in figure 7.6. The data matrix is A. Each row is an image. If this were data from a two-dimensional electrode array or an optical sensor, then the electrodes or pixels are strung out as a long row. We keep the order of the rows of the data sequentially, so that we can readily reshape the rows back into a two-dimensional matrix for display purposes. But to singular value decomposition the original position within the row does not matter. Time T is indexed in the sequential rows. Sensor number x is indexed in the columns.

Singular value decomposition breaks this matrix up into a U, S, and V matrix. The first matrix U has columns u_i that are the same length of time as the original data. These are

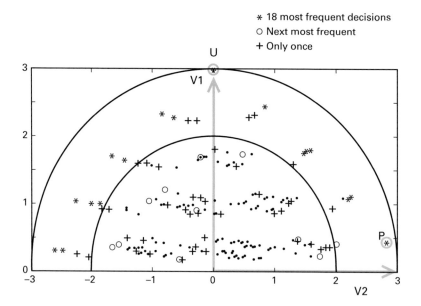

Figure 7.5
Space of decisions, spanned by the first two singular vectors v_1 and v_1. For comparison the unanimous, u, and the presidential, p, decisions are plotted. Modified from [Sir03] with permission.

orthogonal *temporal modes*. On the far right, V is a set of orthogonal *spatial modes* with length x flipped on its side as a transpose. In between is a diagonal matrix of singular values S. U has the dimensions of $T \times T$, V the dimensions of $x \times x$, and S the dimensions of $T \times x$. Note that the number of potentially nonzero singular values are the lesser of T or x to fit on the diagonal within S—the rest of S are zeros. Generally in scientific data collection, you will collect more data, T, than the number of sensors, x. Of course, if you are examining rare events—epidemics perhaps, or phases of the moon—and your data start only when people started recording such phenomena, T might be less than x.

First, I'm going to multiply U by the weightings S. In principal component analysis, this multiplication is the *score*. After each row of U is multiplied against the first column of S, you get a column of $u_1 s_1$. Then, multiplying all the rows of U against the second column of S generates the column $u_2 s_2$, and so on. Thus, each temporal mode becomes weighted by the singular values. If $T > x$, only the first x temporal modes are retained. Now see what happens if we do an outer product between the column $u_1 s_1$ and the first row v_1 of V^T. This generates a matrix with the same dimensions of T rows and x columns. Look carefully at what this gives. The first row in this outer product is literally the first point in time of the first temporal mode weighted by the eigenvalue times the first spatial mode. The second row is the second point in time of the first temporal mode weighted by the same eigenvalue and the same first spatial mode. Every row is the same representation of this eigenimage

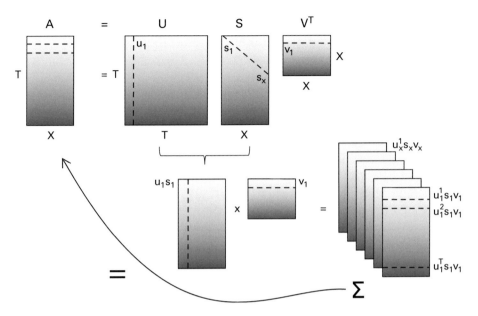

Figure 7.6
The geometry of singular value decomposition.

or eigenvector scaled by the temporal mode. It's a time sequence of the same image if we have two-dimensional data. We will show this in another way shortly.

The second weighted temporal mode, the column $u_2 s_2$, forms an outer product with the second spatial mode, the second row of V^T. We proceed through the x columns of US and form outer products with each of the x rows of V^T. We now have a stack of matrices. Summing this entire stack of the matrices gives us the original A—exactly A. Look at the first row of $u_1^1 s_1 v_1$, and note that there is a stack of rows behind this, that form a linear combination that sums to the first data sample in row 1 of A. This is a linear combination of *all* of the spatial modes. *The data are a linear combination of the modes, and the modes are linear combinations of the data.* The corollary of this is that the modes can contain only what was observed in the data.

What we have done is to find the best representation, on average, for the T images that make up the T rows of matrix A. Let's call this best fit v_1. This first spatial mode maximizes the inner product average of all of the rows, which can be written as

$$\text{maximize } \langle (A, v_1)^2 \rangle_t = \frac{1}{T} \sum_t (A, v_1)^2$$

where (A, v_1) indicates inner product projection, and $\langle \cdot \rangle_t$ indicates average over time t. There is a remarkable set of relations that emerges among the U, S, and V of the singular value decomposition:

$$s_i u_i = (A, v_i)_x$$

$$s_i v_i = (A^T, u_i)_t$$

There is nothing in the u's or v's that was not present in the original data, A.

Now note that the singular values are ordered from large to small, $s_1 \ldots, s_x$. The first singular values are large compared with the later values that are small and eventually meaningless. By the turn of the twentieth century, Schmidt had understood that you could approximate A by using only the top subset of the stack of $u_i s_i v_i$ matrices. This lets you do a crude form of image compression. It also lets you reduce the dimensionality of the space you are working in. Keeping the top two of the nine matrices is what Sirovich did in the Supreme Court example when he showed, in table 7.2, that the top two singular values were much larger than the rest, and figure 7.5 represents a projection of the nine-dimensional justice space in the approximated two-dimensional space following singular value decomposition.

7.4 Static Image Decomposition

Now let's examine a popular image. In figure 7.7 (plate 10), we show the original image, and in the second column, the first four matrices $u_i s_i v_i$. Above the stack of matrices is the partial reconstruction $\sum_{i=1}^{4} u_i s_i v_i$. The image is 124×180 pixels, so there are 124 singular values s_i. If we sum the entire stack to create a full reconstruction, $\sum_{i=1}^{124} u_i s_i v_i$, we get our original image. Now note the modal coordinates, the $s_i u_i$.[6] The first mode is much larger than the others. Why? Because the mean was left in this image. So let's remove the mean, as shown on the far right column. This image looks a bit odd, but the modal coordinates now reveal the details of the image. You can see the belly of the duckie in the first mode, and how the other modes have an orthogonal look and feel from visual inspection compared with the first mode and each other around the belly. Now look at the plots of $log\ s_i$. Comparing these plots with the mean left in with those with the mean removed, you can see (arrows) the substantial difference that the first mode makes with the mean left in. The mean is the average level of illumination of the image. It is important to decide whether this feature is important to the dynamics of image analysis, and whether it is a substantial feature. Note the *tolerance* in the plots of the singular value spectra in figure 7.7. There are many ways of doing this, but in this low-noise figure, I used the product of

---

6. I am using a terminology that will be more palatable shortly when the u_i values are temporal modes. For now, these are really the modes in the vertical direction of the original image.

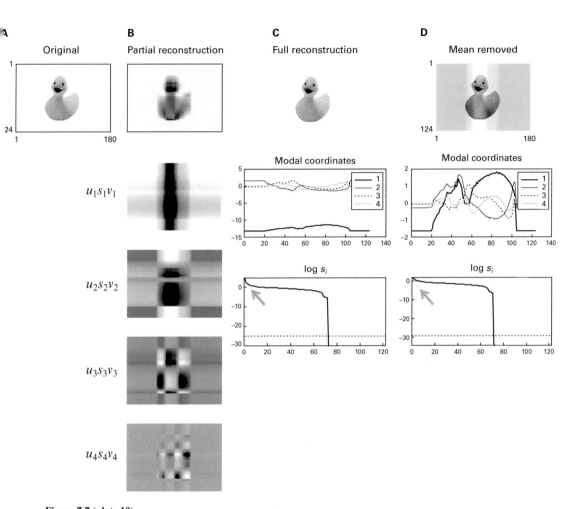

Figure 7.7 (plate 10)
Singular value decomposition of a popular image in A. In B are four matrices $u_i s_i v_i$ and the sum of these matrices is shown at the top. C shows a full reconstruction with the mean retained in the image. D shows the image with the mean removed. The arrows point to the difference that the large first mode contributes to the spectrum of singular values s_i. The horizontal red lines indicate a *tolerance* (red lines) based on the product of the largest of the singular values times the computer machine precision—anything below the tolerance line is computationally meaningless.

the largest singular value times machine precision of the computer. Anything below this line is computationally meaningless. Regardless of whether the mean is left in or out, it is clear that about half of the $u_i s_i v_i$ matrices can be thrown away—they are multiplied by s's so small that they are meaningless in the reconstruction of the figure.

Next, is the subject of *energy*. If we were studying a dynamical system, such as a velocity field in turbulent fluid, then the square of the singular values would be related to the kinetic energy of the system (through $\frac{1}{2}mv^2$). The static image used here has no intrinsic energy, but we can use the relative sum of the squares of the singular values,

$$\text{Energy} \ \sim \ \sum_{i=1}^{m} s_i \bigg/ \sum_{i=1}^{n} s_i$$

as a *probability* that the first m out of the total n components account for the energy or variance in the image. For the case when the mean was removed, the first four modes shown account for 90% of the image. When the mean is left in, the first mode, the mean, accounts for 99% of the image, rendering such energy considerations meaningless.

One last thing to consider: The modes are coherent structures. That is, they reveal features that go up and down together. This is why such techniques are useful in financial market analysis—it gives the industries and stocks that are moving together. In our use in neural data, it will show us which neuronal elements are firing together. So let's make this more concrete with an example of dynamic spatiotemporal image analysis.

7.5 Dynamic Spatiotemporal Image Analysis

A very instructive example of the use of singular value decomposition is in the recognition of faces. The so-called rogue's gallery was solved by Sirovich in a series of papers [SK87, KS90], using a database of volunteer students who lent their faces to computational image analysis fame.[7]

In figure 7.8, we show on the lefthand column the first six facial images from this database. The left column in the figure is literally the data matrix if you were to string out each 128×128 pixel image into a single row of 16,384 image intensities. The data matrix is 6 rows by 16,384 columns. Shown on the top row are the six spatial vectors v_i, or eigenimages, reshaped back into their 128×128 locations. The first mode looks like a face, and the others a bit less so. But the first spatial mode is not any of the faces present in the data matrix. It is the image that is the best statistical representation of all of the faces, and computationally would give the largest set of inner products were one to project each of the original faces onto a single face. It is the best linear combination of the original faces such

---

7. There is an unpublished but widely available 2003 tutorial on image analysis by Sirovich that also employs the rogue gallery, and is highly recommended.

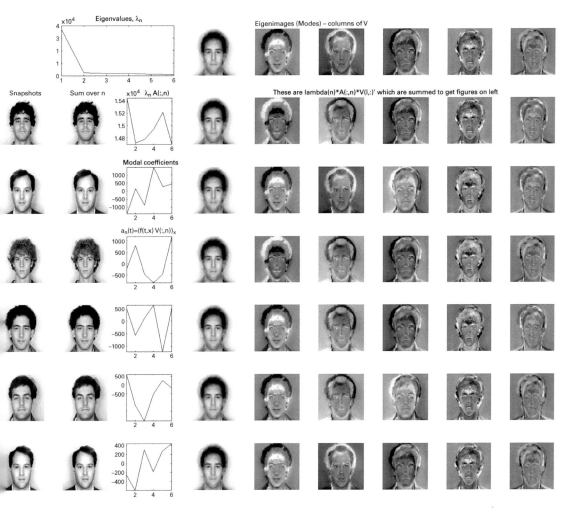

Figure 7.8
The rogues' gallery. Database from [SK87] and [KS90]. The six faces selected are the same as from Sirovich's unpublished tutorial from 2003.

that these inner product projections are maximized. Each subsequent spatial mode along this top row is orthogonal to the other images. The eigenvalue spectrum is shown rapidly decreasing after the first one—the subsequent modes after the first largely add detail. But, of course, this is the detail that makes each face distinct. In the third column are the temporal modes. Note that the first one is very large (multiplied by 10^4). Forming the outer product of this first temporal mode with the first spatial mode gives the time sequence of the first temporal mode in the fourth column. This is the same as the first $u_i s_i v_i$ matrix on top of the stack of matrices shown in the lower right of figure 7.6. Each image is a subsequent row in this matrix. Forming the outer product of the second temporal mode with the second spatial mode gives the second $u_i s_i v_i$ matrix in this stack, and is the fifth column of figure 7.8. If we sum the entire stack of $u_i s_i v_i$ matrices, each corresponding row sums to the original image in the data matrix, shown as the second column from the left, identical to the first column.

The heuristic value of detailing these faces is that if you understand the geometry, then you completely understand figure 7.9.

7.6   Spatiotemporal Brain Dynamics

Figure 7.9 is a spatiotemporal analysis of voltage-sensitive dye imaging of wave pattern formation in cortex. In this figure, we show data from [HTY$^+$04]. The left column are the first 6 snapshots of the images from a 12×12 photodiode array sampling at 1,000 times per second. There are 1,000 images in the data matrix, recorded over 1 second. I show only the first 6 on the left in order to be homologous with the previous face figure, but the $\cdots$ reminds you that this is just the very upper left corner of a huge set of images. Note that the first and second temporal modes oscillate prominently at 10 cycles per second. If you *filtered* by keeping only the first two modes, you would extract the powerful coherent oscillations at 10 cycles per second.

Now focus on the top row in figure 7.9—the spatial modes. The first spatial mode shows a left-right pattern. The second mode is more diagonally oriented. Although there is a lot of structure in the first four modes, by the fifth and sixth modes, there is considerably less structure. This 1 second of data was taken when the cortex was very coherent, and generating a rotational spiral. It takes only a few modes to account for most of the activity. The brain dynamics here can be expressed in only a few dimensions (modes).

If we sum the corresponding rows of the outer product matrices as discussed earlier, we get the second column. Here, all 1,000 rows sum to reconstruct each individual original image (actually a row in the data matrix). It really is the same as the faces.

You can put anything through such a *proper orthogonal decomposition*—Supreme Court judge decisions, voltage-sensitive dye images, frequencies. So let's do spatial frequencies. We perform a spatial Fourier transform on the snapshots along the left column of figure 7.9. We use wave number instead of frequency in the transform. Let's look at a simple example of this in figure 7.10. We start with a vertical grid pattern. Shifting the zero spatial frequency

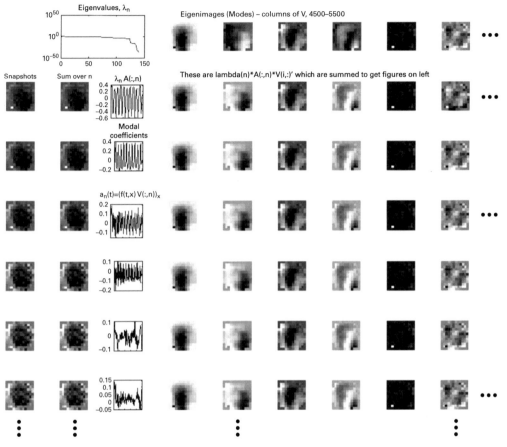

Figure 7.9
Spatiotemporal analysis of brain optical imaging dynamics in the same fashion as shown for faces in figure 7.8.

to the center (crystallographers do this when they present their data), we get a dot at the center, and two dots along the x-axis reflecting the spacing of the grid lines. If I now remove the mean, the point at the center disappears (second column). Now create a checkerboard in the third column. The spatial frequencies form a star pattern, and if I remove the mean in the last column, the center point again disappears. The spatial Fourier transform must be done in the original two-dimensional space and arrangement of the original image (or it would make no sense). But after the transform is done, you can string out the two-dimensional plot of the transform just as was done for the image amplitudes, and perform a principal orthogonal decomposition as we did for justices, faces, and voltages.

One more technical piece before we return to the brain data: Again, the phenomenal contributions of Lawrence Sirovich come to the fore. I just presented a small piece of a set

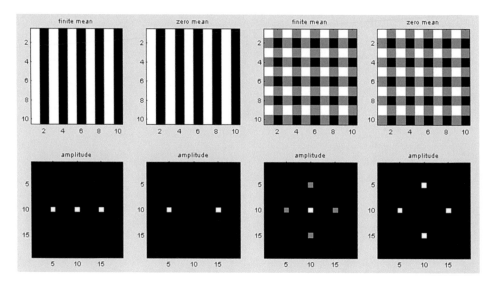

Figure 7.10
Spatial Fourier transform examples.

of optical imaging data—1,000 snapshots over 1 second. But brain dynamics take place over more than 1 second, and each year our imaging cameras, and multielectrode arrays, get larger and faster. Whatever computer you have, at some point you will not be able to solve the singular value decomposition because you run out of memory or your finite lifespan weighs on your patience.

In 1987, Sirovich published a series of three papers in the *Quarterly of Applied Mathematics* on turbulence and the dynamics of coherent structures. In the first of these seminal fluid mechanics papers, he introduced an approach called the *method of snapshots* to deal with the computational problems that large data sets can introduce [Sir87]. For a data matrix, A, where the rows are time, $t = (1, \ldots, T)$, and the columns are space, $x = (1, \ldots, N)$, then for a singular value decomposition,

$$A(t, x) = USV'$$

where $'$ indicates transpose to avoid confusion. U is $T \times T$, S is $T \times N$, and V is $N \times N$. Then we will almost always find that $T \neq N$. For typical laboratory experiments, even for the newest cameras or electrode arrays, $T \gg N$. So we can get rid of T by left multiplying by A':

$$A(t, x)'A(t, x) = VSU'USV'$$

$$= VS^2V'$$

$$A(t, x)'A(t, x)V = VS^2$$

which is an eigenvalue equation where V is the matrix of eigenvectors of $A'A$ and S^2 is the eigenvalue matrix. And since $N \ll T$, it is a much easier eigenvalue equation to solve now that T is removed. Once V is solved, one can calculate S and then determine U from

$$U = A(t, x)VS^{-1}$$

If $N \gg T$, a comparable solution for U as an eigenvector matrix of AA' can be found.

Using the same data as the detailed study of the dynamics of the cortical wave patterns shown in the previous chapter [HTY+04], we examined the principal orthogonal decomposition of such dynamics in both the amplitude and spatial frequency domain. Figure 7.11 shows a sequence of such dynamics as the cortical network progresses from chaotic, through spiral and plane wave dynamics, to finally dissolution through chaotic disorganization. The first set of segmental eigenmodes corresponds to chaotic activity, the second set to spiral wave dynamics, and the third set to chaotic activity. The modes are plotted both as amplitude and spatial frequency modes. Note the trajectory plots in the third column from the right. These three-dimensional plots are created by plotting the first three temporal modes against each other. On the far right are plotted the global eigenmodes calculated for the entire data sample. These latter calculations are the ones that most require the use of the method of snapshots discussed earlier.

Modes can help in filtering and denoising complex data. In their study of dynamic brain imaging [MP99], Mitra and Peseran show how using the largest modes can denoise such data effectively, as shown in figure 7.12 (plate 11). On the left, they show the average spectra from the first twenty modes, summed much as we summed the largest modes in figure 7.7. On the right side of this figure is a snapshot of our spiral cortex data, below that the spectrogram of the original data set, and below that the spectrogram from a reconstruction of the largest six modes. The evolution of the dynamics can be seen much more clearly in the modal filtered version.

There is a very interesting set of modal decompositions of visually excited turtle cortex from Senseman and colleagues [SR99]. This is a creative work, whose modal trajectories we have adopted successfully in this book. The researchers found something remarkable: The first mode accounts for almost all the dynamics, regardless of whether they use diffuse stimulation with different colors or white light, or spot stimulations. Turtles are not the smartest creature in the pond. But nevertheless, one does speculate whether findings of the large first spatial mode dominating their dynamics, under several conditions, might be related to having left the mean amplitude in their optical images. A reevaluation of such data might prove interesting. But Senseman and colleagues' use of the modal trajectories led to the following.

Look again at the three-dimensional trajectory plots from the modal coefficients in figure 7.11. Such state space reconstructions enable a reconstruction of the images to be performed (the modal coefficients and the eigenvalues determine the weightings on the spatial modes). But they also permit local linear fits to the dynamics in phase space so that a new trajectory,

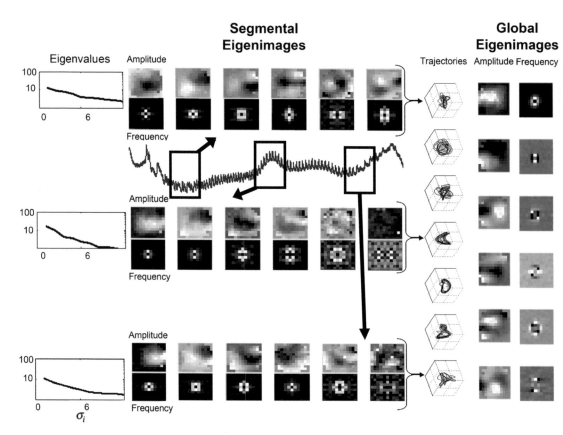

Figure 7.11
Dynamics from a 7-sec oscillatory episode in figure 6.19. At the left side are shown the eigenvalues corresponding to the first six spatial eigenmodes based on the photodetector amplitudes. Below the amplitude eigenmodes are shown the corresponding spatial frequency eigenmodes. These three sets of eigenmodes correspond to seconds 1, 4, and 7 from the episode. The column of trajectory plots from each of the 7-sec episode shows the relationship between the first three temporal modes, u_i. Note the simplification of dynamics during seconds 4–6. On the right side are global amplitude and spatial frequency eigenmodes calculated from all 7,000 images collected. Gray scale proportional to scaled eigenmode amplitudes. Modified from [SHW07] with permission.

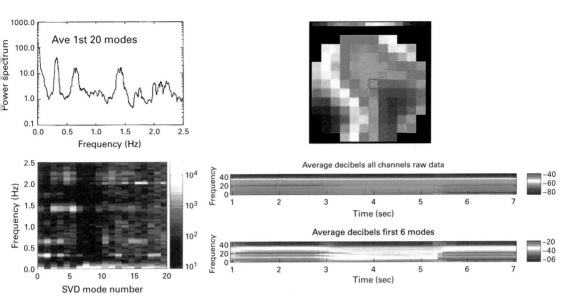

Figure 7.12 (plate 11)
On left is an example of modal filtering with multitapered average spectrum from fMRI dataset, and below that, the spectrogram from individual modes. Reproduced from [MP99] with permission. On the right is an example of our spiral cortical images, with unfiltered and modal-filtered (retention of first six modes) spectrograms. The modal filtered version shows the dynamical evolution much more clearly.

close to previous trajectories, can have a prediction performed based on the nearby trajectories of previous close states. It is this approach that enables a predictive controller to be designed based on such empirical mode decomposition. One can develop an ensemble (unscented) Kalman filter approach by scattering points about such local trajectories and iterating them forward in time. How to choose the modes for the axes in such state space modeling remains an open question. What I show in figure 7.11 are the largest three modes. Picking the best fitting modes in sequence instead of the largest ones is straightforward. This remains an open problem as of this writing.

These dynamics do something that the Wilson-Cowan equations never do. They *evolve*. Wilson-Cowan equations asymptote. Real brains don't asymptote. They do, however, change their activity patterns based on the history of their activity. We are, after all, accommodating learning machines.

If we calculate the equivalent energies of these dynamics, and ask how many modes are required to account for 90% of the energy of the system, we obtain a plot similar to that of figure 7.13. These modes that account for the majority of the energy can be equated to an effective dimension for the dynamics, d_{KL} [Sir89], and we see that those effective dimensions decrease during the episode, then increase. The dimension, or effective complexity of the dynamics, evolves. Later (chapter 12), we will show how this can be accounted for, and it will involve the lower-scale metabolism of ion dynamic such as potassium.

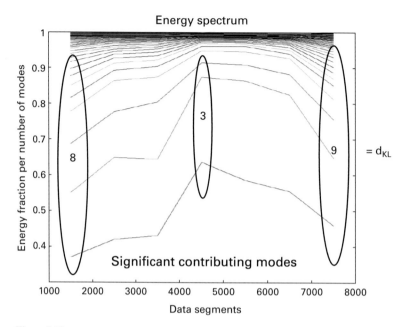

Figure 7.13
Energy spectrum of consecutive data segments from data in figure 7.11. The number of modes required to reach $\geq 90\%$ of each 1 second's total energy is indicated. Reproduced from [SHW07] with permission.

In figure 7.13, the energy estimates are statistical averages over segments of data. Perhaps one can define an instantaneous effective dimension, D_{eff}, as follows. If the inner product of the data, A, with the spatial modes is (A, v_i), then

$$(A, \mathbf{v}_i)^2 = s_i^2 a_i^2, \quad i = 1...n$$

then we can define $D_{eff} = min(m)$ such that $E(m) \geq 0.9$, where

$$E(m) = \frac{\sum\limits_{i=1}^{m} s_i^2 a_i^2}{\sum\limits_{i=1}^{n} s_i^2 a_i^2}, \quad m < n$$

Instantaneous dimensions are effectively the probability that a given snapshot can be accounted for by a given set of spatial modes [SHW07]. As such, they are practical metrics for real-time control. We can also use them to watch the evolution of snapshot to snapshot, as shown in figure 7.14 (plate 12). The frequency-mode representations, and the frequency-based D_{eff}, are smaller and more compact than amplitude. This makes considerable sense. These waves alternate between rotating and not rotating. If they purely rotated, one would

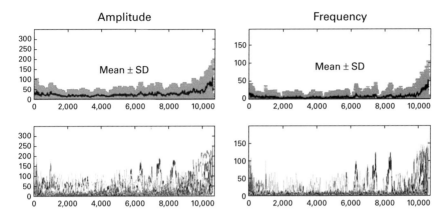

Figure 7.14 (plate 12)
Effective instantaneous dimension shown for each resampled data set, below, and as mean and standard deviation, above, for amplitude and for spatial frequency. Reproduced from [SHW07] with permission.

want to go into a rotating frame of reference that would simplify the representation drastically [BBK03]. An alternative is to use a radially symmetric basis [RFPSS05]. Because we are dealing with brain patterns where the dynamics switch from rotational to irrotational states, I have chosen to stay with a fixed frame of reference. But the reader must keep in mind that there may be better basis representations.

One must be cautious in carrying the analysis from fluid dynamics too far with brain dynamics. Fluids seem simple compared with brains. Their interactions are all nearest neighbor. Brains are nonlocally connected, as reflected in the Wilson-Cowan equations. One technique used frequently in fluid dynamics is to take the basis, such as principal orthogonal decomposition modes, and insert them into the fundamental equations of the fluid dynamics. Then one projects these equations through an inner product onto each basis function mode—a Galerkin projection [HLB96]. For the brain dynamics presented here, the experimental observations suggest that there may be a significant degree of translational symmetry. One observes spirals that drift and wander; they do not seem locked to any particular place in the imaging field. In translationally symmetric cases, a Fourier mode basis may be employed. Especially unclear has been how to handle a Galerkin projection in the setting of the nonlocal connectivity [Lai05] assumed universal in brain network topology. Freestone and colleagues [FAD+11] have recently shown fascinating progress along these lines.

Another issue with brains is that after activity, they are refractory. This means that superposition breaks down. If we employ a linear-mode decomposition technique, then some compensation for the fact that the available brain that can be reexcited changes with system dynamics has not been employed (Tim Lewis, personal communication, 2008). Such a modification of the analyses presented in this chapter would likely be fruitful.

Data

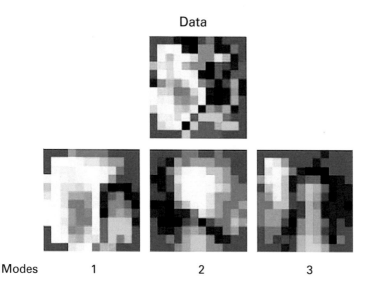

Modes   1   2   3

Figure 7.15 (plate 13)
Mode decomposition of a typical snapshot of spiraling cortical dynamics. Reproduced from [SHW07] with permission.

We can project snapshots of raw data onto individual spatial modes, and watch these as animations. Figure 7.15 (plate 13) shows such an example. The raw data are a spiral rotational wave. The first three spatial modes shown are activated in sequence as they pass the rotating wave off from one to another. (An animated movie of these dynamics is archived with [SHW07].) In fluid mechanics one can take a Fourier basis and insert it into the Navier-Stokes equations, or insert into the equations the modes that have been calculated from data. One result is that the modes interact in a peculiar way. These modes are supposed to be independent. But if they really were independent, the modes would just die out. It is the *interaction* between the modes, and in fluid the energy flow between the modes, that keeps turbulence going [SMH05b, SMH05a]. *In the brain, it would be the interaction between the modes, the basic patterns, that keeps brains persistently active.*

The modes from such brain activity are building blocks of activity. Perhaps they are an alphabet, the kind of dynamical alphabet a piece of brain can use in different combinations to produce a variety of dynamics. If so, then the grammar is contained in the dynamics that link the succession of combinations of such modes. In drumheads, for instance, a reasonably small set of prominent vibrational modes, depending on how you hit the drum, recombine to produce the characteristic sound of the drum [Kac66]. The brain, in a sense, is a crystal. It is arranged in a complex lattice, whose connectivity is not just neighbor to neighbor. So brains are a complicated crystal in terms of the nonlocal connectivity. Further work with the types of empirical mode decompositions discussed in this chapter that includes the nonlinear connectivity and nonlinearities of the brain promises a deeper understanding of brain dynamics.

Do we need to chose between the arrogance of our fundamental models of the brain and the ignorance of our empirical models? In principle, a fundamental model should be more compact for a complex system, and it will always do one thing an empirical model can never do—handle the previously unobserved and unexpected. Fortunately, one can run these different models in parallel. In fact, given the Kalman framework we have presented so far, one can readily compute the prediction errors for each model at each step, and rely more heavily on the one that is more accurate at any point in time. If the computational overhead is practical, running such models in parallel helps us hedge our chances of keeping track of a system that may, or may not, be well accounted for by any one model as it travels through the various regimes of its dynamics.

But for any model, empirical or fundamental, we must formally address their inadequacies. This is rigorously done in the next chapter.

Exercises

7.1. Let's explore singular value decomposition a simple static matrix (figure 7.16):

```
smiley =        [0 0 0 0 0 0 0 0 0
                 0 0 1 0 0 0 1 0 0
                 0 0 0 0 0 0 0 0 0
                 0 1 0 0 1 0 0 1 0
                 0 0 1 0 0 0 1 0 0
                 0 0 0 1 1 1 0 0 0
                 0 0 0 0 0 0 0 0 0
                 0 0 0 0 0 0 0 0 0
                 0 0 0 0 0 0 0 0 0] ;
```

• Perform a singular value decomposition on this matrix using `[U,S,V] = svd(z);` and extract the singular values from S by `lambda = diag(S)`. Plot the log of the singular values.

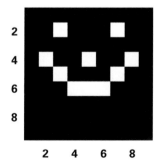

Figure 7.16
Simple smiley matrix.

Which ones are large enough to be useful? A nice way to guess at the important ones is to estimate a tolerance such as

```
tolerance = max(size(S))*eps(norm(S))*...
                    ones(1,length(lambda));
```

• Now form the partial matrices from figure 7.6 using

```
for i = 1:length(lambda)
  A(:,:,i) = lambda(i) * U(:,i) * V(:,i)';
end
```

Plot these individual matrices. How many do you have to sum in order to get a reasonable approximation to your original smiley face?

7.2. Now let's take a sequence of three images:

```
f(:,:,1) =       [0 0 0 0 0 0 0 0 0
                  0 0 1 0 0 0 1 0 0
                  0 0 0 0 0 0 0 0 0
                  0 1 0 0 1 0 0 1 0
                  0 0 1 0 0 0 1 0 0
                  0 0 0 1 1 1 0 0 0
                  0 0 0 0 0 0 0 0 0
                  0 0 0 0 0 0 0 0 0
                  0 0 0 0 0 0 0 0 0] ;

  f(:,:,2) =       [0 0 0 0 0 0 0 0 0
                  0 0 1 0 0 0 1 0 0
                  0 0 0 0 0 0 0 0 0
                  0 0 0 0 1 0 0 0 0
                  0 0 0 0 0 0 0 0 0
                  0 1 1 1 1 1 1 1 0
                  0 0 0 0 0 0 0 0 0
                  0 0 0 0 0 0 0 0 0
                  0 0 0 0 0 0 0 0 0] ;

  f(:,:,3) =       [0 0 0 0 0 0 0 0 0
                  0 0 1 0 0 0 1 0 0
                  0 0 0 0 0 0 0 0 0
                  0 0 0 0 1 0 0 0 0
                  0 0 0 0 0 0 0 0 0
                  0 0 0 1 1 1 0 0 0
                  0 0 1 0 0 0 1 0 0
                  0 1 0 0 0 0 0 1 0
                  0 0 0 0 0 0 0 0 0] ;
```

which represent the images in figure 7.17.

Figure 7.17
Sequence of three simple images.

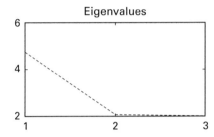

Eigenimages (modest) — columns of V

Figure 7.18
Eigenvalue spectra and the three spatial modes.

• First, let's make a data matrix out of these images. Do this by stringing out the images as a set of row vectors, and consider each image (row) as a sequential point in time. One way to do this is to write:

```
[rows,cols, NumberOfSnapshots] = size(f)
for i = 1: NumberOfSnapshots
  ff(i,:) = double( reshape(f(:,:,i),1,[]) );
end
f=ff;
```

• Now you can perform your singular value decomposition exactly as you did earlier with the static image in exercise 7.1, and form the partial reconstruction matrices.

• Plot the eigenvalue spectra, and plot the three spatial modes. You should get something that looks like figure 7.18.

• Plot the three temporal modes.

• Now plot the outer products of the temporal modes and the spatial modes. Each row of the summed partial matrices should sum to give you each of the original three images.

• But more subtly, note how the values of the temporal modes determine how each of the spatial modes is weighted to form the various forms of the smile or frown in these images. You should get something that looks like figure 7.19.

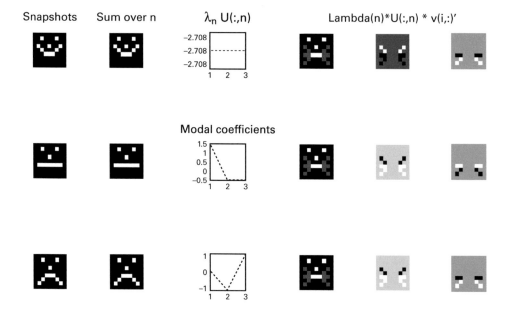

Figure 7.19
Singular value decomposition.

7.3. The canned routine svd in Matlab or Octave is great for small data sets, and miserable for larger ones. I will now give you a routine that is very useful for neuronal data sets. It is the code required for solving Sirovich's *method of snapshots* [Sir89], and I will present it here with sufficient annotation so that you can follow the logic:

```
%Test to see which Dimension is the largest
    [T,N] = size(f);
    if N >= T
        %ff' = (ASV')(ASV')' = (ASV')(VS'U') = AS^2A'
        %ff'U = AS^2
        %and the size of ff' is only the length
        %of the number of snapshots ^2
        [U,SS] = eig(f*f','nobalance');
        S = sqrt(SS);
        lambda = diag( real(S) );
        [lambda,IX] = sort(lambda,'descend');
        lambdamatrix(:,1) = lambda;
        S = diag(lambda);
        %Only retain significant eigenvalues
        Retain = length(find(S > 10*sqrt(eps)));
        LeftoverS = S(1:Retain,1:Retain);
        S = LeftoverS;
```

```
      for i = 1:Retain
         Atemp(:,i)=U(:,IX(i)); %reorders lambda, U
      end
      U = Atemp;
      V = (inv(S) * U' * f)';
   else%N > T
      %f'f = (ASV')'(ASV') = (SV')'U'(ASV') = VS'U'ASV'
      %f'fV = VS^2
      [V,SS] = eig(f'*f,'nobalance');
      S = sqrt(SS);
      lambda = diag( real(S) );
      [lambda,IX] = sort(lambda,'descend');
      lambdamatrix(:,1) = lambda;
      S = diag(lambda);
      %f = ASV'
      %S^- 1A^-1f = V'
      %if f is TxN dimensional, and N >> T,
      %then only first T eigenvalues
      %of S, and first T rows of V' (cols of V) are significant
      %Only retain significant eigenvalues
      Retain = length(find(S > 10*sqrt(eps)));
      LeftoverS = S(1:Retain,1:Retain);
      S = LeftoverS;
      for i = 1:Retain
         Vtemp(:,i)=V(:,IX(i)); %reorders lambda and V
      end
      V = Vtemp;
      clear Vtemp
      U = ( f * V * inv(S) );
   end%
```

This algorithm reduces the problem to the smallest dimension in the data matrix, and allows you to handle much larger data sets for a given amount of computer memory. It can also be much faster even when you have sufficient computer memory to handle the svd decomposition directly.

• Use this method of snapshot algorithm to calculate the singular value decomposition, and see whether the results are the same as before.

7.4. You can perform the same calculation in spatial frequency space. To do this and center zero frequency on the origin, consider the following lines of code:

```
[rows, cols, page] = size(f);
spatialft_f = zeros(rows,cols, page); %preallocate memory
for page = 1:max(page);
  demeaned_f(:,:,page) = f(: ,: ,page) - mean(mean(f(: ,: ,page)));
  spatialft_f(:,:,page) = abs(fftshift(fft2(demeaned_f(:,:,page))));
```

Modal coordinates

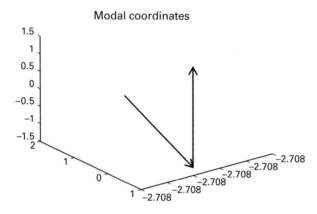

Figure 7.20
Trajectories of temporal modes.

```
end % page
f = spatialft_f; %do in K space
```

You need to do the spatial Fourier transform in the arrangement of the original smiley face images. Then you can reshape the transformed data into the data matrix as done in exercise 7.2, and perform singular value decomposition on this matrix. Examine the nature of the spatially transformed smiley face images. Can you see the differences in structure expressed in the spatial frequencies?

7.5. Last, let's explore the trajectories of the temporal modes plotted in the coordinates of the singular value decomposition. Take the first value of each of the three modes and plot this as a point in space, and then link it to the second, and the third. These are actually the original temporal data transformed by the *matrix times circle equals ellipse* into the space of the new coordinate system in v. In figure 7.11, such a trajectory formed the plots that resembled limit cycles and horseshoes when the dynamics were plane waves or spirals, respectively. For this simple example, your results look like the plot in figure 7.20. But if you had data that exhibited low-dimensional dynamics, and if the patterns recurred, then a local fit to nearby points would give you a predictive model for the next set of values. This is how you can use such empirical models for tracking and prediction.

A

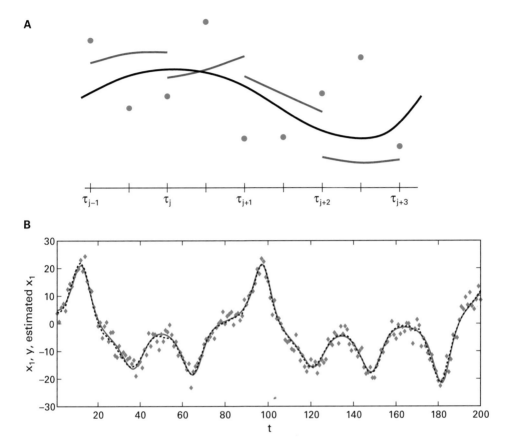

B

Plate 1
Multiple shooting schematic for solving inverse parameter estimation shown in A, and an actual example of noisy Lorenz system is shown in B. Reproduced from [VTK04] with permission.

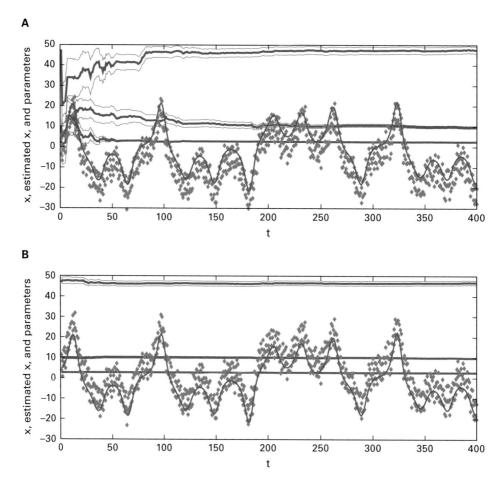

Plate 2
A shows the first pass of the unscented Kalman filter through the noisy x output (dots) of the Lorenz system. The three parameters being estimated are shown as solid lines, with thin line confidence limits gradually converging from left to right. The values at the right are then used to initiate another pass at parameter estimation, shown in the lower panel, B. Although parameter estimation improves a bit more, the upper panel demonstrates the potential power of this technique to work with one pass in real-time applications. Modified from [VTK04] with permission.

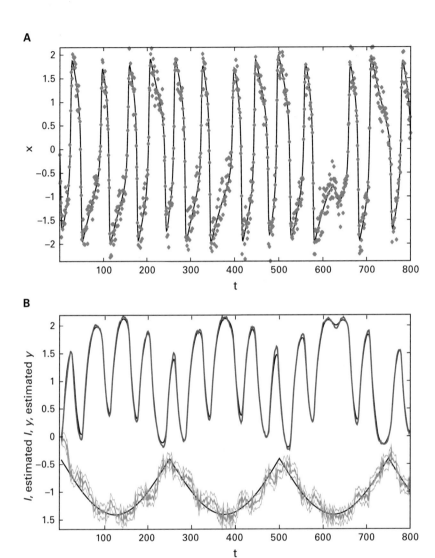

Plate 3
Variable and parameter tracking in the Fitzhugh-Nagumo equations. A shows the underlying trajectory (solid line), the noisy measurements (dots), and B shows the reconstruction (red) of the unmeasured variable y (black, nearly overlapping) in the upper half of the panel, and the applied current $I(t)$ and its estimation as a tracked parameter (magenta) with 1 standard deviation confidence limits (thin magenta lines). Reproduced from [VTK04] with permission.

A

B

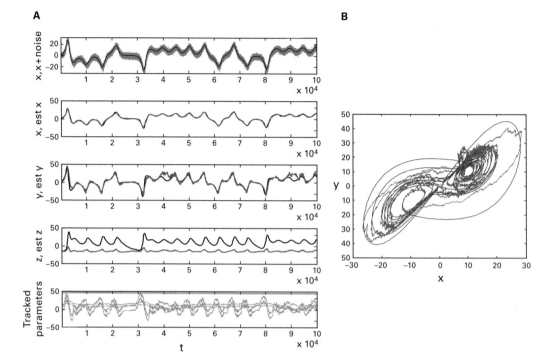

Plate 4
Results of reconstructing the Lorenz equations using the unscented Kalman filter from measuring only noisy x variable and estimating variables and parameters. In this simulation, the integration time step is 0.0001, $\mu = 10$, $r = 46$, and $b = 8/3$. A shows plots of reconstructed Lorenz variables. Upper panel is true (black) and noisy x variable (blue), second panel is true (black) and reconstructed x variable (red), third panel is true (black) and reconstructed y variable (red), fourth panel is true (black) and reconstructed z variable (red), and lower panel shows three reconstructed parameters with confidence limits. B shows true (black) and reconstructed (red) Lorenz x versus y trajectories. Note the difficulties with the z variable—similar troubles are seen in synchronizing Lorenz equations through this variable [PC90].

Same slice

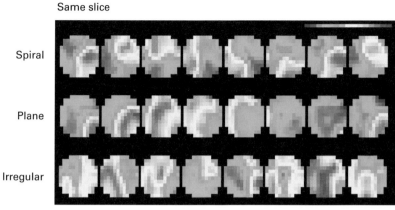

Spiral

Plane

Irregular

Frame interval 15 ms

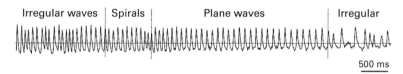

Irregular waves | Spirals | Plane waves | Irregular

500 ms

Plate 5
From a single oscillatory episode, a progression is seen of chaotic irregular, to spiral, to plane, and back to chaotic waves. Reproduced from [HTY+04] with permission.

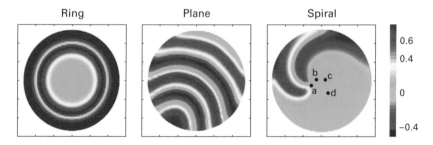

Ring Plane Spiral

0.6
0.4

0

−0.4

Plate 6
Simulation of ring, plane, and spiral waves with the Wilson-Cowan equations. Reproduced from [HTY+04] with permission.

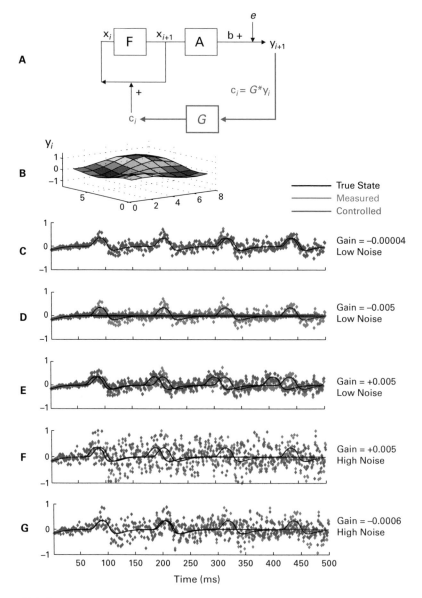

Plate 7

(A) Schematic of dynamical system F, which acts on state variable vector x, and observable function A, which produces output vector b, to which measurement noise e is added, generating output observable y. When direct proportional control is used, control function G is employed, which generates the control vector c. This control vector is directly "injected" into the neuronal elements, adding vector c to the activity variable u (nothing is added to the unobserved variable a or to the parameters of the system). (B) An instantaneous snapshot of the 8×8 output y during a rotational spiral. (C) The output of the observable y under conditions of small negative gain and low noise, from a single element of the grid in B, shown as a time series without (blue) and with (red) control applied. The underlying state vector activity element u_i is plotted as a thin black line. In contrast with C, D shows outputs under conditions of higher negative gain and low noise, E shows positive gain and low noise, and F and G show high noise conditions with positive and negative gains, respectively. Low noise corresponds to 1 to 5, and high noise corresponds to 10 to 14 on an arbitrary scale. Note that under high noise conditions, the application of direct proportional control as illustrated in the schematic renders the system dynamics unstable and destroys the rotational spiral regardless of positive or negative feedback. Reproduced from [SS08] with permission.

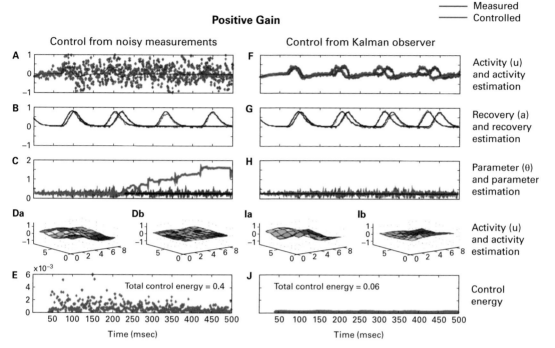

Positive Gain

True state
Measured
Controlled

Control from noisy measurements

Control from Kalman observer

A

F

Activity (u) and activity estimation

B

G

Recovery (a) and recovery estimation

C

H

Parameter (θ) and parameter estimation

Da   Db   Ia   Ib

Activity (u) and activity estimation

E   J

Total control energy = 0.4

Total control energy = 0.06

Control energy

Time (msec)   Time (msec)

Plate 8
Comparison of the use of positive gain proportional control calculated directly from high noise observations, left column A–E, and from the Kalman observer, right column F–J. (A) Output variable y_i from single element of grid, is the same as in figure 6.21F. The application of positive feedback gain at this high level of noise results in instability of the rotational oscillation, with loss of both the underlying oscillation u_i (black line) and measured observation y_i (red). (B) Without control being applied, the Kalman observer can accurately estimate (blue) the unobserved variable a (black). With this level of noise, when control is applied, not only is the observable a oscillation lost (red), but the true system state loses stability and the wave is destroyed after two oscillations (black). (C) Without control, despite the high noise level, the parameter θ (black) is accurately tracked (blue). When the stability of the dynamics is lost during control, the parameter θ is no longer accurately estimated (red). Da shows the variable y, and Db the estimated underlying activity $\hat{u}$. The snapshots in Da and Db are taken from the last frame in the 500 msec simulation, and the full simulation can be viewed as Supplementary Movie 1 from [SS08]. In E is shown the energy (sum of squared voltage) of the control vector c_i, and the total sum of all control vector energy for the 500 msec is 0.4 (arbitrary scale). F shows the estimated observable $\hat{y}$ without (blue) and with (red) control derived from the Kalman observer. (G) With Kalman observer control, the unobserved recovery variable a is now well tracked and the rotational spiral is stably increased in frequency, and in H, the parameter θ is now well tracked. Ia and Ib show the experimental variable y and estimated underlying variable $\hat{u}$. In J is shown that the control energy required is markedly reduced, with total control energy 0.06 (arbitrary scale).

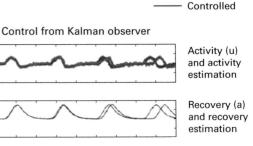

——— True state
——— Measured
——— Controlled

Negative Gain

Control from noisy measurements   Control from Kalman observer

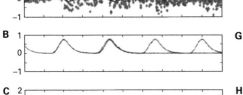
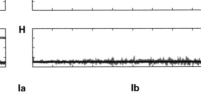

A 1 / 0 / −1   F   Activity (u) and activity estimation

B 1 / 0 / −1   G   Recovery (a) and recovery estimation

C 2 / 1 / 0   H   Parameter (θ) and parameter estimation

Da   Db   Ia   Ib

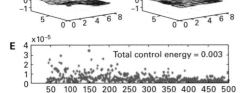
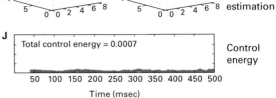

Activity (u) and activity estimation

E ×10⁻⁵ 4 / 3 / 2 / 1 / 0   J   Control energy

Total control energy = 0.003   Total control energy = 0.0007

50 100 150 200 250 300 350 400 450 500   50 100 150 200 250 300 350 400 450 500

Time (msec)   Time (msec)

Plate 9
Comparison of the use of negative gain proportional control calculated directly from high noise observations, in left column A–E, and from the Kalman observer, in right column F–J. (A) Output variable y_i from
single element of grid, same as in figure 6.21G. The variables in this figure are similar to the description for
figure 6.23. Note that with control derived from the Kalman observer (right column), negative gain in high
noise conditions stably slows the frequency of the spiral, and requires considerably less control energy than
the unstable case where direct proportional control is calculated from the noisy observations (left column).

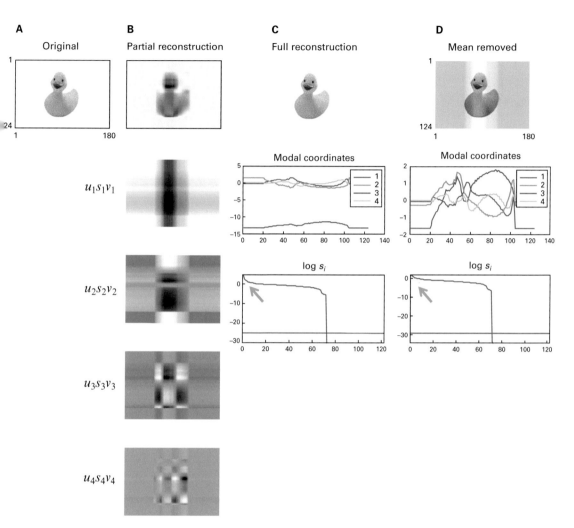

A
Original

B
Partial reconstruction

C
Full reconstruction

D
Mean removed

$u_1 s_1 v_1$

$u_2 s_2 v_2$

$u_3 s_3 v_3$

$u_4 s_4 v_4$

Modal coordinates

Modal coordinates

log s_i

log s_i

Plate 10
Singular value decomposition of a popular image in A. In B are four matrices u_i s_i v_i and the sum of these matrices is shown at the top. C shows a full reconstruction with the mean retained in the image. D shows the image with the mean removed. The arrows point to the difference that the large first mode contributes to the spectrum of singular values s_i. The horizontal red lines indicate a tolerance (red lines) based on the product of the largest of the singular values times the computer machine precision—anything below the tolerance line is computationally meaningless.

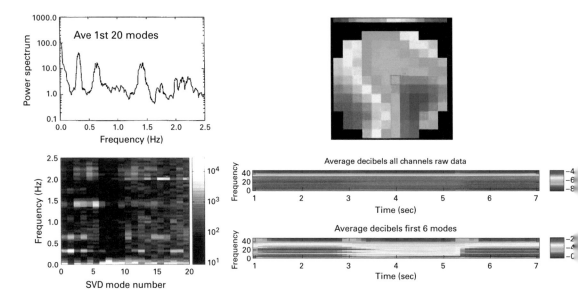

Plate 11
On left is an example of modal filtering with multitapered average spectrum from fMRI dataset, and below that, the spectrogram from individual modes. Reproduced from [MP99] with permission. On the right is an example of our spiral cortical images, with unfiltered and modal-filtered (retention of first six modes) spectrograms. The modal filtered version shows the dynamical evolution much more clearly.

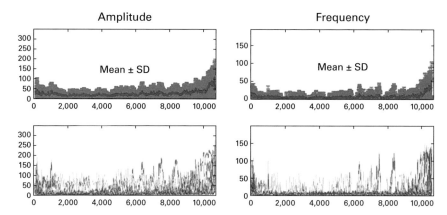

Plate 12
Effective instantaneous dimension shown for each resampled data set, below, and as mean and standard deviation, above, for amplitude and for spatial frequency. Reproduced from [SHW07] with permission.

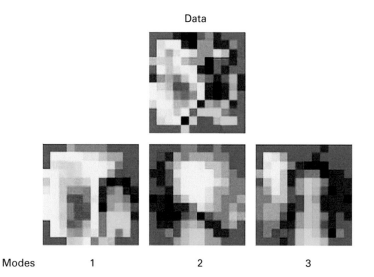

Plate 13
Mode decomposition of a typical snapshot of spiraling cortical dynamics. Reproduced from [SHW07] with permission.

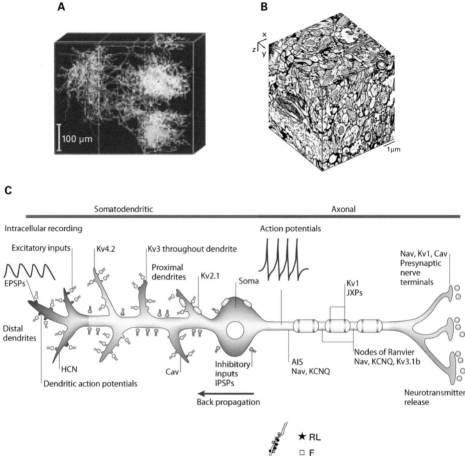

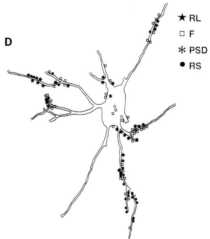

Plate 14
Complexity of neurons and their networks. A shows the reconstruction of six individual interneurons in the hippocampus, whose dendrites are shown in red, and axons in yellow. Reproduced from [GDR+05] with permission. B shows an electron micrograph reconstruction of a 350-μm reconstruction of a small cube of mammalian neuropil into which one must envision the neurons from A are embedded. Reproduced from [BD06] with permission. C shows an inhomogenous distribution of voltage-gated ion channels in a neuron. Reproduced from [LJ06] with permission. D shows an inhomogenous distribution of synapse type on a neuron. Reproduced from [LHJ95] with permission.

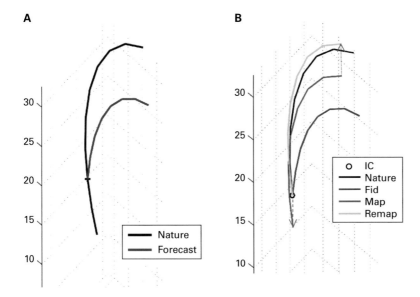

Plate 15
A illustrates imperfect model trajectory. B shows mapping strategy to bias the initial condition, IC, in order to improve the trajectory. Fid indicates the fidelity paradigm. The bias applied to the initial condition in mapping (Map) is corrected following iteration (Remap). Reproduced from [TP07] with permission.

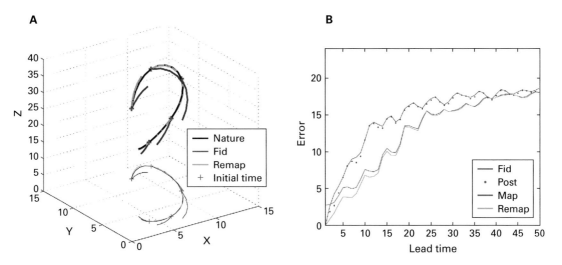

Plate 16
Sequential application, A, of the mapping-remaping (green) versus the fidelity (Fid) paradigm in the two forms of the Lorenz equations as described in the text. On the right are shown the sequential errors between the two methods. Standard post-processing (Post) to remove bias. Reproduced from [TP07] with permission.

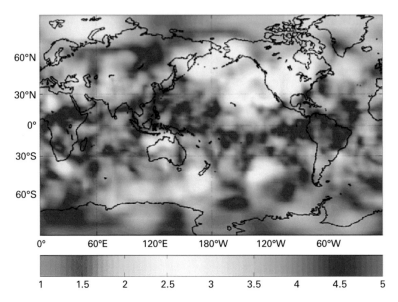

Plate 17
Bred vector dimension for wind speed across the world. Reproduced from [PHK+01] with permission.

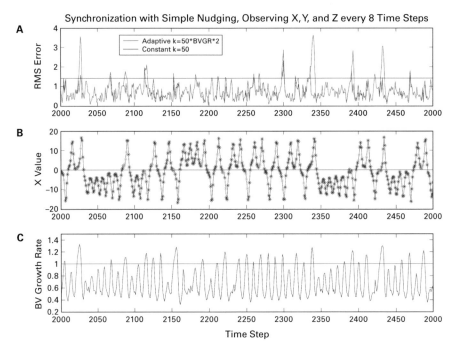

Plate 18
Nudging synchrony control for Lorenz 1963 equations using simple nudging along the axes of the original variables (blue), versus along the bred vector directions (red), shown in panel A. Panel B shows the Lorenz x in the slave system, along with the color-coded value of the magnitude of the singular value for the bred vector with blue being small, green being medium, and red being large in amplitude. Panel C shows the logarithmic bred vector growth rate, each averaged over 8 time steps. This growth rate is predictive of impending desynchronization. Reproduced from [YBL+06] with permission.

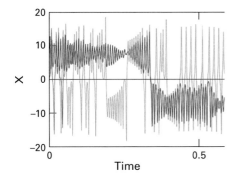

Plate 19
Two identical auxiliary slave systems, in pink and blue, driven by a dissimilar drive system in green. The auxiliary systems synchronize rapidly to each other. Modified from [YBL+06] with permission.

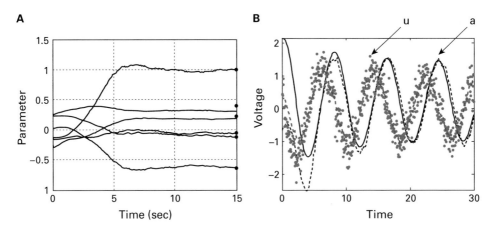

Plate 20
A shows that when tracking a heterogeneous network, the ensemble Kalman filter can be used to let the parameters float and find average values that converge close to the actual mean parameter values in the network being tracked (filled circles). B shows that, from one of the grid points, the u and a variables are closely tracked. Reproduced from [SS09] with permission.

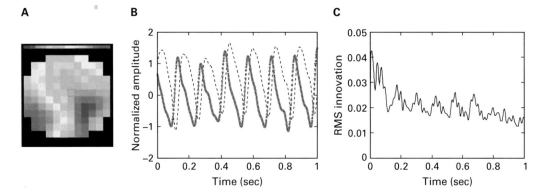

A   B   C

Plate 21
A snapshot of voltage-sensitive dye data from a sequence of thousands of snapshots collected at 1,000 samples per second (A), the consensus set tracking of a single Wilson-Cowan node (B), and the root mean square (RMS) convergence of the tracking errors (C). Reproduced from [SS09] with permission.

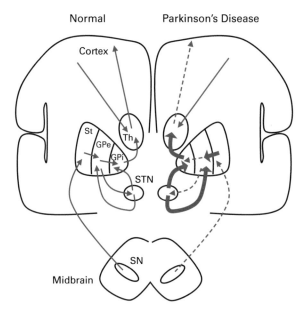

Plate 22
An extremely simplified schematic of network imbalance in Parkinson's disease. Excitation is shown in red, and inhibition in blue. The normal brain (left) is contrasted with the brain in the Parkinson's disease state on the right, where thickened (thinned) lines indicate an increase (decrease) in excitation (red) or inhibition (blue). St, striatum; GPe, globus pallidus externa; GPi, globus pallidus interna; Th, thalamus; STN, subthalamic nucleus; SN, substantia nigra. I have made no distinction between indirect and direct pathways, and customized this for the purposes of the discussion within this chapter. For a more complete and detailed description of this anatomy, see [OMRO+08].

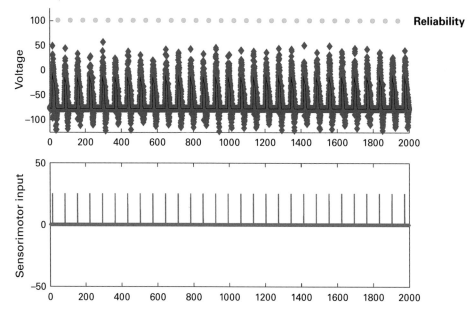

Plate 23
Normal response of TC cell to periodic sensorimotor stimulation. Voltage in reduced TC cell model (thin black line, upper panel), with substantial added noise to serve as a noisy observable (blue markers, upper panel), and sensorimotor input (red pulses, lower panel). Each time a sensorimotor pulse is reliably transmitted, a marker (green) is placed above the successfully transmitted spike. This cell is 100% reliable. Reproduced from [Sch10] with permission.

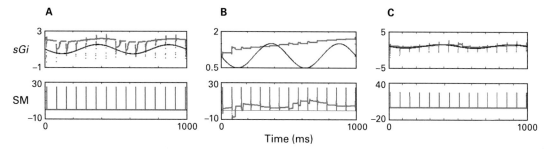

Plate 24
Tracking and estimating parameters as a function of process noises Q. Estimates of synaptic current from GPi (*sGi*, upper), and sensorimotor input (SM input, lower). Process noise parameters are (A) *sGi* $Q = 30$, *SM* $Q = 0.01$, (B) *sGi* $Q = 0.01$, *SM* $Q = 30$, and (C) *sGi* $Q = 10.0$, *SM* $Q = 0.01$. The algorithmic incorporation of such process noise can be explored in the code archive with [SS08]. Reproduced from [Sch10] with permission.

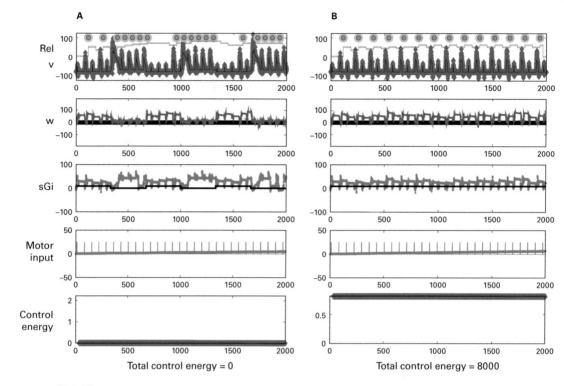

Plate 25
(A) Uncontrolled reduced TC cell dynamics in the Parkinsonian state with fluctuating current from GPi (sGi). (B) Perfect DBS stimulation filling in the troughs in the fluctuating current from GPi. Top panels show noisy observable voltage (blue symbols), reliability as piecewise continuous plots without (blue) and with (red) control on, the green circles are the timing of SM spikes, and the smaller red circles are transmitted spikes. The second panels show estimated w (red). The third panels show real (black) and estimated (magenta) synaptic current from GPi (sGi, estimated values multiplied by 10 for discriminability from the true values). The fourth panel shows real (red) and estimated (magenta) motor input (we are deliberately not trying to reconstruct motor input in the reconstruction through Q ratio adjustment). The bottom panel shows the running control energy, the squared value of the control signal at each time point, and the total sum of squares given as total control energy. Reproduced from [Sch10] with permission.

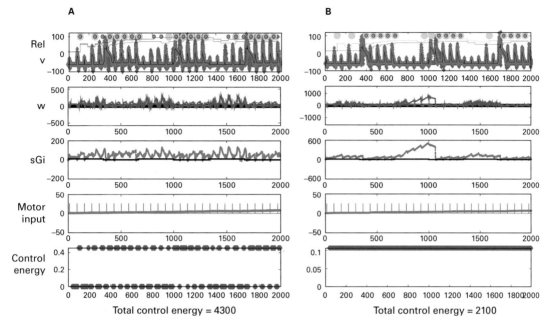

Plate 26
Feedback control scenario based on turning on and off deep brain stimulator based on a running average of the estimated T-current availability shown in A. B shows a scenario where a constant amount of deep brain stimulation is simply added to the fluctuating Parkinsonian GPi output. No adjustment of GPi constant stimulation comes close to the reliability achieved with the closed loop feedback scenario shown in A. Symbols as in figure 10.13. Reproduced from [Sch10] with permission.

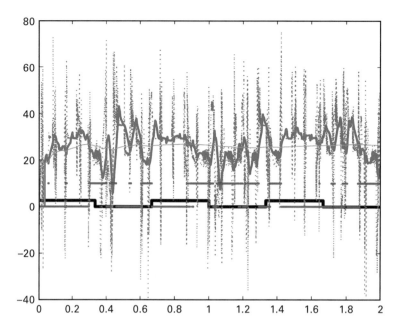

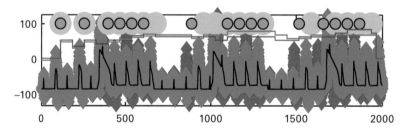

Plate 27
Estimated GPi input to TC cell (blue dotted line), and smoothed short-term moving average of this GPi input (blue solid line). We take a long-term moving average of this current (magenta line) as an adapting threshold to tell when the more instantaneous GPi input is fluctuating up or down. Crossing below the threshold determines when to turn the control on (red). The actual GPi fluctuations are shown (black lines). In the lower panel are the results with control off (blue markers), and on (red markers), and the uncontrolled (green markers) and controlled (magenta markers) spikes transmitted. The running reliability of the TC cell is plotted as a piecewise continuous line for uncontrolled (blue line) and controlled (red line) scenarios. Reproduced from [Sch10] with permission.

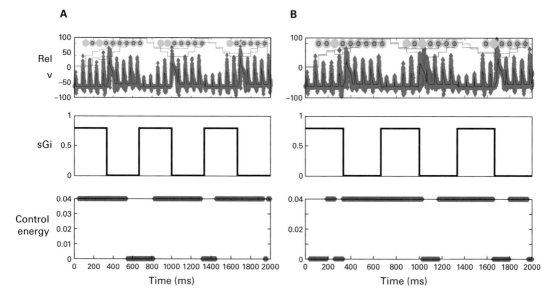

Plate 28
Control of a TC reduced-cell model using reliability as a control parameter. A shows a threshold of turning on GPi stimulation when reliability is <0.9. B shows a different strategy, using an inverse approach. In B, control is turned on when reliability is >0.5. Note that the relevant reliability in both examples is the controlled (red) piecewise continuous line in the upper panel (the blue reliability line is the uncontrolled state shown for comparison). The inherent delays in employing the moving average of reliability can be exploited so that inverse reliability control can be more reliable than when using a more intuitive strategy based on turning on stimulation when reliability falls. Reproduced from [Sch10] with permission.

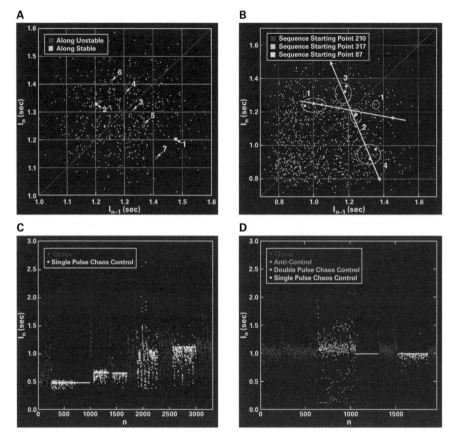

Plate 29
Controlling chaos in the brain. A shows the identification of a motif that shows an approach to the line of identity where sequential intervals, I_n and I_{n+1}, are equal. B shows multiple repetitions of this motif. C demonstrates that by directing the system to fall near the contracting direction (manifold), that the chaotic intervals could be regularized. D shows that by pushing the system away from the contracting direction, that we could prevent regularization. Reproduced from [SJD+94] with permission.

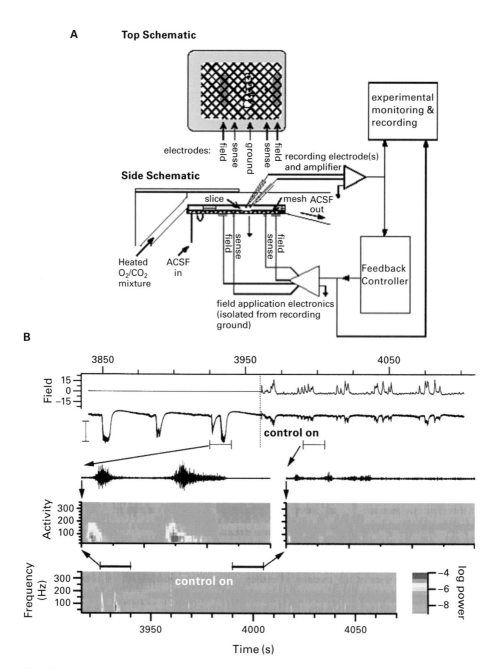

Plate 30
Adaptive feedback control of neuronal activity. Schematic of experimental chamber and electronics shown in A. In B are results demonstrating that a proportional control law could effectively suppress seizure activity in the brain slice. Reproduced from [GNWS01] with permission.

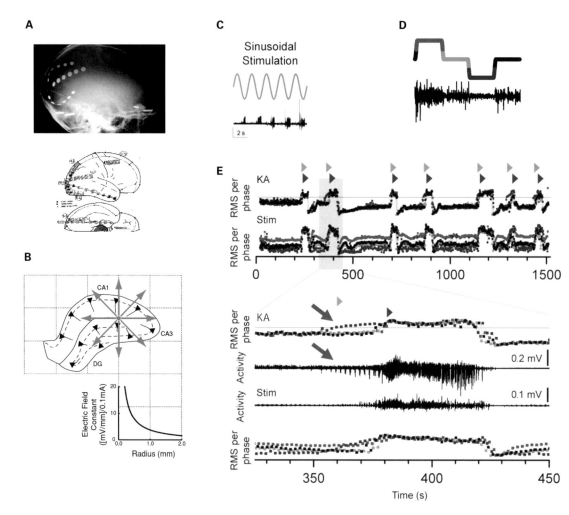

Plate 31

A shows an example of occipitally placed recording depth electrode in human epilepsy patient. Reproduced from [SSKW05] with permission. B is schematic of an axially placed stimulating electrode for epilepsy control in experimental hippocampal seizures. C shows experimental verification in animal models of sinusoidal entrainment of hippocampal neurons with low-frequency sinusoidal electrical fields. Using DC electrical fields of alternating polarity, in D, we demonstrated in E that by stimulating the contralateral side to the kainic acid (KA)–injected side with subthreshold alternating pulses, that the network demonstrated progressively increasing responses to these small stimuli in the seconds leading up to the start of the seizure (red arrows). B through E are reproduced from [RGW+03] with permission.

8 Model Inadequacy

All Models are Wrong.
—Leonard Smith, 2006[1]

8.1 Introduction

All models are wrong. The simplest physical system, a pendulum, if started with a particular initial condition and let go free, will soon diverge from the trajectory you calculate from the simple model of the pendulum that we all learned as school children. The easiest way to solve the pendulum is for small perturbations, where the small-angle approximation makes the calculation simpler. Such boring dynamics. Drive a pendulum such that it can reach the unstable fixed point standing upside down, and the trajectories can be fully chaotic and thoroughly unpredictable.

So if a simple pendulum model is difficult to use to track the motion of the physical device, can we anticipate that our neuronal, or cortical models, will be anything but abysmal in reflecting reality?

The answer is yes, and no.

The harsh reality is that if you really wanted to use a model of a pendulum to track a real pendulum, you would need to fit parameters as well as variables. Friction must be included to account for damping. And this dual-estimation problem of variables and parameters is one that we have already dealt with in chapter 5. Since our model of a pendulum is a *good* one, if we incorporate it into our nonlinear Kalman framework, it will do much better than an empirical model might at short-term prediction of the dynamics. This is one hallmark of a good model for our purposes—that it can be used in our recursive predictor-controller framework to accomplish accurate dynamical prediction.

But this is a book about neural control. Certainly, not only are the Hodgkin-Huxley equations, Fitzhugh-Nagumo equations, and Wilson-Cowan equations wrong, but the network

---

1. Second International Seizure Prediction Meeting, Bethesda, 2006.

A

B

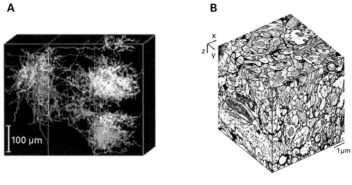

C

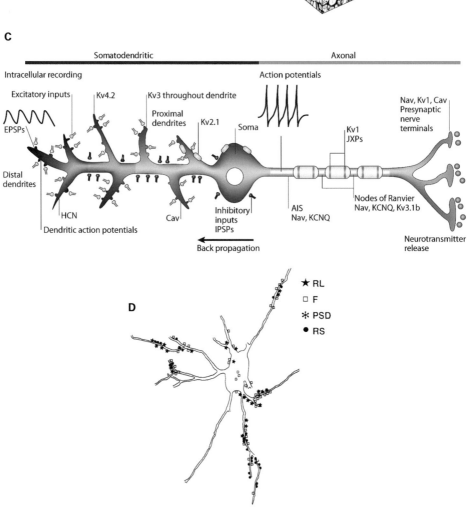

Figure 8.1 (plate 14)
(Caption on facing page)

topologies we used to account for network dynamics in chapter 6 are wrong. In fact, we know almost nothing about the true network topologies of any small piece of mammalian brain. There have been heroic efforts to define connection topologies electrically cell-pair by cell-pair [MW84, MW86]. Such issues have spawned intense research into what many call the *connectome* problem, and many researchers are intensively trying to fill this gap by, for example, attempting to define the exhaustive connectivity of cubic millimeter volumes of brain using electron microscopic and machine learning algorithms [BD06, Seu09].

The complexity of single neurons is mind-boggling (figure 8.1A). The complexity of networks of neurons is both mind-boggling and mind-generating (figure 8.1B). Viewing the morphologies from figure 8.1 (plate 14) brings up a limits of knowledge question. When we needed superb models, such as of aerospace vehicles, NASA or Boeing were able to pour resources into creating models of spacecraft or airframes that were so accurate that their model-based control systems functioned superbly. But rockets and airplanes are simple compared with the neurons in figure 8.1. Not only are these mammalian neurons complicated by their topology (figures 8.1A, B), but within each neuron, the distribution of the dynamical ion channels [LJ06] (figure 8.1C) and ligand-based receptors [LHJ95] (figure 8.1D) are very inhomogenous. Would any amount of finite research effort permit the full modeling of the features in Figure 8.1, panels A, C, and D, into panels B? No one can say at present. But skepticism over the accuracy of our large-scale biophysical models of brain [Mar06] is warranted at the present time. And if such models were to be verifiably accurate in their details, fitting their variable values and parameters to data would prove intractable with current techniques. The latter statement is diplomatic. One must worry that we may face fundamental limits to knowledge in encountering the extraordinary complexity of the brain [Sco95].

All of this is not intended to instill despondency and early termination of the activity of reading this book. But it emphasizes the paramount importance of addressing *model inadequacy*. And it underscores the need for the methods presented in this book in dealing with the ever-growing complexity of computational neuroscience models.

8.2   The Philosophy of Model Inadequacy

All of the control theoretic frameworks so far discussed—linear, extended, and ensemble Kalman filtering—focused on *noise* as a source of error. They did not implicitly formalize

Figure 8.1 (plate 14)
Complexity of neurons and their networks. A shows the reconstruction of six individual interneurons in the hippocampus, whose dendrites are shown in red, and axons in yellow. Reproduced from [GDR+05] with permission. B shows an electron micrograph reconstruction of a 350-μm reconstruction of a small cube of mammalian neuropil into which one must envision the neurons from A are embedded. Reproduced from [BD06] with permission. C shows an inhomogenous distribution of voltage-gated ion channels in a neuron. Reproduced from [LJ06] with permission. D shows an inhomogenous distribution of synapse type on a neuron. Reproduced from [LHJ95] with permission.

that the models were bad. Nevertheless, every control engineer recognizes that the uncertainty in such frameworks lump into the noise terms the uncertainty in the models
themselves.

But even the very act of sampling systems introduces model error. Natural and manmade
systems generally express their dynamics in continuous time. Most digitized implementations of control frameworks sample such continuous systems discretely, and then iterate a
discrete model to project such dynamics forward in time. *All discrete predictor-corrector
implementations of continuous systems are by definition approximations to continuous truth.*

Certainly with respect to the nervous system, we typically sample data at time scales
relatively long with respect to the underlying dynamics. While collecting data, we are not
iterating the models. While iterating the models, we are not sampling data. Although clearly
making a case for the parallel implementation of such roles, the point is to underscore that
such iterative and discrete computation misses a lot.

Another fundamental issue is that models tend to be formed at particular temporal and
spatial scales of the hierarchical organization of a natural system. For neural dynamics,
we generally assume that we can focus on cellular membrane voltage and spike dynamics,
without delving down to the level of quantum indeterminacy or quarks [BK86]. In terms
of time, we think that milliseconds count, rather than femtoseconds.[2] Our models *cannot*
account for scales of dynamics below the resolution of the model. We typically lump all such
fine scale errors into the noise terms in Kalman filtering applications. But this ubiquitous
scale dichotomy between nature and model is at its essence another ubiquitous source of
model inadequacy.

We will, in chapter 12, explicitly model lower scales into neuronal models, and show
examples of their necessity. But one can always ask how the inadequacy of the representation
of even lower scales diminishes the veracity of our models. And the more scales we add,
the more complex a tracking model becomes. In the end, we will need to trade off model
complexity against model tracking accuracy; there will always be a point at which adding
more complexity, even valid complexity, results in a worse model fit. Such a fact is, however,
very good news for our hopes to run such neuronal tracking models in real time.

8.3   The Mapping Paradigm—Initial Conditions

The first technical approach we will examine focuses on initial conditions of model iteration.
A consequence of starting an imperfect model out with a given set of initial conditions is
that it will map these starting points to the wrong place. So a practical approach is to change
the initial conditions in order to achieve a more perfect trajectory.

---

2. In certain neuronal applications, such as dynamic clamp [DBNW07], or in waveform spike sorting [MPIQ09],
the temporal scale of several tens of microseconds is commonly employed.

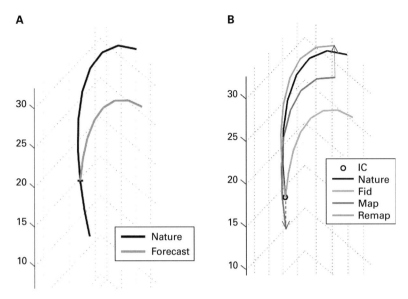

Figure 8.2 (plate 15)
A illustrates an imperfect model trajectory. B shows mapping strategy to bias the initial condition, IC, in order to improve the trajectory. Fid indicates the fidelity paradigm. The bias applied to the initial condition in mapping (Map) is corrected following iteration (Remap). Reproduced from [TP07] with permission.

An important subtlety here is that, faced with an imperfect model, we ordinarily consider that the first thing to do is to improve the model. The next thing to do is to get the initial conditions closer to nature. But bringing initial conditions closer to nature might worsen the results when the model is flawed. *The goal is to get the trajectory to end closer to the truth, not to get the initial conditions closer to the truth.* Toth and Peña argued in 2007 [TP07], that fidelity to the truth in initial conditions in fact is applicable only when the model is perfect. Perhaps only in highly constrained scientific or engineering experiments might this be reasonable.

A schematic of the trajectory produced by an imperfect model is illustrated in figure 8.2A (plate 15).

Attractors in dynamical systems theory are the asymptotic set of points to which all trajectories are, well, attracted. The physical issue involved with figure 8.2A is that the difference in dynamics between nature and the imperfect model implies that the attractor of nature is *different* from the attractor of the imperfect model. The *model drift* encountered then occurs because the initial condition, albeit close to nature, was not close to the equivalent position in the model system.

The fix for this deviant initial condition is elegantly simple: Map the natural initial condition close to the equivalent model initial condition, iterate the model, and then remap the model trajectory result back to the natural attractor coordinates. The idea is schematized with mapping vectors (magenta arrows) in figure 8.2B. Such a mapping vector is the asymptotic difference between nature and the model, and is not at all appropriately mapped by assuming that expanding the random errors of uncertainty will appropriately capture this mapping. A simple unoriented expansion of covariance inflation (expansion of the sigma points) would capture the mapping, but at the expense of almost pure uncertainty in essentially every direction but the mapping.

Toth and colleagues create a sample system by (of course, since they are working on meteorology) modifying the Lorenz 1963 model that we introduced in chapter 5 in the following way. They take the Lorenz equations from equations (5.4),

$$\dot{x} = -\sigma x + \sigma y$$

$$\dot{y} = -xz + rx - y$$

$$\dot{z} = xy - bz$$

translate the attractor in the z direction with $z = z + 2.5$, and change the σ variable from 10 to 9 to alter the dynamics. They keep $r = 28$ and $b = 8/3$ for both models.

One of the very useful aspects of the mapping paradigm is that one can look at historical data (whether weather or neurons), or run models for a while, and derive the mapping vector as the difference in the mean values of the two attractors: nature versus model.

Figure 8.2 was actually plotted from these two Lorenz systems. In figure 8.3A are shown the sequential steps of the fidelity paradigm (blue) and the remapped predictions (green), and the prediction errors for such sequential steps for these procedures are shown in figure 8.3B (see plate 16).

Note in figure 8.3B that if you keep using any of these models, the forecast errors eventually asymptote to the same values. The parameters used in both of the Lorenz systems were in the chaotic range. You can shadow a true trajectory of a chaotic system for only so long—the dynamics through sensitive dependence of initial conditions destroys information, and eventually you are just guessing the mean value of the system. This mathematical observation fits with our efforts to predict the weather. Beyond a time horizon of about two weeks, our detailed weather predictions are useless. We think that the weather is chaotic because of such observations, and it is one of the reasons why so much of the numerical meteorology literature uses chaotic systems to bench test prediction algorithms.[3] One of the characteristics of stationary nonlinear dynamical systems, even if chaotic, is that although their short-term predicability is limited, their long-term statistical averages such as mean (or dimension, or entropy) are rock steady.

---

3. Is this applicable to nervous systems? We seem adapted to be able to mirror and predict what our acquaintances are going to do in the short term. We generally have no idea what they will do two weeks from now. But following this analogy any further is not sensible.

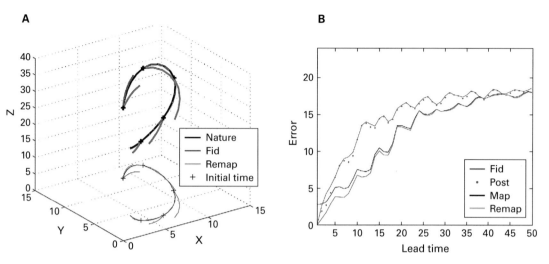

Figure 8.3 (plate 16)
Sequential application, A, of the mapping-remaping (green) versus the fidelity (Fid) paradigm in the two forms of the Lorenz equations as described in the text. On the right are shown the sequential errors between the two methods. Standard post-processing (Post) to remove bias. Reproduced from [TP07] with permission.

What if the natural system is not stationary with time?[4] Then the mean between nature and model, the mapping vector, will change as a function of time. The solution is to apply this mapping paradigm adaptively by keeping a running average of the offset between measurement and predicted model state, the innovations $y - \hat{y}$ (chapter 2). Averaging a set of innovations lets you estimate and update the mapping vector at some reasonable frequency with this time-dependent mean difference [TP07].

The mapping paradigm requires some cautions. Toth and colleagues show examples where they assume that they know the dimensions and variables of both nature and the model. This is not actually necessary for implementing such a mapping paradigm. You always have measurements of nature and your prediction of the state of nature, and you can always create a mapping vector by taking their *difference*. But at the time of this writing, the full implications of applying this elegant mapping vector to cases where we explicitly understand that the state space, variables, and dimension of the natural system are unknown remains unclear to the author.

Toth and colleagues avoided augmenting their state vector with model offsets. Their mapping paradigm was calculated *outside* of their Kalman filter. They did this to achieve simplicity, yet there is no reason to presume that model inadequacy lives outside of the state

---

4. We will define nonstationarity as time-dependent parameters. In the case where the entire model itself varies with time, one might envision *switching* models [Mur98]. But the more general problem of changing attractors has no general solution.

of a system. Very likely, a given model fits nature better at some times than others—that is, the model errors are correlated with state. Toth and colleagues were aware of an earlier work that attempted to address these issues, but its complexity and difficulties seemed to be substantial. This deep work brings us to the heart of a powerful abstraction of model inadequacy, which we focus on next.

8.4   The Transformation Paradigm

[W]e treat the forecast model as a black box, that does not yield the true time evolution of the atmosphere, and we attempt to use this black box in conjunction with observations to account for model bias in the state estimation.
—Baek et al., 2006 [BHK$^+$06]

Baek and colleagues [BHK$^+$06] sought to explicitly incorporate model inadequacy into an ensemble Kalman filter framework. They focused on model errors that have no random component, and call this offset *model bias*.

Their approach is to augment state vectors with bulk model error terms. They estimate these error terms dynamically with their ensemble Kalman filter, seeking to *parameterize* the forms of model bias.

Baek and colleagues assume that the model is a finite state representation of an infinite dimensional system. They assume that the model equations are not precisely known, and they expressly ignore natural dynamics at finer scales than the model.

Let's assume that there is a true state, denoted by x^t, for time n, of a natural system whose dynamics, F^t, can be described as

$$x^t_{n+1} = F^t(x^t_n)$$

Let's further assume that there is a model system and state, denoted by x^m, whose dynamics F^m are governed by

$$x^m_{n+1} = F^m(x^m_n)$$

Almost always, the dimensions of $x^t_n > x^m_n$, but for now we will assume that they are the same. There are three general bias models:

Bias Model I

Let's assume, in the fidelity paradigm of Toth et al. [TP07], that ideally we place the initial condition of the model very close to the initial condition of nature, so that $x^m_{n-1} \approx x^t_{n-1}$. Then the *bias* is

$$b^t_n = F^t(x^t_{n-1}) - F^m(x^t_{n-1})$$

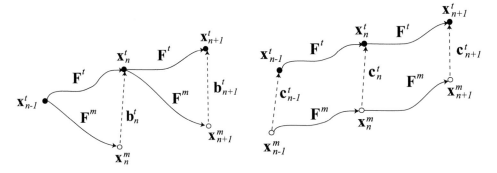

Figure 8.4
Bias model I is shown in A. Because of the differences in true, F^t versus model F^m dynamics, the bias, b, must be accounted for. B shows the contrasting geometry from bias model II. Reproduced from [BHK$^+$06] with permission.

as illustrated in figure 8.4. If the bias is constant in time, its dynamics are expressed as

$$x_n^t = F^m(x_{n-1}^t) + b_n^t$$

$$b_n^t = b_{n-1}^t$$

The constant model for b is one that we have encountered previously, in chapters 2 and 5, and we can treat it as a parameter with trivial dynamics in an ensemble Kalman filter. This model assumes that the best forecast is produced when the fidelity paradigm is operative, that is, you can place the initial condition very close to that of nature, $x_{n-1}^m \approx x_{n-1}^t$. Indeed, we assume that following correction through $x_n^t = F^m(x_{n-1}^t) + b_n^t$, our observation function is

$$y = A(x_n^m + b_n^t) + \varepsilon_n$$

where ε represents random measurement error.

   Bias model I, by itself, is not as good as the mapping paradigm described by Toth and Peña [TP07], which leads us to bias model II.

Bias Model II

Let's assume that our model system has a different attractor from nature. We expect that this is always true. Put your best and most perfect initial condition measured from fidelity with nature, and no simple additive drift such as bias model I can account for the discrepancy.
   We seek a transformation of coordinates from nature to the model, as

$$c_n^t = F^t(x_{n-1}^t) - F^m(x_{n-1}^m)$$

$$= F^t(x_{n-1}^t) - F^m(x_{n-1}^t - c_{n-1}^t)$$

The geometry of bias model II is brilliantly illustrated in figure 8.4B reproduced from [BHK$^+$06]. The system dynamics can now be approximated, if we assume a constant model for c:

$$x_n^t = F^m(x_{n-1}^t - c_{n-1}^t) + c_n^t$$

$$c_n^t = c_{n-1}^t$$

Assuming a constant c is equivalent to [TP07], but in general, one can employ nontrivial dynamics on c, as discussed in detail in [BHK$^+$06]. As before, we assume that our observation function is now

$$y = A(x_n^m + c_n^t) + \varepsilon_n$$

It is now a logical step to consider bias model III.

Bias Model III

In bias model III, we combine bias models I and II to produce, for static models b and c,

$$x_n^t = F^m(x_{n-1}^t - c_{n-1}^t) + b_n^t + c_n^t$$

$$b_n^t = b_{n-1}^t$$

$$c_n^t = c_{n-1}^t$$

Parameterizing Model Inadequacy

We now create vectors of parameters that correspond to the model bias schemes just described.

Assume that the evolution of state x, dx/dt represents a series of N ordinary differential equations that prescribe the evolution of system state. Then the most general form would be

$$\frac{dx}{dt} = F(x + c) + b$$

where

$$b = [b_1, b_2, \ldots, b_N]^T$$

$$c = [c_1, c_2, \ldots, c_N]^T$$

where T indicates transpose. We now augment the state vector, v, to be used in the ensemble Kalman filter, with

$$v = [x, b, c]^T$$

Finally, Baek and colleagues [BHK$^+$06] employ covariance inflation as we used in chapter 2, which they find can stabilize the Kalman filter in the presence of model error.[5]

Figure 8.5A–C illustrates the results of numerical simulation of the Lorenz 1998 model [LA98] subjected to model biases I, II, and III. Interestingly, the use of type III bias parameters always gives the best results, whether the true model bias is type I, II, or III. One reason to understand this is, for instance, in the case of type I bias. In figure 8.5D, note that using type III parameterization, rather than type I parameterization, decreases the distance from the true trajectory for true type I bias.

Using the augmented state vector v leads to a covariance matrix for the errors that expands from x^2, from chapter 5, to $(2x)^2$ for bias model II, and $(3x)^2$ for bias model III. Such augmentation dramatically increases the computational overhead when Cholesky decompositions of such a matrix are required in the calculation of sigma points.

The convergence or *settling* time for such algorithms can be large. Baek and colleagues demonstrated that this settling time is strongly influenced by the magnitude of the covariance inflation parameter. Another finding was that if they placed spatial diffusion constraints on the rate at which the model bias parameters could change, this improved the settling time. For all of the results in figure 8.5, the tuning of covariance inflation is critical to accuracy. All of this tuning was performed exhaustively offline. It's important to note in figure 8.5 that, to some degree, increasing covariance inflation can compensate for a less adequate bias model.

Another issue that the author raises is whether Baek's insertion of the model error parameters within the augmented state vector is the best place for them. The ensemble Kalman filter will tend to find mean values for the distribution of ensemble particles or sigma points. But such values might not be the best reflection of the model errors. It remains an open question whether optimizing type I, II, or III model error parameters might best be performed in the manner of Toth and Peña [TP07], outside of the Kalman filter. Such optimization might be far more computationally efficient, since the covariance matrices would be substantially smaller.

The complexity of the calculations in Baek et al. [BHK$^+$06] increased dramatically for type I or type II parameterization. But the complexity was tremendously increased for type III. One would, if possible, want to pick just type I or type II, but the results show that type III, if the covariance inflation is tuned properly, is always numerically best.

Baek and colleagues never addressed how one might approach tuning covariance inflation in real time. We will address this next.

---

5. In their implementation, their covariance inflation is proportional to the trace of the state covariance matrix times the identity matrix, rather than by an arbitrary small constant times the identity matrix as we employed in chapter 5. Their formulation provides a degree of adaptability as the trace of the covariance matrix changes dynamically with time.

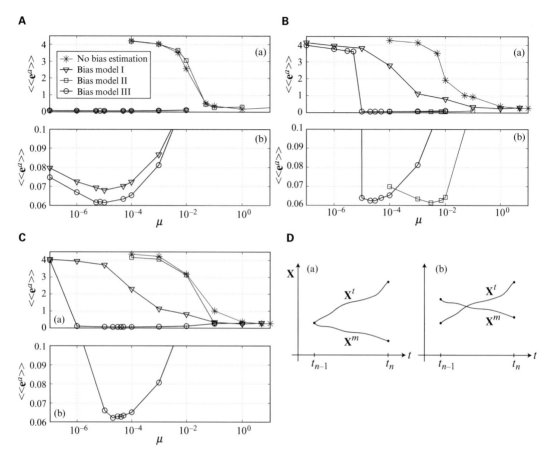

Figure 8.5
Employing the three model error parameterization schemes for type I model bias (A), type II model bias (B) and
type III model bias (C) The ordinates are forecast error, while the abscissa for A–C are covariance inflation. Note
that type III model bias, for appropriate tuning of covariance inflation, always gives the best results. One reason
for this is illustrated in D, where in true type I bias, the overall error is less when model parameter c is used.
Reproduced from [BHK+06] with permission.

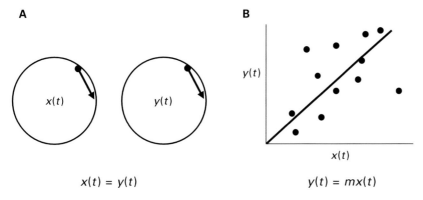

Figure 8.6
Schematic of two systems in synchrony in A, and inferring synchrony through correlation in B.

8.5   Generalized Synchrony

These findings lend credibility to the suggestion that a synchronization principle is also fundamental to the relationship between the brain and the external world,
—Duane et al., 2006 [DTW06]

Consider for a moment what a Kalman observer system is doing. It is a dynamical system that, if perfect, would emulate the system it hopes to track. It was recognized several years ago that control systems seek to *synchronize* with the natural system. In an airplane cockpit, the model of the airframe dynamics seeks to perfectly synchronize with what the aircraft is actually doing.

Synchrony implies that two systems, $x(t)$ and $y(t)$, do the same thing at the same time: $x(t) = y(t)$ (figure 8.6A). For identical systems, synchronization can be exact. For non-identical systems, they can come close, but never exactly synchronize.

Synchrony is traditionally inferred from correlation (in time) or coherence (in frequency). Figure 8.6B shows that, for noisy data, we often try to demonstrate that a linear functional relationship relates two systems to each other when they are synchronizing. For linear systems (which do not exist), this is reasonable.

It came as a surprise to the dynamics community when Lou Pecora and Tom Carroll demonstrated in 1990 that chaotic systems could synchronize [PC90]. In figure 8.7, we show one of their examples for two sets of chaotic equations. Although the state of the system on the right starts far away from the drive system on the left, it rapidly synchronizes to it. There are some serious subtleties to chaotic synchronization. Both systems remain chaotic. But if you perturb the driven system, there are no longer expanding directions. All small perturbations stably return to the same trajectory, which is the same as the drive system that enslaves it. This is an important condition for all synchronization—it is stable to perturbation. Another way to envision this is that when synchronized, such driven systems

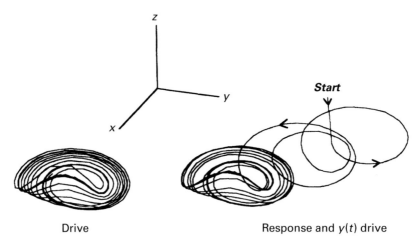

Figure 8.7
Chaotic synchronization of Rössler equations. Reproduced from [PC90] with permission.

forget their initial conditions and adopt the dynamics of the driving system. In neuroscience applications, we rarely apply stability tests to infer synchrony, relying on correlation in time, although there are exceptions [FGS03].

What if the systems were not identical? We never expect systems to be identical. And in this book, we never anticipate that they have linear dynamics. In 1995, Nikolai Rulkov and colleagues [RSTA95] defined what they termed *generalized synchronization* for dissimilar coupled nonlinear systems. Figure 8.8 shows examples of the difference, as coupling (C) is increased, between identical and nonidentical nonlinear systems as they synchronize.

When we defined linear correlation in figure 8.6B, we defined a *functional relationship* between two systems. For nonidentical systems, a straight line is no longer adequate to describe the relationship. The function is nonlinear.

Defining metrics to tell when such dissimilar nonlinear systems were or were not in synchrony really depends on whether you can map from one system's state to another. If there was a 1:1 correspondence, then mathematically a homeomorphism might be present. Such 1:1 correspondence could be inferred through testing for *continuity*. Furthermore, if the functional relationship preserved the orientation within such continuity functions, there might be differentiability of the function—a *diffeomorphism* between the two systems. A detailed set of mathematical and statistical tests emerged to quantify generalized synchrony [PCH95]. Another way to quantify such synchronization was through mutual prediction, and this was applied to neural networks and single neurons in mammalian spinal cord circuitry in [SSC+96].

Duane and colleagues [DTW06] proposed that data assimilation through Kalman filtering can be envisioned as a synchronization problem. Without a controller, the coupling between

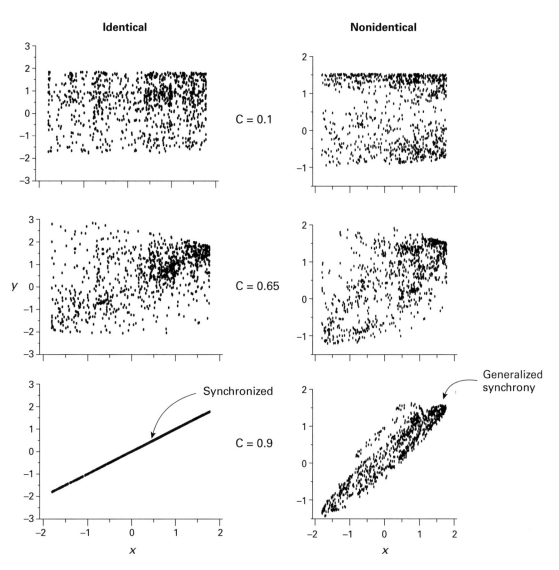

Figure 8.8
Contrast between identical and nonidentical nonlinear systems as a coupling parameter, C, is increased. Identical (left) and generalized (right) synchronization patterns are illustrated. Reproduced from [SSC+96] with permission.

nature and the observing model is one way, and the situation is similar to the coupling for nonlinear systems described in [PC90] or [RSTA95]. They also speculated that some of the psychology of Carl Jung on *synchronicities* [JP55a] might fit within such an observing synchrony framework (we defer further discussion of Jung until chapter 13).

They examined a simple example of tracking a system where the attractor switched from one set of Lorenz 1963 equations to another

$$\left. \begin{aligned} \frac{dx}{dt} &= \sigma(y - x) \\[2mm] \frac{dy}{dt} &= rx - y - xz \\[2mm] \frac{dz}{dt} &= xy - bz \end{aligned} \right\} \;\rightarrow\; \left\{ \begin{aligned} \frac{dx}{dt} &= zy - bx \\[2mm] \frac{dy}{dt} &= rx - y - xz \\[2mm] \frac{dz}{dt} &= \sigma(y - z) \end{aligned} \right.$$

In figure 8.9, they illustrate an experiment where they use a Kalman filter to track the above switching system. It does a reasonable job, except that at the point of switching

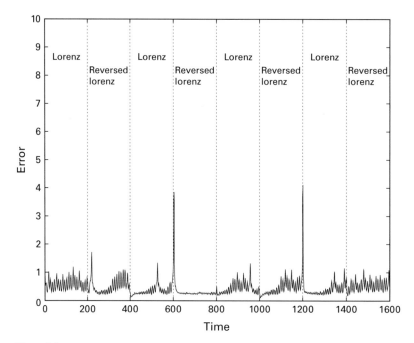

Figure 8.9
Kalman filter loses track of a system during rapid changes in dynamics. During such time there are tracking catastrophes. Increasing covariance can help mitigate against this. Reproduced from [DTW06].

(dotted lines), there are tracking *catastrophes* as the filter looses track of the system, and it takes finite time to reacquire the trajectory. These desynchronization bursts are seen when the trajectory of the coupled system blows off of the synchronization manifold that had previously defined generalized synchrony.

One way to decrease the effect of these blowouts is to increase covariance inflation so that the tracking ensemble is broad enough to not lose the trajectory. And a strategy to deal with anticipating such blowouts is considered next.

8.6   Data Assimilation as Synchronization of Truth and Model

It seems unlikely that a group of undergraduate students gathered during their summer vacations under a grant to foster research would produce a seminal work at the cutting edge of control theory, but the work of Yang and colleagues appears to be that rare event [YBL+06].

Using a coupling function, G, of the difference between a variable, α, and its observed measure, α_{obs}, was termed *nudging* by Hoke and Anthes [HA76]. Yang and colleagues [YBL+06] proposed nudging the Lorenz 1963 equations as follows:

$$\frac{dx_s}{dt} = \sigma(y_s - x_s) + \delta_x k(\bar{x} - x_s)$$

$$\frac{dy_s}{dt} = rx_s - y_s - x_s z_s + \delta_y k(\bar{y} - y_s)$$

$$\frac{dz_s}{dt} = x_s y_s - bz_s + \delta_z k(\bar{z} - z_s)$$

where s refers to the *slave* system being driven by nature, k is a coupling constant (the inverse of which is a time scale of the coupling), $\bar{x}$, $\bar{y}$, and $\bar{z}$ are the observations of nature, and the $\delta_{x,y,z}$ are indicator functions equal to 1 when the nudging is in the direction of a given variable, and zero otherwise. Simple nudging through the x variables of two Lorenz systems was shown by Pecora and Carroll [PC90] (see figure 8.7).

Yang and colleagues [YBL+06] wished to observe a noisy Lorenz system, and sought to synchronize to it using nudging. So they set up their coupling to use all combinations of variables: x, y, xz, yz, xy, and xyz.[6] In figure 8.10, we show the errors of such coupling. There are some notable things to comment on. First, coupling through all variables is not necessarily as good as coupling through only one or two variables. The efficacy of coupling depends on coupling strength. These issues open a new dimension to Kalman observers— that picking the variables to observe, and the strength of the coupling gain on the observable

---

6. Coupling through z alone is often unsuccessful in the literature for the Lorenz 1963 equations.

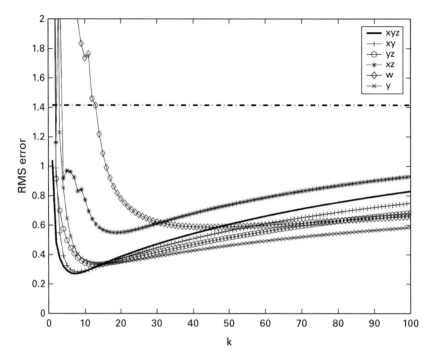

Figure 8.10
Synchrony error versus coupling strength for a variety of coupling variables in the Lorenz 1963 equations. Reproduced from [YBL+06] with permission.

variables, are factors that will affect the ability of the observer system to track a natural system. Intriguingly, Yang and colleagues found that if the sampling interval was lengthened, increasing the coupling strength would help compensate for this.

The next finding of Yang and colleagues was that as the synchronization coupling strength was lowered below the value that would synchronize the systems, intermittent bursts of desynchronization (tracking catastrophes) would be observed. Such *intermittency* in dynamics is well described. It can be a consequence of the system dwelling close to but not asymptotically on an attractor, when the dramatic excursions are termed *blowout bifurcations* [OS94]. Or, especially attractive to our purposes, such excursions can be found in attracting systems when noise is intruduced, and this has been termed *bubbling*, especially in the case of synchronizing systems [ABS94].

Where the Wind Blows

We take a brief excursion to examine *bred vectors*. If we took weather data, and measured, say, wind speed velocity from a grid of points above our heads, we can iterate the vector of wind velocities forward through our favorite weather model and see what the forecast is. We could also perturb this initial condition, making sigma points as in chapter 2, and push each of the ensemble points through the forecast model. Let's say that our ensemble of k initial points were spherical in state space about the initial condition. Then, if the atmosphere operator were a matrix, matrix times circle equals ellipse (chapter 7), and we have an ellipsoid. Perform a singular value decomposition on the covariance matrix $B^T B$ of the ensemble of k bred vectors, B, and you can ask which of the expanding directions is most significant [PHK+01].

The largest expanding direction(s) are where the initial state was most sensitive to perturbation. You can define a *bred vector dimension* ψ from

$$\psi(\sigma_1, \ldots, \sigma_n) = \frac{\left(\sum_{i=1}^{k} \sigma_i\right)^2}{\left(\sum_{i=1}^{k} \sigma_i^2\right)}$$

where if each column of B were normalized to 1, then $\sum_{i=1}^{k} \sigma_i^2 = k$. The key to this computation is that the number of singular values, σ_i, that are significantly larger than zero will dictate an effective dimension between 1 and k. So let's look at what this implies for the winds of the world in figure 8.11 (plate 17). San Diego, California, is easy to predict. It's the same every day. Miami, Florida, is (with respect to wind speed) chaotic. So let's breed these vectors in a control context.

From Wind to Control

We will breed with the slave system (the observer), since the natural system can be considered unknown. This is the case for nearly every control process we will encounter with the nervous system. The bred vector, calculated from sigma points surrounding the state estimate applied to the model system, predicts where the system will be growing. One can use nudging in the direction of the bred vectors to correct this expected blowout/bubbling. In figure 8.12A we present the calculations from [YBL+06], showing how nudging from the bred vectors gives better synchrony with a true system than simple nudging. Note that the substantial excursions from synchrony, the peaks in panel A, occur as intermittent desynchronous bursts. But bred vectors give us far more than that. Look at panel B in figure 8.12. The color coding indicates the magnitude of the singular values of the largest bred vector, superimposed on the Lorenz x variable. Blue is small, green is medium, and red is the largest in magnitude. The large-magnitude bred vectors tell you when the system is about to blow out of synchrony in panel (see plate 18).

All of this is well and good if you know the equations for nature. You don't. But you do know the equations of your observer model.

If you couple (unidirectionally) a natural system to a model system, at sufficient levels of coupling, you may develop a functional relationship, ψ, between the drive and response system, as schematized in figure 8.13. For unidirectional coupling, the response system

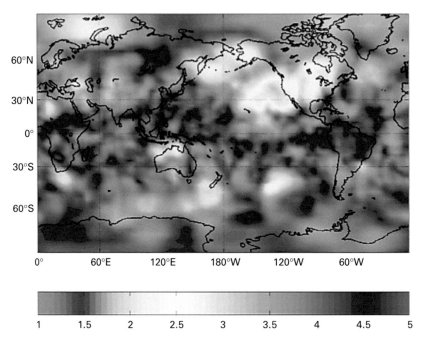

Figure 8.11 (plate 17)
Bred vector dimension for wind speed across the world. Reproduced from [PHK+01] with permission.

always gives you information on the drive system, so you have some drive system predictability from observing the slave system before it is enslaved. Once it is sufficiently
coupled, the drive system can enslave the slave system. This means that a stable functional relationship can develop. At this point, the existence of the function means that
you not only can predict the drive system, but the drive system can predict the response
system. The prediction becomes bidirectional. If you were using a method such as continuity [FSGS01] to test the existence of a functional relationship, you would see that for
a given set of points, ϵ, in the response system, they all came from a given place, δ, in
the drive system [PCH95, NPS04]. Our ability to detect such functions always has limits. In real life, the function will always have a certain uncertainty or thickness, and we
can define the limits to the detection of generalized synchrony in such settings [SBJ+02].
Nevertheless, our ability to detect synchronization (and the absence thereof) in settings
of increasing nonlinearity and noise is improving, and further discussion can be found in
[NPS04, SSKW05, NCPS06].

 Key in this picture is that the stability of the functional relationship means that the slave
system must *forget its initial conditions*. So by analogy, if you started two copies of the
slave system from different initial conditions, and coupled them strongly enough to a single
drive system, you would find that they synchronized *to each other*. Since other than the

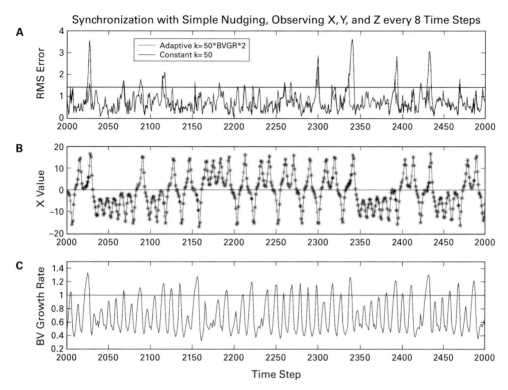

Figure 8.12 (plate 18)
Nudging synchrony control for Lorenz 1963 equations using simple nudging along the axes of the original variables (blue), versus along the bred vector directions (red), shown in panel A. Panel B shows the Lorenz x in the slave system, along with the color-coded value of the magnitude of the singular value for the bred vector with blue being small, green being medium, and red being large in amplitude. Panel C shows the logarithmic bred vector growth rate, each averaged over 8 time steps. This growth rate is predictive of impending desynchronization. Reproduced from [YBL+06] with permission.

initial conditions, the slave systems you know, and can be identical, they will identically synchronize as schematized in figure 8.14.

As you might imagine, the conditions to see paired slave or *auxiliary* systems synchronize are not quite as strong as between the driving system and the slave systems. In 1996, Pyragas [Pyr96] defined the use of auxiliary systems to infer the possibility of generalized synchronization. He defined a *weak* and *strong* case for generalized synchronization, and considered the identical synchronization of the auxiliary systems as an expression of weak synchronization. Nevertheless, seeing such synchrony is a necessary, if not sufficient condition, for the existence of generalized synchronization between a natural system and these auxiliary systems. The situation is schematized in figure 8.15. Interestingly, Pyragas also examined the case of one completely different system (Rossler) driving two identical auxiliary systems (Lorenz). The onset of generalized synchronization can be seen at sufficient

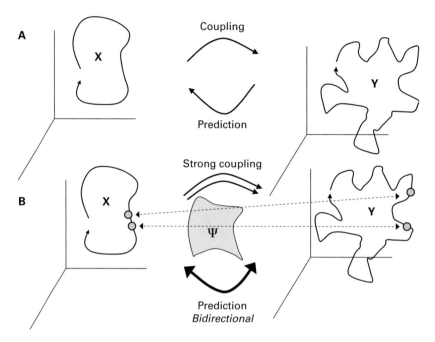

Figure 8.13
A shows a drive system, X, driving a response system, Y. The coupling goes from left to right, and the prediction goes from right to left for weak coupling. In B, as coupling is increased, a functional relationship develops, ψ, after which predictability is bidirectional.

levels of coupling in the synchronization of the auxiliary systems. We have also examined the stochastic synchronization of auxiliary systems; they can be observed to synchronize from a purely stochastic drive [SSC+96]. This stretches the bounds of what we envision as the functional relationship between slave and drive. The slave system has information from the stochastic drive system, but cannot predict the stochastic drive system because it is random. In contrast, at sufficient levels of coupling, the slave system is predictable from the drive systems [SSC+96]. Such a situation does establish a formal relationship between drive and response systems, and the auxiliary systems do tell you *something* about the uncharacterized drive system. This situation is further discussed in [YBL+06]. You don't have to know everything about a natural system to have your tracking be useful.

Now back to control.

If one sets up a Lorenz system driver, and a set of matched copies of auxiliary systems with different parameters from the drive and with different initial conditions from each other, one can observe an example of such auxiliary systems synchronizing (figure 8.16, plate 19).

Note the time needed to observe synchronization. This points to a fundamental difference between synchrony and typical Kalman frameworks. Kalman filtering predicts the best guess

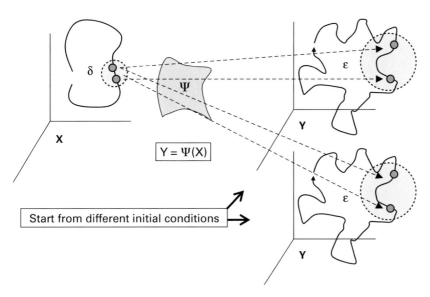

$$Y = \Psi(X)$$

Start from different initial conditions

Figure 8.14
Example of driving two copies of response systems. Starting from different initial conditions in the response systems, they will identically synchronize with each other under conditions when generalized synchronization with the drive system develops in each. If the drive system dynamics is unknown, we may be able to observe the auxiliary response systems synchronize with each other.

at the present state. A bred vector approach tells you what is about to happen a short time into the future. Kalman filters are a bit of a *nowcast* [DTW06].

And finally, Yang and colleagues [YBL+06] offer a prescription for adaptive adjustment of observation models in anticipation of an increase in forecast errors. If one were either to calculate the amplitude of the bred vector singular values (figure 8.12B), or average the values for several recent time steps (figure 8.12C), such an increase in local instability of the synchronization could predict the desynchronization bursts shown in figure 8.12A. The results pave the path for what we have not had up until now—a way to adaptively tune *covariance inflation* on the fly. Yang and colleagues also suggest tuning the coupling strength or gain on the fly. All of this can be proposed based on calculating bred vectors on the fly using auxiliary systems. One can envision spawning auxiliary systems, with random perturbations, at periodic intervals.

But how to inflate covariance with respect to bred vectors remains an open issue. In Yang et al. [YBL+06], the focus is on the direction of greatest expansion. In certain dynamical systems, actually in probably *every* real natural system, the number of expanding and contracting directions will not be constant as a system evolves. The mathematical term for this is non-hyperbolic. In such systems, the *thin* contracting directions of a set of initial conditions may cause the propagated model trajectories to miss the real trajectories entirely.

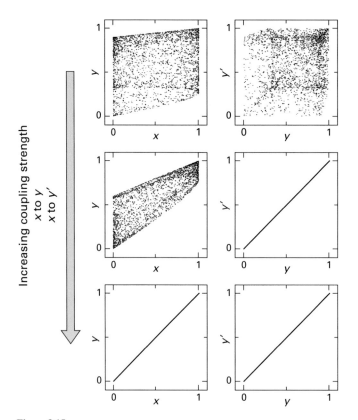

Figure 8.15
As coupling is increased from top to bottom, a drive system, x, will synchronize with a response system, y. In addition, copies of the response system, y and y', will synchronize to each other, but at a lower coupling than that required to see synchrony between the driver and response systems. Reproduced from [Pyr96] with permission.

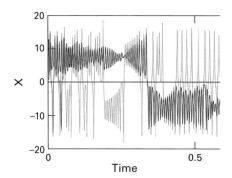

Figure 8.16 (plate 19)
Two identical auxiliary slave systems, in pink and blue, driven by a dissimilar drive system in green. The auxiliary systems synchronize rapidly to each other. Modified from [YBL$^+$06] with permission.

One might need to inflate such initial conditions along the thin or contracting dimensions [DY06]. Reconcilling these two tales of inflation will require additional research, but they both point to the same conclusion. Use enough anticipatory inflation to keep close to real trajectories.

And if auxiliary systems were run for only one time step, what are they equivalent to? They are our sigma points from previous chapters—each is a perturbed copy of our slave system. In addition to calculating the mean of the iterated sigma points, one can use the singular value decomposition of the covariation of the iterated sigma points to estimate whether the system is going to be prone to impending instability or not. If so, then add more covariance inflation to try to hold onto the system for the expected wild ride to come. There is more data in the ensemble Kalman filter than has been previously appreciated.

Fusion - Synchronous Nudging and Kalman Filtering

In a fusion of the concepts discussed in this chapter, one can also add a control system on an observer system to help it synchronize to a natural system. The natural system would be uncontrolled in this pure data assimilation scenario. Why do this? Because this is a way to adaptively estimate both state and parameters of the natural system dynamically over time.

Discussions of such adaptive control of the synchoronization of an observer system can be found in [Hua06] and [DYK07].

If we consider a dynamical system

$$\dot{x} = f(x)$$

Where x is an augmented state of varibales and parameters. Then following [Hua06], we can design an adaptive feedback controller for an observer system

$$\dot{y} = f(y)$$

by adding a control term, u, as

$$\dot{y} = f(y) + u$$

where the control law

$$u = -k(y - x)$$

and coupling strength for positive k can be adaptively estimated as

$$\dot{k} = \lambda(y - x)^2$$

for small λ.[7]

---

7. It is probably best to split the variables, x, from the parameters, p, in such systems as $\dot{x} = f(x, p)$, and for the observing system, as $\dot{y} = f(y, q) + u$. One then develops a separate evolution term for the parameters as $\dot{q} = -\beta(y - x)f(y)$, where the β are a vector of arbitrary constants [DWC09].

If we knew x, then for low noise and uncertainty in model and system, such an adaptive nudging scheme would serve to synchronize the observer to the natural system. Of course, this is a chapter about ubiquitous uncertainty and, of course, we never know x.

It was [DWC09] who suggested fusing the above adaptive nudging scheme with an unscented Kalman filter estimation of x, $\hat{x}$, as

$$\dot{y} = f(y) - k(y - \hat{x}) \tag{8.1}$$

Deng and colleagues [DWC09] go further to apply the approach in (8.1) to both the Fitzhugh-Nagumo and the Hindmarsh-Rose neuronal models. In this application, they assumed compete knowledge of the underlying equations, and demonstrated that this *combined* approach (adaptive nudging and unscented Kalman filter) to parameter estimation was extremely robust to added noise. The further application of their combined approach with the concepts of model uncertainty described in this chapter remain open problems begging for exploration.

8.7   The Consensus Set

Finally in our brief tour of model inadequacy, we will return to cortex and real data. In our implementation of the Wilson-Cowan equations in chapter 6, we assumed that each node in these equations had the same parameters. Such an assumption is unreasonable for brain, no matter how homogeneous or isotropic we imagine it to be.

In their most general form, the Wilson-Cowan equations can be written as follows, where for each (i, j) grid point, we have

$$\dot{u}_{ij} = c_{ij1}u_{ij} + c_{ij2}a_{ij} + c_{ij3} \sum_{i'j'} e^{-d} H(u_{i'j'} - c_{ij4})$$

$$\dot{a}_{ij} = c_{ij5}u_{ij} + c_{ij6}a_{ij} \tag{8.2}$$

where the sum is taken over all other nodes $i'j'$ in the network, d denotes the distance between grid points ij and $i'j'$, and H denotes the Heaviside function or a smoothed version (such as a steep sigmoid) if a continuous equation is desired. The u variable represents instantaneous excitation, and the a variable represents recovery or adaptation. Biologically meaningful parameters ordinarily require $c_{ij1}, c_{ij2}, c_{ij6} < 0$ and $c_{ij3}, c_{ij5} > 0$.

We will assume that the true system is heterogeneous, and vary the parameters for each node by 10%. For a 16×16 grid, we have $256 \times 6 = 1,536$ actual parameters. Fitting such a multiplicity of parameters would be unfeasible. One way to address this is to let each parameter float in the ensemble Kalman filter, and see if they will settle down and converge to optimal values that are the best local mean fits for the grid in question. In [SS09], we show that this is feasible, as shown in figure 8.17 (plate 20). But this demonstration is a

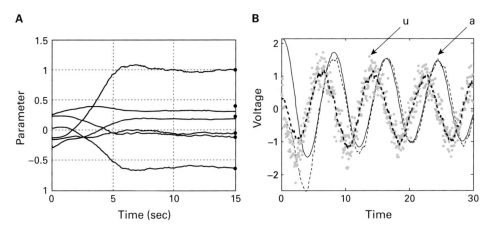

Figure 8.17 (plate 20)
A shows that when tracking a heterogeneous network, the ensemble Kalman filter can be used to let the parameters float and find average values that converge close to the actual mean parameter values in the network being tracked (filled circles). B shows that, from one of the grid points, the u and a variables are closely tracked. Reproduced from [SS09] with permission.

bit unrealistic, to say the least. We used knowledge of both u and a, and we used the same model (with heterogeneity) to both generate the data and to assimilate it.

Before going further, we need to deal with symmetries (see also discussion in chapter 6). Symmetries in equations are important to address in any Kalman filtering assimilation of data. The reconstruction algorithm cannot distinguish between variables that are functions of each other. Consider the task of determining $a(t)$ and $c_1, \ldots, c_6$ in equations (8.2) from the knowledge of $u(t)$ alone (we drop the ij subscripts for simplicity). Any experimental data u that is consistent with $a(t), c_1, c_2, c_3, c_4, c_5, c_6$ is consistent with $a(t)/\gamma, c_1, \gamma c_2, c_3, c_4, c_5/\gamma, c_6$ for arbitrary nonzero γ, meaning that the dual estimation problem is underdetermined. Applying the ensemble Kalman filter directly to this problem results in parameter drift. To achieve data assimilation in this context, we must force γ to assume a unique value.

Now we apply such symmetry-less observation equations to real data from the cortex. We will employ voltage-sensitive dye imaging data from [HTY+04] (see chapter 6), and seek a consensus set in our assimilation model. The results are shown in figure 8.18 (plate 21). There are some important subtleties in these results. The consensus set is the one that predictively tracks the data the best. *It is not going to reengineer the brain and reveal the physiological parameters that correspond to our tracking model.* The parameters in figure 8.18B are physically meaningless. But the convergence of the tracking is excellent in figure 8.18C.

Simultaneous with our work in brains, came the results of Cornick and colleagues in fluids [CHO+09]. They were studying Rayleigh-Bénard convection experiments. In these

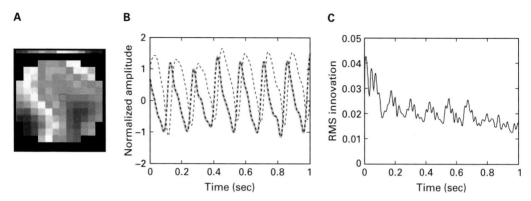

Figure 8.18 (plate 21)
A snapshot of voltage-sensitive dye data from a sequence of thousands of snapshots collected at 1,000 samples per second (A), the consensus set tracking of a single Wilson-Cowan node (B), and the root mean square (RMS) convergence of the tracking errors (C). Reproduced from [SS09] with permission.

experiments, a thin depth of fluid is heated uniformly from below by a plate, and convection rolls develop. The physics is described by the Navier-Stokes equations, but they used a simpler approximation to these equations, the Boussinesq equations, to assimilate the data. They employed an ensemble Kalman filter, and use a tiling of the spatial domain [OHS+04] to locally approximate the system. Their results [CHO+09] are shown in figure 8.19. There are several important points to note about these results for our purposes. First, in case the reader was wondering whether directly inserting data into an assimilation model might be a good shortcut to using the Bayesian assimilation of a Kalman filter, the forecast errors are always worse (figure 8.19). Use a filter. The next point is that the parameters in their tracking model that best tracked the experiment were *not* the most physical ones, but the ones that best reduced tracking error. This is exactly what we concluded above regarding the tracking of brain data [SS09].

How do you know how much to tile data? Well, if you can get away with just one tile, as we used in figure 8.18, that will increase the efficiency of calculations. The next step is to break up the spatial field to see if tiling the space will increase accuracy. The tiling scheme in [OHS+04] is elegant. Obviously, the ultimate tiling would be to fit each cell of, say, a Wilson-Cowan network to assimilate each channel of data—a very computationally expensive proposition. The answer is probably straightforward. Start with one large tile, and increase the number of tiles until you hit a minimum in tracking error. The more tiles, the harder it is to fit parameters. Such tiling optimization has not, to our knowledge, been well explored in applications.

The current research that is driving the study of model inadequacy is coming primarily from branches of science where the natural systems, meteorology and fluid turbulence, are both spatially and dynamically very complex. They share complexity features with our

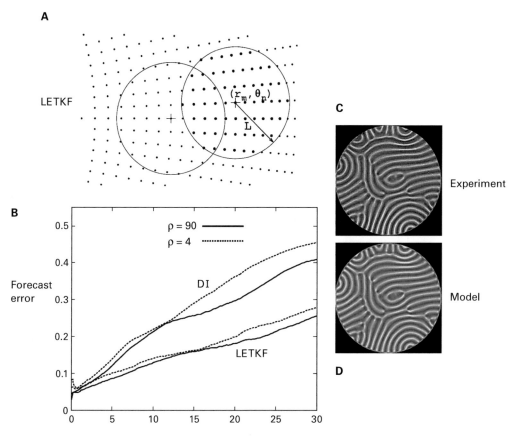

Figure 8.19
The local tiling of Raleigh-Bénard convection experiment (local ensemble transform Kalman filter), and forecast errors using the Boussinesq equations for data assimilation. In the forecast error plot, direct insertion of data, DI, is always worse than using the ensemble Kalman filter; p is the sparseness of sampling. Reproduced from [CHO⁺09] with permission.

observational problems in the nervous system. We have the delightful prospect that although our colleagues in meteorology will long be stuck simply observing and forecasting weather, we should be able to use their findings to develop feedback control schemes for dynamical diseases of the brain.

Exercises

8.1. Let's replicate the model from section 8.3 [TP07]. Use the Lorenz 1963 equations for your data assimilation model, setting $\sigma = 10$, $r = 28$, and $b = 8/3$.

$$\dot{x} = -\sigma x + \sigma y$$

$$\dot{y} = -xz + rx - y$$

$$\dot{z} = xy - bz$$

Now modify the equations that you want to track by translating them in the z direction with $z = z + 2.5$, and change the σ variable from 10 to 9 to alter the dynamics. Keep everything else the same.

• You need to calculate the mapping vector. Run both models for a while, so that initial transients die down, and observe their means. It should not matter what the initial conditions are, but you might best consider using several were this a real application. For now, if you use the same initial conditions each time for both systems, that is enough for starters.

• What are the errors, the innovations $(y - \hat{y})$, in tracking for the model with and without using the mapping paradigm? How does it compare with figure 8.3B.

• Now set up the data-generating model so that the dynamics drift. You might want to change $z = z + 2.5 * t / T$, where t is the time, and T is the total time of the simulation, and do this *only* in the underlying system that you want to track (i.e., assume that you do not know that this drift is present when you do the tracking). Set up a mapping paradigm as before, using a long simulation to generate the mean mapping vectors. Now try calculating a running average of the innovations $(y - \hat{y})$ in order to estimate the mapping vector. Can you improve the tracking errors by doing this?

8.2. Now set up the same scenario as in exercise 8.1 again. Use the altered Lorenz equations as nature, and use the original Lorenz equations as your data assimilation model. Leave the time-dependent drift out for now (the t / T term).

• Implement Bias Model II from section 8.4. Augment the state variable by a vector of b's, one for each variable x, y, z to serve as drift offsets. In principle, this will give very similar results to the mapping paradigm. Creating these uncertainty parameters is a more adaptable strategy than using a fixed mapping paradigm, but the increase in size of the covariance calculations Pxx will slow things down. Try a range of covariance inflation for several runs, and plot the mean tracking errors.

• Implement Bias Model III by further augmenting the state variable with a vector of c's, one for each variable x, y, z to serve as variable offsets. Now your Pxx will be 9 times larger. But the offset should fit the offset z variable well. Try a range of covariance inflation for several runs, and plot the mean tracking errors.

8.3. Again, set up the same scenario as in exercise 8.1. Use the altered Lorenz equations as nature, and use the original Lorenz equations as your data assimilation model. Leave the time-dependent drift out for now (the t / T term). Don't apply the mapping paradigm or the bias models as in exercises 8.1 or 8.2.

• Your sigma points are philosophically each a set of perturbed model states arranged in a spheroid. If we arrange them as columns in a matrix B, they are equivalent to the perturbed wind vectors of section 8.6. When you iterate them, they form an ellipsoid. Recall matrix times circle equals ellipse from chapter 7. Form a covariance matrix of such an iteration and perform a singular value decomposition as suggested in section 8.6.

• Can you plot the largest singular value along with the tracking error at each step?

• Try using the largest singular value at each step, or averaging it for several steps, to adaptively adjust covariance inflation. Can you find a strategy that decreases tracking errors?

9 Brain-Machine Interfaces

Predicting is more advantageous than simply reacting.
—Simone Schütz-Bosbach and Wolfgang Prinz [SBP07]

9.1   Overview

No aspect of neural engineering has captured the popular imagination as much as brain machine interfaces. From the medical grail of producing thought-powered prosthetics for the disabled, the military grail of thought-powered ordinance, to the entertainment grail of thought-powered fantasy games, brain machine interfaces are on the minds of many people.

Our advances in brain-machine interfaces have been almost entirely through the use of empirical *black box* statistical models to solve the decoding problem between neuronal recordings and motor movements. Some of the world's best minds in neuroscience and statistical theory have produced a set of powerful solutions to this *decoding problem*. Yet the results so far could have been produced without knowing anything much about the brain that produces the signals beneath our electrodes.

The best of the generalized statistical models employed for such decoding problems observes the brain activity, trains a model by fitting data, and uses that model fit to decode future data. No one has yet, to the author's knowledge, produced fundamental and biophysically based models that can observe and track such brain circuitry *without data fitting*, and create recursive predictive models. Are we at the point where our computational neuroscience of, say, the motor system, is sufficiently advanced that we can begin to shed some light inside the black box? Perhaps soon. Let's examine this intriguing field.

9.2   The Brain

We begin our tour of brain-machine interfaces by introducing, in figure 9.1, a schematic of the brain. The motor and sensory sheets of the brain have highly localized functions. By this we mean that if you stimulate while following along their length, you can get motor

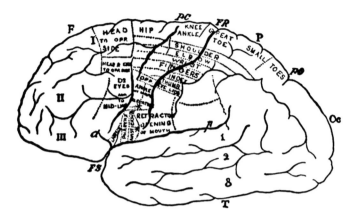

Figure 9.1
Schematic of topological layout of the motor *homunculus* of the brain. The figure represents William Gowers's
mapping [Gow01] onto the human brain of Victor Horseley's findings from the stimulation of macaque [BH90a]
and orangutan [BH90b] primates. Reproduced from [Gow01].

responses, or sensory responses, respectively, that reflect a 1:1 spatial mapping of the body.
Figure 9.1 shows that this mapping is distorted with respect to body size, and we call this
small humanoid map the homunculus. If one were to map tones in the auditory area, you
would get a tonotopic map of frequencies. If you map the visual field in the primary visual
cortex, you would get a spatial map of the visual field. It is the primary motor and sensory
regions of the brain that have been most carefully recorded from, in an effort to construct
brain-machine interfaces.

9.3   In the Beginning

The study of brain-machine interfaces really begins in 1982 with the seminal work of
Apostolos Georgopoulos [GKCM82]. In these experiments, Georgopoulos studied neurons
recorded from the primary motor cortex of the brain, and recorded single neurons while a
monkey played a game with a manipulandum. The monkey's task was to first center the
manipulandum over a light in the center of the board. Once held there, the task was to next
move the device over one of eight possible lights arranged at 45-degree angles in a circular
field. If this was done correctly, the monkey received a juice reward. An example of the
trajectories of an experiment are shown in figure 9.2C.

The response of a single cell during trajectories in eight directions are shown in figure 9.3.
There is clearly a strong directional preference in the firing rates associated with movement
in preferred directions, and these firings increased prior to movement onset. Interestingly, in
directions away from the preferred one, firing would decrease and at times stop at movement
onset and in the period immediately afterwards. Georgopoulos fit these firing rates to a cosine
tuning function:

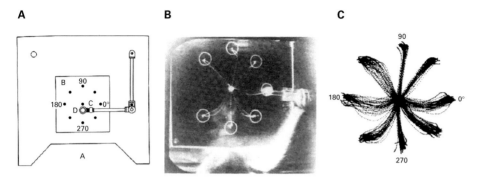

Figure 9.2
Schematic of experiment (A), picture of the monkey performing an experiment (B), and trajectories overlapped at the end of an experiment (C). Reproduced from [GKCM82] with permission.

$$z_k = h_0 + h_p \cos(\theta_k - \theta_p)$$

The firing rate of a given neuron, z_k, was modeled as a baseline rate, h_0, plus a cosine function of the difference between the movement direction, θ_k, and a preferred direction for that neuron, θ_p (with an adjustable constant h_p). From the trigonometric identity that none of us could remember as school children,

$$cos(a - b) = cos(a)cos(b) + sin(a)sin(b)$$

one gets

$$z_k = h_0 + h_1 sin(\theta_k) + h_2 cos(\theta_k)$$

because the preferred direction, θ_p, for each neuron is a constant and can be lumped into the constants h_1 and h_2.[1]

Georgopoulos found that when a given movement in a particular direction was made, most of the cells that fired had directional preferences close to the direction of movement. This work suggested that were one to have recorded from these cells at the same time, a sum of the vector angles of the cells that fired would predict the angle of the arm movement. *This finding is the foundation of all brain-machine interfaces that we presently have, and are about to discuss.*

In 1986, Georgopoulos published the three-dimensional version of this experiment [GSK86]. Again, summing the population vectors for multiple cells was highly predictive

---

1. Be very careful in assuming that because the cosine function fits these data, that the neurons are little cosine generators. In further work from Georgopoulos and colleagues [AG00], they explored a variety of uni- and multimodal functions. A generalized form of circular distributions, the von Mises function, can often fit these tuning curves better, and represents a sum of Bessel's functions. The reader should think of cosines as a convenient way to represent a unimodal continuous peaked function in representing a neuron's preferred direction.

A

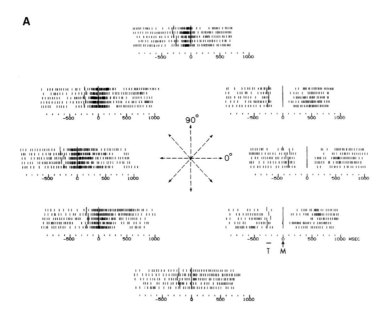

B

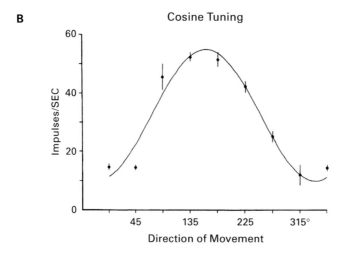

Figure 9.3
Raster plots showing the sequential firing times of spikes from a single cortical cell for different directional movements in A, and the fit of these firing rates to a cosine tuning curve in B. Reproduced from [GKCM82] with permission.

A   **B**

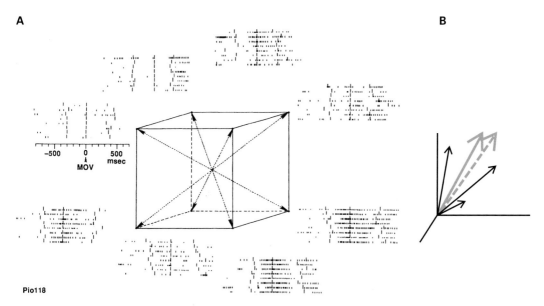

Pio118

Figure 9.4
Three-dimensional version of the arm trajectory experiment. A single cell firing in response to direction in A, and a three-dimensional plot illustrating how individual cell cosine tuning vectors (black arrows) can be summed through the population model to reflect a direction in three-dimensional space (gray solid arrow) that closely approximates the true movement direction (gray dashed arrow). Image in A is reproduced from [GSK86] with permission.

of the actual trajectory that the arm made within the eight possible vertices of the cube, as shown in figure 9.4.

9.4   After the Beginning

Within several years of Georgopoulos's work, our technology advanced to the point that placing arrays of microelectrodes offered the possibility of recording from more than one cell at a time, and extracting their spike times in real time.

In 2002, John Donoghue's group published a paper that examined less constrained movements than we showed above, and with simultaneous recordings from multiple neurons [SHP+02]. In these experiments, a target appeared at multiple positions on the screen, and a monkey needed to trace the trajectory using a manipulandum. Figure 9.5 shows a combined trajectory of the hand movement (green) as well as the brain-firing reconstruction (blue).

In contrast to the population vector approach, Serruya and colleagues [SHP+02] created a linear filter that converted neuronal firing to hand position. They established a firing rate matrix for k neurons, n, each observed over fifty time bins as

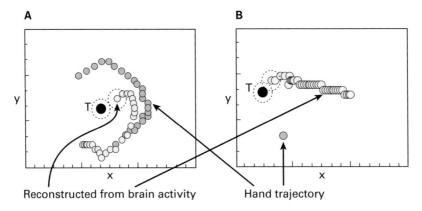

A

B

y

y

T

T

x

x

Reconstructed from brain activity   Hand trajectory

Figure 9.5
Target, T, hand trajectory, and reconstruction of hand trajectory from neuronal activity given filter $\hat{f}$. A shows the hand and reconstructed trajectory. B shows a trajectory where the monkey did not move its hand. Reproduced from [SHP+02] with permission.

$$R = \text{firing rates} = \begin{pmatrix} n_1^{t-49} & \cdots & n_k^{t-49} \\ \vdots & \ddots & \vdots \\ n_1^t & \cdots & n_k^t \end{pmatrix}$$

where

$$Rf = x$$

and x is the hand position. The filter, f, converts firing rates to position. Here, hand position is modeled as a linear combination of firing rates. The *best* filter for this, $\hat{f}$, is the solution of an equation that we solved in chapter 1, with equations (1.28). It is an over-determined problem, which we solve with least squares

$$R^T R \hat{f} = R^T x$$

$$\hat{f} = (R^T R)^{-1} R^T x$$

$$R \hat{f} = R(R^T R)^{-1} R^T x$$

Schematically, these equations are shown in figure 9.6. Each row of the matrix R, representing a collection of neurons $1 \ldots k$, forms an inner product with each column of the filter, $\hat{f}$, producing two numbers: a pair of coordinates x and y. Least squares finds the best filter given the data to accomplish this task. This simple linear filter solves a stationary system.[2]

---

2. A Wiener filter solves for the optimal filter for a stationary system where the information being destroyed by noise or damping is exactly balanced by the incoming information such as our spikes [Gel74]. You will see Wiener filters being frequently applied to such linear problems, but rarely will the underlying requirements for the existence of the filter be fulfilled.

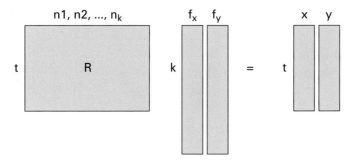

Figure 9.6
Schematic of linear filtering of spike times.

Serruya and colleagues then made a peculiar observation. Sometimes the monkey didn't move its hand, but nonetheless was able to rapidly move the cursor to the target using brain activity alone (figure 9.6).[3] It's the serendipitous findings that are so often the most important.

Two years later, this same group demonstrated that both hand velocity and position could be independently encoded or decoded from these same experiments [PFHD04]. There is something very subtle here. In Georgeopoulos's experiments, the task was highly constrained. The monkey moved along prescribed trajectories that were really quite different from each other and algorithms that attempted to extract these different directions could be very successful. In the pursuit task, where the directions were part of a random walk, the situation is far more complex. Adding more features to the model for which qualitative features in hand movement are encoded, can give improved performance of such a filter. But how many features can be extracted from neurons firing in the motor cortex? The question is a deep one.

Is the motor cortex thinking? It is probably not thinking nearly as much as it is serving as a final common pathway [She06] for decisions made elsewhere in the brain. Of course, I don't know that this is true. One does suspect from research in recent years that whatever projects to the motor cortex may well read its output activity in order to form a feedback loop with it. Such efference copies have long been considered in motor movement. They might be very useful to a part of the brain attempting to drive the motor cortex using predictive coding [Bal97]; but they also are very convenient for people trying to interact with the brain using interfaces in the twenty-first century. The point that I want to emphasize is that just because one can find representation of particular features within the primary motor cortex does not necessarily imply that those neurons are using those properties themselves. But it is nice for us that these representations are there.

---

3. It turns out that giving the brain visual feedback, closing the loop in contrast to the open-loop experiments of Georgeopoulos, was critical to have observed this (we will return to this issue in section 9.7)

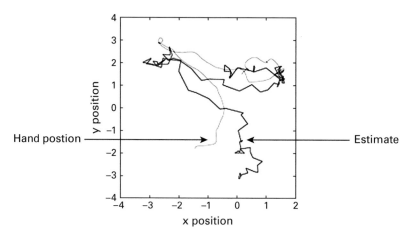

Figure 9.7
Two-dimensional hand position and estimate. Reproduced from [PFHD04] with permission.

Using position and velocity, Paninski and colleagues [PFHD04] demonstrated that they could improve their reconstruction of hand position using the data from Serruya et al. [SHP+02]. They had an interesting approach—assuming that hand kinematics were input random variables of position and velocity, and that neural activity was the response. Their reconstruction of trajectory, as shown in figure 9.7, is compelling in interesting ways. One cannot help but keep asking oneself, on inspection of such reconstructions, what qualities are "good," and which are not. It is clear from this undoubtedly exemplary example that the correlation between the reconstruction and reality is better than the actual error in position. Recall from chapter 8 that the model and reconstruction are here shown in a different space from reality, which is much more complex. Examine the geometry of model inadequacy in figures 8.2 and 8.4. It is interesting to speculate how adapting some of the model inadequacy formalism from chapter 8 might improve some of the position errors in figure 9.7. But speculation on decreasing position errors in [PFHD04] is not the reason to introduce this remarkable work.

One of the most remarkable aspects of [PFHD04] was that by randomly picking a small subset of the recorded neurons, the researchers achieved incredibly high reconstruction accuracy, as shown in figure 9.8.

This is very different from the Georgopoulos result where only the neurons that gave significant direction tuning were considered in the analysis. Here, using a linear filter of position and direction, any sampling of twelve random cells gave excellent reconstruction accuracy. Were the investigators lucky in positioning their electrodes near the cells? Not at all. Such results imply that very high percentages of the neurons recorded within the region of the primary motor cortex have representations of both the position and velocity of the hand. What this means is that you can go into, say, the hand region of the motor cortex,

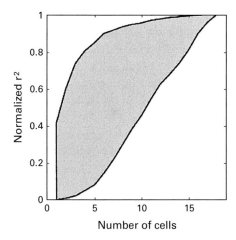

Figure 9.8
Range of reconstruction accuracy versus number of randomly selected neurons. Reproduced from [PFHD04] with permission.

grab a dozen or so neurons—it does *not* matter which ones—and reproduce the results of [SHP⁺02] shown in figure 9.6.

But why are such higher-order qualities of movement represented in Sherrington's *final common path* for motor movement? We think that at some point, the thinking is done, and motor systems are carrying out instructions generated upstream. Perhaps the motor instructions are more complex than just pulling levers of muscles (in primates, much of the output from primary motor cortex is flowing to spinal cord interneurons for further processing rather than directly to motoneurons). But also it may well be that a degree of feedforward and feedback flow of information to and from motor cortex is essential in this process. In [PFHD04] it was apparent that, for velocity tuning, there was a clearly predictive element emerging about 100 msec prior to motor action. One can look for higher-order derivatives of position as well (acceleration, jerk, snap, crackle, etc.), and find a (decreasing) representation of these quantities in such motor cortex neurons (see [WGB⁺06]).

There seems to be three possibilities for these representations: (1) they are just there for the convenience of twenty-first–century scientists wishing to construct brain-machine interfaces; (2) they are just there and the nervous system wastes energy transporting this information because it is too expensive to remove (the informational appendix of the motor system); or (3) they serve a purpose other than direct transmission of motor movement. As in other paradigms of anticipation found in nervous systems, whether the stimulus expectancy of primates [RGDA97], or the omitted stimulus response of ants and bees [RG05], it is hard to write this book without thinking speculatively on the structure of the nervous system's possible predictor-corrector structures.

In 2006, using the same linear filter for position from [SHP$^+$02], the Donahue group showed that their original monkey experiments taught them how to translate this paradigm to humans [HSF$^+$06]. They first demonstrated this in a patient with a high spinal cord injury. He was unable to move his arms or legs, and required a ventilator to breathe. They used the same microarray used in the [SHP$^+$02] experiments, and implanted it in the small bump in the motor cortex affectionately known as the *hand knob* by those who map such things (figure 9.9C). An important difference between training such disabled people and neurologically intact monkeys is that the patient cannot be trained by moving his or her hands. In figure 9.9D a technician is seen providing cursor training early on in such tasks.

By watching the technician and *mirroring* in his mind, he learns to perform these tasks with his mind. He can often perform these tasks while chatting with the technician about tangentially related things. A rich theory in psychology helps us weave a network of brain mechanisms that underlie such phenomena. The discovery of *mirror neurons* by Rizzolatti and colleagues [dPFF$^+$92] has opened up a vista originally speculated on by William James [Jam90]. A century later, we have divided our nomenclature into instantly predictive coding of sensory phenomena by motor actions through generative models [KFF07], and future prospective coding of sensory events anticipated in a future time [SBP07]. Perhaps James's speculations filtered through mirror neurons and their implications render the observations of [HSF$^+$06] plausible but no less remarkable. We will return to the broader implications of mirror neurons in chapter 13.

The relative ease of training and the ability to perform tasks with modest degrees of attention are rather different from the extracranial experience with brain-machine interface implementations [WMVS03]. Nevertheless, with sufficient training, scalp-based electro-encephalograms (EEGs) can perform accurately once the subject (and algorithm) masters the learning curve [WM04]. Recent work with subdural sensing electrodes suggests that an intermediate location between inside the cortex and outside the head may provide an accurate signal for decoding without transgressing the cortex itself [SKM$^+$07, MSF$^+$10].

9.5 Beyond Bins—Moving from Rates to Points in Time

A brain-machine interface can respond only as fast as the signals being processed can be calculated.[4] In both [SHP$^+$02] and [HSF$^+$06], a 50-msec time bin was used to count spikes and form an average rate. We now look at efforts to minimize this by examining the timing of individual spikes themselves.

We begin with a rodent experiment. O'Keefe and Dostrovsky noted that, in addition to other complex sensory and arousal states that caused firing in hippocampal neurons, certain

---

4. We leave out for the moment the significant time lags introduced by the human brain between the presentation of a visual cue and the perception of such an event by the brain, which can typically take more than 100 msec.

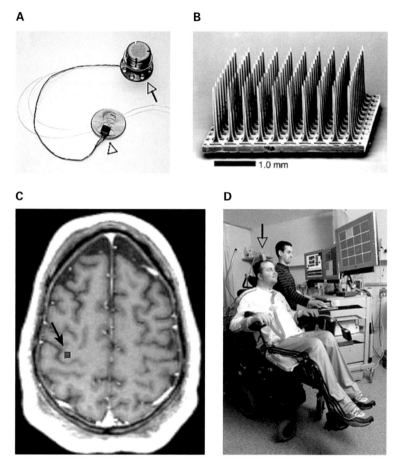

Figure 9.9
The microfabricated Utah array is shown in A and B; the positioning of the array on the *hand knob* of the motor cortex is shown in C; and the training of a quadriplegic subject with this implant with a technician is shown D. Reproduced from [HSF[+]06] with permission.

neurons were particularly sensitive to spatial orientation of the animal and position in space [OD71]. Later, it was observed that the firing of place neurons often occurs in preferred phases of what is called *theta* rhythm (generally 4 to 12 Hz in animal EEGs).[5] In fact, the phase that such place cells fire often *precessed* in relation to the phase of theta rhythm, as the animal entered the place-receptive field and moved through it [OR93].

---

5. A source of continual annoyance to clinicians is that human EEG "theta" rhythm is rather rigidly defined as 4 to 7 Hz. But brains are not standard clocks, and a more functional banding of neural frequencies, especially in animal behavioral studies, is often used.

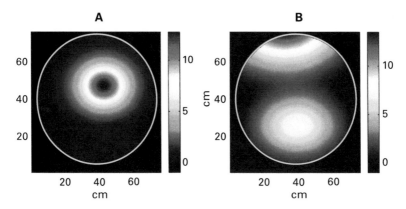

Figure 9.10
Inhomogenous Poisson model fit to firing of individual place cells within an open field arena. The gray scale is spikes per second, as indicated in the intensity bars. A shows a unitary receptive field, and B shows a split receptive field with two distinct regions where the place cell firing peaks. Reproduced from [BFT+98] with permission.

A Poisson process is a single-parameter random process in which the rate parameter, λ, is the expected number of events within a unit of time, and the probability of seeing a given number of events, k, within a unit length of time is proportional to $e^{-\lambda}(\lambda)^k/k!$. If the rate λ is itself time or position dependent, then the Poisson process is *inhomogeneous*.

Can one encode the continuous position of the animal, x, and construct a conditional probability for neural spiking, z, as $p = (z|x)$?

In 1998, Emery Brown and colleagues [BFT+98] used an inhomogenous Poisson process to fit to place cell firing and theta phase data as shown in figure 9.10. These maps represent the initial *encoding* process.

One then takes this encoding and uses it to observe the neural spiking and, in a Bayesian framework, *decodes* the posterior probability of position $p(x|z)$. [BFT+98] solved the problem of taking a point process of spike events and estimating the continuous position of the animal. The original Kalman filter discussed in chapter 2 estimated continuous variables from measuring continuous observables, so some adaptation was required.

Figure 9.11 compares a linear model and the estimates with the Bayes filter. In the Bayes filter, the underlying model is that of a random walk. This *generative model* has the effect of placing a huge prior constraint on each next estimate of the animal's position. Rats don't fly. The linear model contrasted in figure 9.11 had no such constraint.

One might then ask whether the simpler linear model might have done much better using a constraining generative model. Perhaps a simple linear Kalman filter, using spike rates instead of point processes, might perform admirably well given kinematic constraints and dynamics.

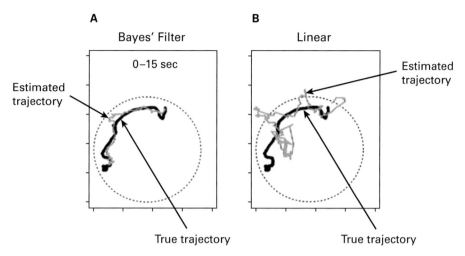

Figure 9.11
True and estimated trajectories using Bayes' filter (A) and a linear filter (B). Reproduced from [BFT⁺98] with
permission.

9.6   Back from the Future

In [WGB⁺06] the authors took the data analyzed in [SHP⁺02] and [PFHD04]. They rean-
alyzed it using kinematic constraints and a generative model of limb dynamics, and used
continuous measures of spike rates rather than point process observations of spike times.
And they applied the linear Kalman filter as we have described in chapter 2.

   They examined two tasks: a *pursuit* tracking task where the target moves in a random
walk, and a *pinball* game where the target appears at random locations and the goal is to
hit it. The kinematic variables will be position, as well as derivatives of position (velocity,
acceleration, jerk, snap, crackle, etc.).

   Linear models of hand kinematics include

$$z_k = h_0 + h_x v_{x,k} + h_y v_{y,k}$$

where z_k is the firing rate of a cell at time k, and the h's are linear functions relating the
velocities v_x and v_y at time k to the neuron firing. Position would be linearly encoded as

$$z_k = f_0 + f_x x_k + f_y y_k$$

where the linear functions f relate position at time k to firing rate from the coordinates of
hand position (x, y). [WGB⁺06] will also include higher-order derivatives in such linear
functions (especially acceleration).

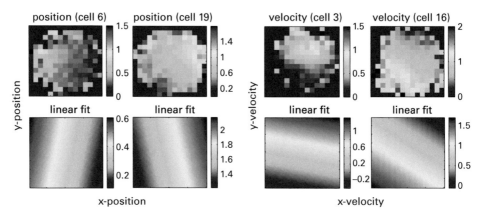

Figure 9.12
Cell firing frequency versus position or velocity as a function of position, with gray scale calibrated as shown for firing frequency. Below are the best linear fits. Reproduced from [WGB$^+$06] with permission.

Figure 9.12 shows examples of kinematic data that was encoded into their linear functions for pursuit tracking. Hand position is plotted against the intensity map of cell firing frequency.

The first thing to discuss about figure 9.12 is that fitting linear models to receptive fields that are circular is problematic. Kalman filtering works when the models of the dynamics are better than random guessing. The authors describe these fits as *crude but reasonable* [WGB$^+$06]. One might have said that the models are *not quite entirely unlike* the data.[6] Clearly, these fits can be improved with radial functions, and we will return to this issue later in this chapter. But for now, the point is that these linear fits capture *something* from these data that is useful, and the power of the Kalman filter is that it optimizes the uncertainty of this inadequate model.

Are these single neurons? [WGB$^+$06] actually used any waveform that crossed a given threshold. So their firing rates undoubtedly included the possibility of multiple single neurons, and combinations of neurons. Isolating single neurons would require spike sorting, and matching multiple waveform features of individual spikes recorded at high sampling frequencies (see, for example, [SDW02]). The literature on *single-cell* spike rates and timings from electrode arrays is full of examples where a critical view of the quality of single-cell isolation can be raised. In [WGB$^+$06] this issue is acknowledged. As for the model fits discussed earlier, whatever these signals are, single versus multiunit recordings, they *are* sufficient to permit hand kinematic estimation. On the other hand, this issue raises the important question of how detailed the data need to be to function in a brain-machine interface. Might local field potentials or local EEGs suffice? Do we need to use microfabricated electrode arrays that corrode over merely months in typical applications, rendering

---

6. A paraphrase of Douglas Adam's expression from his *Hitchiker's Guide to the Galaxy* [Ada79].

them useless in their present instantiations beyond a year or two? Might we be able to use recordings from the subdural or epidural space to achieve adequate brain-machine interface accuracy [SKM$^+$07, MSF$^+$10]?

Also important to note is the sparseness of neurons recorded by the arrays used in [SHP$^+$02], [PFHD04], and [WGB$^+$06]. There should be an order of magnitude more neurons from considerations of neuronal packing density, or from the contrast between extracellular and intracellular recordings. Shoham and colleagues [SOS06] call this the *dark matter* problem in the mammalian brain. Most of the signal recorded by microelectrode arrays may not be neuron firing. Most neurons may be quietly listening most of the time, rather than firing. We have been taught to be spike-centric—the mammalian cortex may not be.

With all of these cautionary issues in mind, let us assume an augmented state of position, velocity, and acceleration:

$$x_k = \begin{bmatrix} x \\ y \\ v_x \\ v_y \\ a_x \\ a_y \end{bmatrix}$$

where k is an index of time such that $k = t/\Delta t$. Δt is the time over which the spikes are binned to derive the firing rates, 50 to 75 msec. Assume that the observations are firing rates z, such that for C neurons at time index k

$$\mathbf{z}_k = \begin{bmatrix} z_1 \\ z_2 \\ \vdots \\ z_C \end{bmatrix}$$

and Z_k represents the history of measurements

$$\mathbf{Z}_k = [\mathbf{z}_1, \mathbf{z}_2, \ldots, \mathbf{z}_k]^T$$

Next assume that the a posteriori probability of system state x_k, conditioned on measurements z_k, is

$$p(\mathbf{x}_k | \mathbf{z}_k)$$

and now assume two Markov assumptions

$$p(x_k | x_{k-1}, x_{k-1}, \ldots, x_1) = p(x_k | x_{k-1})$$

where the hand kinematics really depend on only the most recent values of position, velocity, and acceleration, and

$$p(z_k|x_k, z_{k-1}) = p(z_k|x_k)$$

which tells you that neurons are forgetful of their past spikes, so you can forget them also in your formula relating neuronal spiking to hand kinematics. For now, assume that this relationship is instantaneous—we will change this ridiculous assumption shortly.

Now recall Bayes's rule from chapter 1:

$$P(x, z) = \text{Joint Probability} = P(x|z)P(z) = P(z|x)P(x)$$

which leads us to the central formulation in [WGB$^+$06] and related literature

$$p(x_k|\mathbf{z}_k) = \kappa \ p(z_k|x_k) \int p(x_k|x_{k-1})p(x_{k-1}|z_{k-1})dx_{k-1}$$

This is nothing more than Bayes' rule. Let's dissect it. $p(x_k|z_k)$ is the posterior probability, the *decoding* estimate of hand state x_k from neuronal firing z_k that we seek to find. $\kappa = 1/p(z_k)$, and will end up normalizing the total probability on the right to ensure that it sums to 1. $p(z_k|x_k)$ is the *likelihood*—it is estimated from the actual data used to determine encoding such as in figure 9.12. $p(x_{k-1}|z_{k-1})$ is the previous posterior probability at time index $k - 1$, used to establish the initial conditions for the new *temporal prior*, which is $p(x_k|x_{k-1})$. Integrating $p(x_k|x_{k-1})p(x_{k-1}|z_{k-1})$ is the same as integrating the equations of motion for this dynamical system. It propagates the hand state at time $k - 1$ to the state at time k, $p(x_k)$. If the likelihood and temporal prior are both unimodal Gaussian functions, then we can use the original Kalman filter to solve this problem recursively.

The likelihood model is $z_k = H_k x_k + r_k$, where H is a linear observation function matrix, and r_k is drawn from a Gaussian probability distribution. The temporal prior is $x_{k+1} = A_k x_k + q_k$, where A is a linear matrix of the equations of motion, and q_k is drawn from a Gaussian probability distribution.[7] Wu and colleagues [WGB$^+$06] are careful to include a full covariance matrix, R, which is CxC in their noise structure for the C neurons. This is important. One knows immediately from results such as those shown in figure 9.8 that there was much interdependency and redundancy in representation of information among these neurons. The full covariance structure of R represents this covariation among each individual neuron, and allows for dependency. This is important to improving the accuracy of their results.

Using the Kalman filter equations from chapter 2, the central recursion is

$$\hat{x}_k = x_k^- + K_k(z_k - H\hat{x}_k^-)$$

The observations z_k are compared with the estimated observations $H\hat{x}_k^-$ and their difference is weighted by the Kalman gain K_k.

---

7. If the reader is using this discussion to decode Wu et al. [WGB$^+$06], I apologize for switching notation to maintain consistency with earlier chapters in this book.

The likelihood model, $z_k = H_k x_k + r_k$, is their model of the *neural code*. H is the brain. Is this a good physical representation of the dynamics of the brain? You do not need to know much about the brain other than how to get into it to write this down. This is an empirical fit of a statistical model. But the counter-question would be: Would all of your knowledge of the brain gathered over the past century enable you to do a better job in decoding its signals? Perhaps not now. But at some point, our increasing understanding of neuronal dynamics should enable us to consider whether we can do better than the present state of the field, which uses *black box* statistical models as described here.

It takes time for sensory input to travel through our nervous system to be perceived, and similarly, it takes time for motor intention to travel through our nervous system and muscle actuators to result in hand movement. Wu and colleagues [WGB⁺06] explored a range of time lags, l, for neuronal firing data:

$$\mathbf{z}_k = \begin{bmatrix} z_1 \\ z_2 \\ \vdots \\ z_C \end{bmatrix} \rightarrow \pm \{l_j\}$$

and found that an optimal time lag was about 140 to 150 msec (similar to the optimal lags found by Moran and Schwartz [MS99b]).

The results of this decoding are shown in figure 9.13. Are these reconstructions good? Again we see the observation, that "we observed decoding results with relatively high correlation coefficients that were sometimes far from the true 2D hand trajectory" [WGB⁺06].[8]

And all of this makes a lot of sense. Again, the model is fit into a space much smaller than the brain operates within. All of the model inadequacy is being lumped into the random R and Q distributions from chapter 2. The considerations from chapter 8, in terms of mapping and transforming model results back to the space that nature operates within, seem very pertinent here.

Adding kinematic derivatives beyond one or two did not help (beyond acceleration for pinball tasks, and beyond velocity for pursuit tracking). This is not to say that such variables are not present in the kinematics of the hand. But increasing the complexity of these models comes at a price—one needs to estimate more state variables. There is a trade-off between model complexity and filter accuracy. This is very advantageous for us. If the best filter required a full computational model of the brain running on a high-density computer cluster

---

8. One can try to gain additional accuracy in such calculations by using *target conditioning*—essentially adding an additional boundary condition to the recursive initial value problem that the Kalman filter solves [KP08].

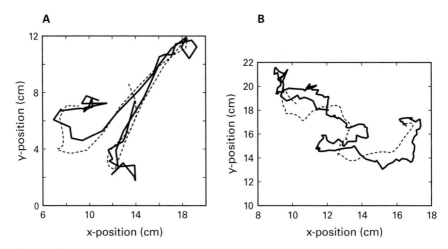

Figure 9.13
True (dashed) and reconstructed (solid) trajectories for pinball task (A) and pursuit (B) Reproduced from [WGB+06] with permission.

of the size of the universe, it would be somewhat impractical to consider using these in the laboratory or implanting them into a person.

Last, Wu and colleagues [WGB+06] compared their results with that of their previously used linear filters and polulation vectors. Their performance with Kalman filtering was better than that when using a linear filter with no explicit generative model of hand kinematics, similar to the findings of [BFT+98]. Their results were particularly better than those from the older population vector encoding. But don't discount such simpler models yet.

9.7   When Bad Models Happen to Good Monkeys

In 2002, Dawn Taylor and colleagues [TTS02] did a remarkable experiment. They used a version of population vector encoding in a three-dimensional center-out task. But they let the monkeys have visual feedback for some of these experiments, allowing their brains to do the work with their hands restrained. They compared these results with those when the monkeys used their hands. And they did one more thing—they retuned their algorithms in case the monkeys changed or adapted their neurons' tuning properties over the period of training.

Figure 9.14 demonstrates the main result. In this *coadaptive* paradigm, where the brain and the algorithm adapt to each other, there is a very significant improvement in accuracy of this neural prosthetic from day to day. This accuracy was linked to an improvement in the neurons' cosine tuning properties for direction for subjects that had visual feedback.

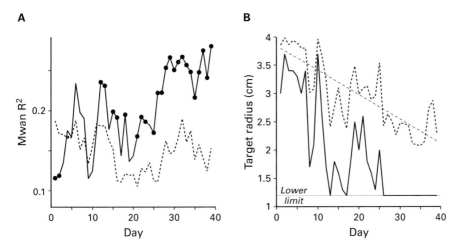

Figure 9.14
A shows the quality of cosine tuning expressed as the average R^2 (coefficient of correlation) statistic for the neurons recorded. Dotted lines are for hand-controlled subjects; solid lines are for brain-control subjects. Significant differences are indicated by solid circle markers. B shows improvement in brain-controlled accuracy by contrasting target size required for 70% accuracy (dotted line is mean; solid line is minimum). Reproduced from [TTS02] with permission.

The coadaptive brain-controlled subjects could achieve accuracy similar to hand-controlled subjects.

The brain can take a mediocre algorithm, and adjust its own tuning curves to adapt to that algorithm and produce better performance. Filters can optimize to the brain activity and motor performance. But *brains can adapt to filters*. Perhaps that is why much of the studies described in this chapter worked well. You just have to give the brain something it can work with. Hammers are not optimized for driving nails, but many of us can become skilled in using them given a chance.

Perhaps the brain has its Kalman filter adapting to our Kalman filters in these experiments. One can immediately see that we should be able to create adaptive controllers that are not at all well suited for the brain. Which types of adaptation are best for the brain's native adaptation capabilities remains a fully open question.

Similarly, this chapter has focused on extracting signals from the motor cortex. It is not at all clear that this is necessary or best. But at the time of this writing, exploring motor control prosthetics from brain signals outside of the motor cortex has not yet been explored.

Much of this book has focused on the nonlinear ensemble Kalman filters. There was an effort to apply such an ensemble filter to brain machine interfaces in 2004 [BRK04]. Brockwell and colleagues recognized the power and flexibility of such methods in that the linear Gaussian models of prior and likelihood models were no longer restrictions. But

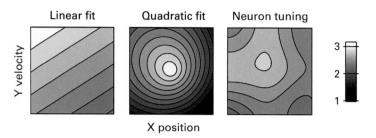

Figure 9.15
Fitting neuronal activity as a function of hand position with linear versus quadratic functions. Brightness is proportional to firing rate. Reproduced from [LOH+09] with permission.

despite the efficiency of their *particle filter*, running thousands of points as initial conditions for such an ensemble filter was not efficient enough to run in real time.[9]

More recently, Li and colleagues employed a sparse ensemble filter—the unscented Kalman filter—in a multineuron brain-machine interface [LOH+09]. There were three important advances in this work. First, they finally got beyond the linear tuning curves, as shown in figure 9.12. Figure 9.15 shows the improvement in fit for neuronal tuning curves for things such as hand position when using quadratic versus linear functions. Second, they used a ten-step autoregressive model for hand kinematics, and although they implemented this filter causally, they used an acausal model within the filter. Third, using the unscented Kalman filter, they could demonstrate real-time performance in both center-out and pursuit tasks. Comparing the performance accuracy with the linear Kalman filter from [WGB+06], they achieved improved accuracy with their tenth-order uscented Kalman filter.

9.8 Toward the Future

In spite of all the impressive developments outlined in this chapter, where is the brain? In between the neuronal output spikes and limb movement for instance, there is a void where our neurophysiology remains uninformative (other than spike tuning curves). This work has often been focused on a decoding paradigm: Neuronal firing is conditioned on arm movement, for instance, and arm movement is conditioned on neuronal firing. If one is building a brain-machine interface, recent work toward a general model that incorporates a hybrid composite of discrete and continuous states, as well as neuronal firing, is likely the most advanced framework currently available [SB07].

---

9. Nevertheless, computer processor and multicore architecture offer progressively improving computational capabilities, and robust particle filtering methods become more practical for real time use with each passing day [BKS07]. As a general filtering design principle, one should consider increasing the ensemble size to the limit that computational overhead permits.

It is tempting to speculate whether we can improve brain-machine interfaces by building models not on fitting data to empirical tuning curves, but rather by building models based on neurophysiology. Such models would incorporate the properties of individual neurons, and their network connectivity, that have have been observed through electrode recordings. Such models would be fused with data—*data assimilation* [Kal03].

We remain far from such a goal at present. But our modeling sophistication, our means for making measurements within the brain, and our control theoretical tools are ready when we are ready.

10 Parkinson's Disease

[T]he unhappy sufferer has considered it as an evil, from the domination of which he had no prospect of escape.
—James Parkinson, 1817 [Par17]

10.1 Overview

Parkinson's disease was first described by James Parkinson in a monograph published in 1817 [Par17]. It is a degenerative neurological condition, and we attempt to condense its neurological manifestations into four key signs: tremor, rigidity, bradykinesia (slowness of movement), and postural instability. But a far better picture of the signs of this condition was captured in William Gowers's short description in his 1901 textbook [Gow01], which flanked his famous sketch of a typical patient walking (figure 10.1):

[T]he aspect of the patient is very characteristic. The head is bent forward, and the expression of the face is anxious and fixed, unchanged by any play of emotion. The arms are slightly flexed at all joints from muscular rigidity, and (the hands especially) are in constant rhythmical movement, which continues when the limbs are at rest so far as the will is concerned. The tremor is usually more marked on one side than on the other. Voluntary movements are performed slowly and with little power. The patient often walks with short quick steps, leaning forward as if about to run. [Gow01]

We have learned a considerable amount about the neurobiology of Parkinson's disease since the nineteenth century, but we do not know how to prevent it, and all of our present-day treatments remain palliative [LHR09]. The discovery by Carlsson and colleagues that a precursor to dopamine in the brain could ameliorate the effects of dopamine depletion [CLM57] led to successful medical therapy of Parkinson's disease with L-DOPA. Early on in the disease, L-DOPA provides the chemical precursor to produce more of the waning dopamine neurotransmitter, but gradually, fewer of these cells remain that can benefit from such a boost in chemical processing. Furthermore, there are two side effects that patients find especially disturbing: (1) the returning symptoms during the wearing-off phase of the drug, producing more radical on-off swings in motor symptoms, and (2) the gradual development

Figure 10.1
A *well-marked case of this disease*, as described by William Gowers. Reproduced from the second edition of his 1901 textbook [Gow01].

of involuntary movements termed *dyskinesias* [SEJP09]. Although drug therapy remains the firstline standard of treatment for patients with Parkinson's disease, the long-term medical side effects of phamacological therapy has kept alive the surgical treatment options for phamacologically intractable Parkinson's disease.

There are three surgical lesion targets in the brain that have been found to reduce the symptomatology of Parkinson's disease effectively: the ventral intermediate nucleus (VIM) of the thalamus, the internal segment of the globus pallidus (GPi), and the subthalamic nucleus (STN). Because patients with Parkinson's disease generally require treatment on both sides of the brain, the efficacy of single-sided lesion treatment was often tempered by the complications of bilateral therapy. One never wants to lesion a brain symmetrically. So lesions can be administered asymmetrically. One also is hesitant to use too large a lesion at any one time, so there was a frequency of having to return to surgery to enlarge a lesion that was not effective enough.

Parkinson's disease is now the disease most widely treated by deep brain stimulation (DBS) [KJOA07]. In 1997, the U.S. Food and Drug Administration approved its use in Parkinson's disease. DBS is effective when used bilaterally, without the symmetrical lesion complications. Stimulation of the same targets that were lesioned produced palliative effects [KPLA99]. VIM lesions and stimulation were well characterized to reduce tremor, but although tremor is a hallmark of the disease, it is not the most disabling symptom for most patients. To better deal with the bradykinesia and rigidity, the GPi and STN became the preferred targets for stimulation. Although STN is now the dominant DBS target, it is unclear whether GPi, with a less burdensome set of cognitive side effects, might be better for some patients [ABH+05]. As our experience with DBS has progressed, a direct comparison of pharmacological versus DBS (bilateral GPi or STN stimulation) for Parkinson's disease found DBS to have advantages in quality of life outcomes despite the risks inherent in surgical treatment [DSBK+06, WFS+09]. Nevertheless, interest remains in lesions, whose effectiveness versus pharmacological therapy has also been shown [VBF+03], and whose long-term medical management and costs are considerably less than for patients who require life-long maintenance of DBS systems [BH06]. An especially compelling study of bilateral subthalamotomy argues the rationale for lesions and demonstrates the apparent safety and risk assessment of symmetric subthalamic nucleus lesions [AML+05]. The issue requires continued debate, as the majority of patients with Parkinson's disease on the planet, and their health care systems, lack the resources for DBS therapy.

The lesion debate notwithstanding, there is no more fruitful arena to consider a radical new approach to neural systems control than DBS technologies for Parkinson's disease. Our present approach has been to focus on high-frequency stimulation (130 Hz) delivered in open loop without feedback sensing. Along with the rise of such empirical DBS therapy, we have developed increasingly sophisticated computational models of the fundamental networks involved in the pathophysiology of Parkinson's disease. And we have developed increasingly sophisticated models of the physics of DBS stimulation to help us understand how DBS interacts with neurons. Let's examine this in detail.

10.2 The Networks of Parkinson's Disease

A great deal of the brain, especially the regions beneath the cortex, is heavily involved with movement regulation. Such areas include the connected set of basal ganglia, portions of the thalamus, and the cerebellum. The cerebellum may well contain as many neurons as the rest of the brain. Coordination and movement require a lot of mind.

In Parkinson's disease, there is degeneration of neurons that use dopamine as a neurotransmitter, which have their cell bodies in the substantia nigra at the upper edge of the midbrain. The decrease in neural output from the substantia nigra causes a disturbance in the network balance of excitation and inhibition, as schematized in figure 10.2. The result is a net increase in inhibition from GPi to thalamus (for a much more detailed discussion of the circuitry, see [OMRO+08]). But the lines and arrows in these static diagrams refer to average firing rate or activity and do not reflect the dynamics that is critical to understand what is happening. In Parkinson's disease the inhibition to the thalamus becomes phasic and oscillates.

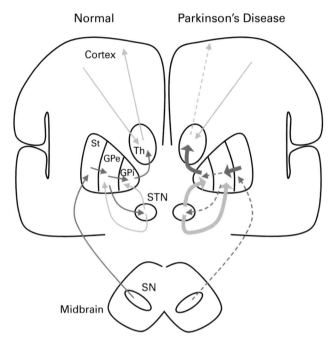

Figure 10.2 (plate 22)
An extremely simplified schematic of network imbalance in Parkinson's disease. Excitation is shown in red, and inhibition in blue. The normal brain (left) is contrasted with the brain in the Parkinson's disease state on the right, where thickened (thinned) lines indicate an increase (decrease) in excitation (red) or inhibition (blue). St, striatum; GPe, globus pallidus externa; GPi, globus pallidus interna; Th, thalamus, STN, subthalamic nucleus; SN, substantia nigra. I have made no distinction between indirect and direct pathways, and customized this for the purposes of the discussion within this chapter. For a more complete and detailed description of this anatomy, see [OMRO+08].

If, as in figure 10.2 (plate 22), you place a thin probe and burn a small hole in the GPi or STN, the symptoms of tremor, bradykinesia, and rigidity decrease.[1] We have understood this improvement in symptoms as resulting from a decrease in the excessive inhibition streaming out of the GPi and into the thalamus. If, instead, you place a thin electrode into these same structures, and stimulate at 130 Hz, you will get an almost identical decrease in the symptoms of tremor, bradykinesia, and rigidity. And this paradox requires explanation.

10.3   The Thalamus—It's Not a Simple Relay Anymore

The thalamus (Latin for *inner chamber*) is a walnut-sized piece of very high-end real estate in the center of the brain. We used to view it predominantly as a relay center for things

---

1. This is done by image- and microelectrode-guided stereotactic placement of a thin probe to the target nucleus within the brain, and the use of a calibrated radiofrequency thermal lesion.

such as sensory information. The thalamus has a variety of relay stations resynapsing touch, vibration, sound, and visual information onto a fresh set of neurons coursing to their respective cortical targets (see schematic in [GS02]). With substantial damage to the thalamus such as that caused by a stroke, the brain loss consciousness [PP07]. Neuroscientists, evolutionary biologists, epileptologists, and philosophers have always gravitated to the human cortex as the object of their fascination. But gradually the reality of a cortex heavily interconnected with a series of relay loops to the deep thalamus has emerged, substantially complicating earlier views [GS02].

Transmission within the cortex tends to be slow, with local conductance velocities in the range of centimeters per second, in contrast to the fast long-range neural connections where many tens of meters per second are commonly observed. We think that the thalamus forms an essential transcortical relay system to help integrate information processing across cortical areas where local conduction speeds would otherwise render us far dimmer.

The reliability of this relaying of motor information through the thalamus is a focus of the following computational theory. Nevertheless, one must ask why, if there is such an intimate relationship between cortex and thalamus, does so much of the medical treatment focus on the lesioning or stimulation of targets many inches beneath the surface of the brain? Recent clinical trials have explored this issue, but so far, the results of these early efforts at superficial cortical stimulation have not been as effective as DBS or lesions [GSAV+09].

10.4   The Contribution of China White

The study of the dynamics of Parkinson's disease was immeasurably helped by the seemingly inherent cravings of humans for narcotics. In the late 1970s synthesizing heroin-like compounds such as MPPP,[2] perhaps with a bit too much heat or acid in the reaction, produced a byproduct of the related MPTP.[3] A solitary report of poisoning from this byproduct quietly appeared in 1979 [DWM+79]. By 1982, a clandestine drug chemist in California was selling this product under the name *china white*, until a bad batch started producing profound Parkinson's disease symptoms in a group of young addicts [LBTI83]. It was quickly discovered that this compound could produce the same symptoms in nonhuman primates [BCM+83]. Both the human and animal data showed that MPTP was very selective in destroying the dopamine-containing neurons in the pars compacta of the substantia nigra—the same substructure that prominently degenerated in human Parkinson's disease. The symptoms and signs of the drug-induced and natural disease were nearly identical, and both diseases responded to treatment with the dopamine precursor drug L-DOPA.[4] The full story of these events is remarkable [LP95, FW02].

---

2. 1-methyl-4-phenyl-4-proprionoxy-piperidine.
3. 1-methyl-4-phenyl-1,2,3,6-tetrahydropyridine.
4. L-3,4-dihydroxyphenylalanine.

The critical benefit from these events was the proof that loss of a small group of neurons in the substantia nigra could produce the triad of tremor, rigidity, and bradykinesia. The primate animal model provided us with a way to dramatically increase both our knowledge of the electrophysiology of the neuronal networks involved, and their potential electrical modulation, to the extent that the following can be described.

10.5   Dynamics of Parkinson's Networks

Prior to 2002, most models of Parkinson's disease were displayed as static diagrams (as in figure 10.2). Nevertheless, the advent of the MPTP primate model [WD03, RFMF+01], and increasingly the recording of neurons from human Parkinson's patients during deep brain surgery [MMJ00, BOM+01], revealed that the neurons within the Parkinsonian networks were strongly oscillatory (thalamus, GPi, and the external segment of the globus pallidus, GPe). In a fundamental experiment, Plenz and Kital [PK99] observed that if they mixed bits of GPe and STN in tissue culture, the cells would spontaneously connect and generate oscillations.

In the schematic of the basal ganglia, fundamental features of *central pattern generators* are seen in the connectivity of GPe and STN. GPe cells inhibit each other as well as STN. STN excites GPe. When STN cells are inhibited, they also demonstrate an exaggerated rebound excitation [BMT+02]. Any simple neuronal membrane will spike if released suddenly from an inhibitory input—*anode break excitation* (chapter 3) [HH52d]. Some of the neurons critical to Parkinsonian dynamics have additional currents that exaggerate such inhibition-induced rebound excitation.

David Terman and colleagues [TRYW02] set out to explain these network effects on the basis of the biophysical properties of the individual neuronal types and their synaptic connectivity. They focused on the essence of what appeared to be the rhythm-generating circuitry, which turns out to also be the targets for both lesioning and stimulation in surgical therapy. (Their schematic is reproduced in figure 10.3.)

Applying typical experimental techniques in brain studies, we can record from single cells. But only in exceptional circumstances do we survey enough pairs of connected cells to develop strong characterizations about the details of the connectivity. There have been heroic experiments where investigators spent years carefully sticking microelectrodes into pairs of neurons to directly sample such connectivity [TM91], but such efforts remain rare. To make up for the paucity of connectivity data, Terman and colleagues [TRYW02] explored a range of connectivities. They had a great deal of intracellular data from networks from brain slice experiments that preserved native network architecture and connectivity to draw on (see [HB05] and references therein).

Three types of network topologies were constructed by Terman and colleagues [TRYW02]: random and sparsely connected, structured and sparsely connected, and structured and tightly connected, as shown in figure 10.4.

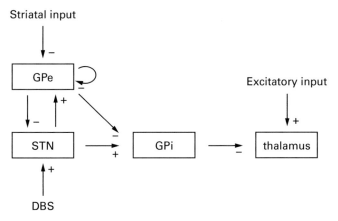

Figure 10.3
Schematic of rhythm-generating structures in Parkinson's disease. Striatal input refers to the outer segments of the basal ganglia that send input to these deeper segments. Excitatory input refers to sensorimotor input to the thalamus that needs to be relayed to cortex. Excitation is indicated by +, and inhibition by −. Reproduced from [RT04] with permission.

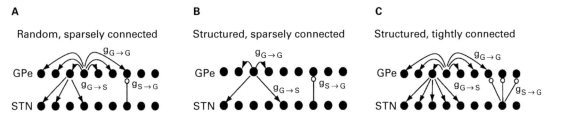

Figure 10.4
(A) Sparse random network. Each STN neuron excites a single random GPe cell, and each GPe neuron inhibits three random STN cells. GPe cells inhibit each other through all-to-all coupling. (B) Sparse structured network. Although more structured than the random sparse network of A, it is designed to avoid direct reciprocal connections between STN and GPe cells. Each STN neuron excites the single closest GPe cell, and each GPe neuron inhibits two STN cells skipping the three closest. GPe cells inhibit two immediate neighboring GPe cells. (C) Tightly connected structured network. Each STN neuron excites three closest GPe cells, and each GPe neuron inhibits the five closest STN cells. GPe cells inhibit each other through all-to-all coupling. Spatially periodic boundary conditions are applied (the network wraps around on itself). Reproduced from [TRYW02] with permission.

To model the STN neurons, and bring out their rebound excitability, Terman and colleagues [TRYW02] model the STN membrane current as a combination of

$$C_m \frac{dv}{dt} = I_L - I_K - I_{Na} - I_T - I_{Ca} - I_{AHP} - I_{G \to S} \tag{10.1}$$

where C_m is membrane capacitance.

The first three currents ($I_L - I_K - I_{Na}$) are the Hodgkin-Huxley leak, rectifying potassium, and fast sodium, which evolution elected to conserve from cephalopods to mammals [HH52d] (chapter 3). In this case, these currents had parameters tailored to the mammalian neurons rather than the squid axon. $I_{G \to S}$ represents the current injected into these cells from the GPe synapses. And now three special currents will be described: I_{AHP}, I_{Ca}, and I_T.

The afterhyperpolarization current, I_{AHP}, is a potassium channel that opens in response to increasing amounts of intracellular calcium. While it is on, it prevents the cell from firing another spike, so it takes an important role in turning off excitation, as well as regulating firing frequency. It's a good way to create resonance frequency ranges within which a cell would prefer to fire. And it's a good way to generate a pacemaker. It is prominent in motoneurons in the spinal cord, where the frequency of discharge must be matched to the characteristics of the muscle fibers they connect to (fast or slow) [KBC99]. I_{AHP} is prominent in the suprachiasmatic nucleus where it helps translate the molecular clock of our circadian rhythms into neuronal firing frequency [CS03]. And I_{AHP} is important in the subthalamic nucleus, helping to make the neurons more sensitive to inputs at motor frequencies [BW99], and helping to create an oscillatory central pattern generator out of the network it is embedded within in the basal ganglia [BMT+02].

The high-threshold calcium current, I_{Ca}, is a representation of what are likely several high-threshold calcium channels in such cells [SBOM00]. High threshold means that they are activated with depolarization. Because the calcium Nernst reversal potential is very positive (even more positive than sodium), these channels in general support regenerative potentials (that is, they boost depolarization already in progress). Bursting cells tend to have such channels.

The low-threshold calcium current, I_T, is prominent in thalamic neurons where following inhibition such neurons rebound burst fire [MB97]. This current activates on hyperpolarization, and then deactivates more slowly than I_{Na} deactivates as the neuron depolarizes. Since the reversal potential for calcium is positive, activating such a current gives the neuron a boost to move its membrane potential away from rest and accentuate the tendency of any neuron to rebound a bit (as anode break excitation). I_T plays a role in the prominent postinhibitory rebound spiking seen in STN cells [BMT+02]. Interestingly, the same T-type calcium currents play a role in the automaticity and pacemaker function in heart cells, and can play a role in certain cardiac arrhythmias [VTA06].

Terman and colleagues used the same currents as in equation (10.1) for modeling GPe neurons, but changed their proportions to better match the firing properties seen in a variety of experimental studies [TRYW02].

In the sparse random network (figure 10.4A), the strength of connectivity (maximal synaptic conductance) was varied from STN to GPe, and within the GPe to GPe network. Increasing the STN to GPe coupling produced a range of behaviors from sparse irregular firing, to episodic bursting, to continuous firing. The episodic regime was qualitatively similar to that reported in the classic study by Mahlon DeLong for cells in the normal primate GPe [DeL71]. One of the counterintuitive features of increasing the GPe to GPe inhibition is that it can increase the spread of activity through rebound. In the modeling of Terman and colleagues [TRYW02], the episodic burst firing is terminated by I_{AHP} as calcium is built up in GPe cells during the high-frequency firing episodes.

In the sparse structured network (figure 10.4B), there was more temporal clustering in the interactions among these cells, reflecting an increase in the topological spatial clustering of the neuronal connections.

In the tightly connected structured network (figure 10.4C), they also wrapped the network around on itself (periodic boundary conditions). This network generated waves that traveled. Initiating such waves required symmetry breaking in the STN-to-GPe and GPe-to-STN footprints. To form a solitary traveling wave, the GPe-to-GPe inhibitory footprint had to be larger spatially than the STN-to-GPe footprint (supporting a Turing instability [CH93, Mur03]). Key here was that, for a wave to be propagated, the STN and GPe cells had to have structured footprints so as to orderly spread activity to cells ahead of the leading edge of the wave [TRYW02].

Traveling waves have been experimentally observed in thalamic slices generating what are called spindle oscillations [KBM95]. Spindle waves have been related to sleep physiology, but the relationship to Parkinson's disease is not presently clear. Nevertheless, whenever there is evidence of temporal oscillations in a neural network, it is worth considering what the spatial structure of those oscillations are [HTY+04]. Especially when networks are sparsely connected (as opposed to all-to-all connected where there is no meaningful spatial structure), considering whether waves might underlie such rhythms is reasonable.

Strong oscillations emerge in the GPe-STN network in Parkinson's disease and in dopamine depletion in experimental animals. Nevertheless, a body of experimental evidence, in human patients as well as MPTP primates (see [TRYW02]), fails to find the sort of highly correlated and synchronized firing that would support the coherent waves predicted in the most structured networks of figure 10.4. The picture emerging from this work is that, among the more sparse networks, the conversion from normal to Parkinsonian dynamics fit well with the schematic in figure 10.5. This schematic illustrates that following a loss of dopamine input to the striatum, a strengthening of striatal input to the GPe, perhaps with a concurrent weakening of recurring inhibitory connections within the GPe, could create a Parkinsonian state.

Now go back to the standard firing rate model of Parkinson's disease from figure 10.2. In this static model, a *decrease* in inhibition to the striatum led to an *increased* activity of STN, which *increased* excitation to GPe and GPi, which *increased* the inhibition to the thalamus. No oscillations arise in the static model, in a disease whose hallmarks are the

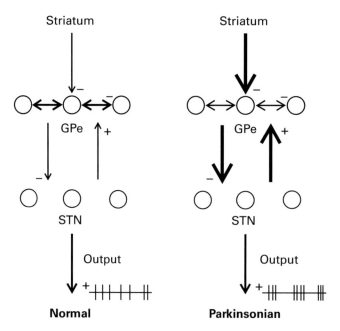

Figure 10.5
Schematic of Terman and colleagues [TRYW02] suggesting how an increase in striatal input and decrease in GPe internal connections would generate the oscillations of a Parkinsonian state. Reproduced with permission.

oscillations both within and outside the brain. In the view of Terman and colleagues, the level of striatal input (inhibitory) is a major regulator of whether oscillations arise from within this network that is naturally wired and poised to oscillate. Increased striatal inhibition effectively strengthened the coupling from GPe to STN. Decreasing intra-GPe inhibition would promote clustered activity within GPe. And inhibition, as is found in other dynamic phenomena in the nervous system such as seizures, plays multifaceted and complex roles [ZCBS06, MM08]. Inhibition in neuronal systems, similar to inhibitory reactants in reaction diffusion systems, is crucial in organizing patterns into clustered as opposed to homogeneous patterns [CH93, Mur03].

10.6   The Deep Brain Stimulation Paradox

Deep brain stimulation of the GPi or STN has almost identical effects on the symptoms of Parkinson's disease as do lesions of the GPi or STN. We have evidence of excessive activity of GPi in Parkinson's disease, but stimulation of STN should further increase GPi activity.

For several years it was therefore assumed that stimulation at the DBS frequencies being used (typically 130 Hz) must have been suppressing activity within the nuclei. This was supported by data showing that recording in the vicinity of the cell bodies within STN, when stimulating STN [BGN+00], or recording within GPi with GPi stimulation [BBBG96], demonstrated a *decrease* in apparent cell firing following nearby high-frequency stimulation. Helping to make further sense of this were the findings, in neurons different from the basal ganglia, that synaptic depression could occur at modest stimulation frequencies while the synapses ploddingly worked to repackage neurotransmitter within synaptic vesicles for release [SLBY98]. These findings were all consistent in explaining why DBS had the same effects as ablative lesions. This assumption was unfortunately wrong.

In computational modeling, Cameron McIntyre and colleagues [MGST04] demonstrated that such DBS stimulation close to cell bodies and axons might preferentially initiate action potentials further out along the axons of such neurons than we have appreciated previously, and furthermore that the cell bodies might not reflect these action potentials. Such findings also remind one of the differing requirements in the curvature of a stimulating electrical potential field needed to initiate firing at the terminal end of a neuron process membrane, as opposed to the middle of a membrane such as an axon coursing near an electrode *en passage* [WGD92]. More recently, optical imaging has also confirmed that action potential initiation may be well beyond the axon hillock, in contrast with what we previously assumed [BIA+04].

The definitive experimental evidence that laid to rest the hypothesis that DBS worked by suppressing neuronal activity and creating a *reversible lesion* was the demonstration by Jerrold Vitek and colleagues that such stimulation led to an *increase* in firing frequency in the nucleus receiving the efferent activity from the nucleus being stimulated [HEO+03] as shown in figure 10.6.

So if increasing the frequency of GPi activity through stimulation of STN had the same effect as burning a hole in GPi, the reason had to lie in the dynamics of the effect of increasing the firing rate of GPi. Note two things about the effect of stimulation on GPi activity in figure 10.6: (1) the frequency is increased, and (2) the neural response becomes less episodic and more continuous. The rest of this chapter seeks to understand and exploit this finding.

Increasingly, we run into examples such as this where you cannot possibly understand the raw data recorded from the nervous system without forming a computational model. The model is necessary to create the equations of motion for the network involved. As with Newton's laws, it does not take a very complex system before even the simplest of nonlinear interactions between elements, whether planets through gravity or neurons through synapses, become impossible to put together from visual inspection and guessing.

Jonathan Rubin and Terman [RT04] set out to ask whether the apparent stimulation paradox outlined here was explainable on the *basis of the biophysical properties of the neurons within these networks*. Their goal was "to demonstrate, with a computational model, why this is actually not contradictory, but rather is a natural consequence of the properties of the cells involved" [RT04]. Fitting empirical models to Parkinson's data might produce an

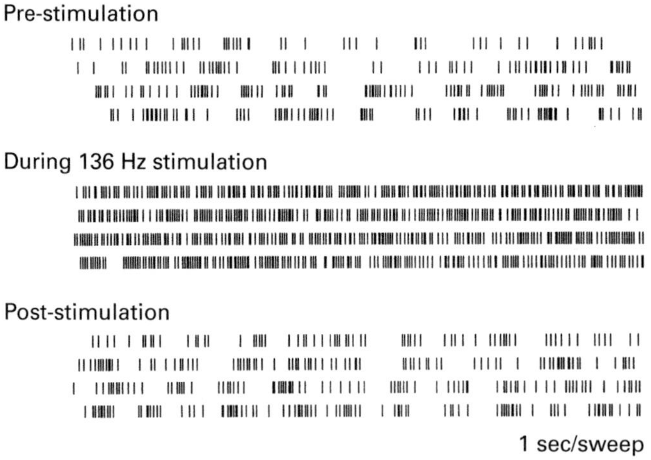

Pre-stimulation

During 136 Hz stimulation

Post-stimulation

1 sec/sweep

Figure 10.6
Demonstration that 130-Hz stimulation of STN in an MPTP primate increased the GPi cell firing rates. Reproduced from [HEO⁺03] with permission.

effective controller, but would be no more insightful than proving that you could fit a model to prove that the neurons did what you had observed them to do. The power of model-based control approaches is that we take the insights from what we have learned about these pieces of brain we are working with, and use those fundamental models to guide our observations and control.

10.7   Reductionist Cracking the Deep Brain Stimulation Paradox

We will now extend our model, from the oscillations between GPe and STN [TRYW02], to the effect of these oscillations on the thalamus through the intermediary way station of GPi. The motor structures within the thalamus, with apology to the vast complexity of this organ [GS02], will be viewed as a structure whose task is to faithfully relay information. The thalamus will have two inputs: GPi and sensorimotor signals (figure 10.3).

Rubin and Terman [RT04] distilled the essence of Parkinson's disease symptoms within the output of GPi. In the normal state, these cells fire irregularly and do not interfere with thalamic information relay. In Parkinson's disease, GPi cells fire bursts of action potentials at the frequency of tremor (3 to 8 Hz). The researchers [RT04] assumed that these burst firing cells will exhibit some degree of synchrony in the pathological state. Their hypothesis was that such clustered firing in bursts would impair the sensorimotor relay properties of the thalamic cells.

The sparse structured network (figure 10.4B) from Terman and colleagues [TRYW02] was chosen based on its results compared with the other topologies examined.

Thalamic cells were modeled with

$$C_m \frac{d v_{Th}}{dt} = -I_L - I_{Na} - I_K - I_T - I_{Gi \to Th} + I_{SM} \qquad (10.2)$$

where $I_{Gi \to Th}$ represents synaptic current from GPi to thalamus, and I_{SM} represents sensorimotor input to thalamus. They are of opposite signs because one is inhibitory and the other excitatory.

The specific structures of these currents are of interest to us now, as we will shortly reduce them as a prelude to our control framework.

The leak current is simple

$$I_L = g_L \cdot (v_{Th} - E_L)$$

where g_L is the maximal leak conductance, v_{Th} is the transmembrane voltage on the thalamic cell, and E_L is the reversal potential at which there would be no leak current when $v_{Th} = E_L$. The sodium current is from Hodgkin and Huxley (Hodgkin:1952d):

$$I_{Na} = g_{Na} \cdot m_\infty^3 (v_{Th}) \cdot h_{Th} \cdot (v_{Th} - E_{Na})$$

except for the use of Rinzel's approximation [Rin85], substituting m_∞ for m in the sodium gating variable, and in the following potassium current equation, substituting $1 - h_{Th}$ for n as the potassium gating variable (h is the sodium inactivation gating variable)

$$I_K = g_K \cdot [0.75(1 - h_{Th})^4] \cdot (v_{Th} - E_K)$$

The T-type calcium current equation is

$$I_T = g_T \cdot p_\infty^2 (v_{Th}) \cdot w_{Th} \cdot (v_{Th} - E_T) \qquad (10.3)$$

where $p_\infty(v_{Th})$ is the T-current gating variable, and w_{Th} is the T-current inactivation variable. In these equations, the reversal potentials for leak, sodium, potassium, and T-current are E_L, E_{Na}, E_K, and E_T, respectively. The inactivation variables follow the Hodgkin-Huxley formalism for first-order kinetics:

$$\frac{d h_{Th}}{dt} = (h_\infty(v_{Th}) - h_{Th})/\tau_h(v_{Th})$$

and

$$\frac{d w_{Th}}{dt} = (w_\infty(v_{Th}) - w_{Th})/\tau_w(v_{Th}) \qquad (10.4)$$

The sensorimotor current, I_{SM}, was prescribed to be either periodic or, at times, random. The symmetry introduced in this equation by the canceling currents $-I_{Gi \to Th} + I_{SM}$ will cause us difficulties, which we will address in section 10.10 (a more complete discussion of symmetries in reconstruction can be found in [SS09]).

These thalamocortical (TC) cells are silent if unstimulated. If stimulated with depolarizing current, they fire progressively faster. However, if they are hyperpolarized, we see progressively more and more intense rebound activity due to the T-current.

We can provide this TC cell with an analog to sensorimotor stimulation by periodically stimulating it. This is a signal that we hope the cell can relay. In figure 10.7A, we start with slow stimulation. The signal is reliably relayed. Now simultaneously we provide the cell with excessive inhibition, such as from an overactive GPi in Parkinson's disease. In figure 10.7B the baseline membrane potential is now more hyperpolarized. With each sensorimotor pulse, the cell rebound spikes because the T-current is deinactivated. This is because hyperpolarization removes I_T inactivation, just as in the sodium inactivation in the Hodgkin-Huxley gating variable h [HH52d]. But removing inactivation is also a relatively slow process. In figure 10.7B, there is sufficient time for this variable to be fully deinactivated. Contrast this with higher stimulation frequencies. In figure 10.7C, the TC cell is reliable, and it remains so in the setting of additional inhibition in figure 10.7D because there is insufficient time for the inactivation to deinactivate.

STN cells were modeled with

$$C_m \frac{dv_{Sn}}{dt} = -I_L - I_{Na} - I_K - I_T - I_{Ca} - I_{AHP} - I_{Ge \to Sn} + I_{DBS} \tag{10.5}$$

Here Rubin and Terman [RT04] added a high-threshold calcium current, I_{Ca}, and the synaptic current is now from GPe to STN, $I_{Ge \to Sn}$. There is also a DBS current, I_{DBS}. The parameters on the STN cells were adjusted so that the cell was spontaneously active, fired at high frequencies when depolarized, and had a less prominent rebound than the TC cell. All of these adjustments were to preserve as much of the qualitative distinction between these cells' firing properties as was observed in prior experiments.

GPe cells were modeled with

$$C_m \frac{dv_{Ge}}{dt} = -I_L - I_{Na} - I_K - I_T - I_{Ca} - I_{AHP} - I_{Sn \to Ge} - I_{Ge \to Ge} + I_{app} \tag{10.6}$$

Input from the striatum was modeled with I_{app}. The GPe cells would continuously fire, initially decrease their firing rate with inhibition, and with more inhibition cluster fire consistent with the results of Terman et al. [TRYW02]. The GPi cells were similarly modeled, except the parameters were adjusted to account for the experimental evidence that these cells tended to fire faster than GPe cells [DeL71].[5]

---

5. Some further discussion about the applicability of DeLong's results [DeL71] to the classification of human Parkinsonian cellular activity can be found in [SSKW05].

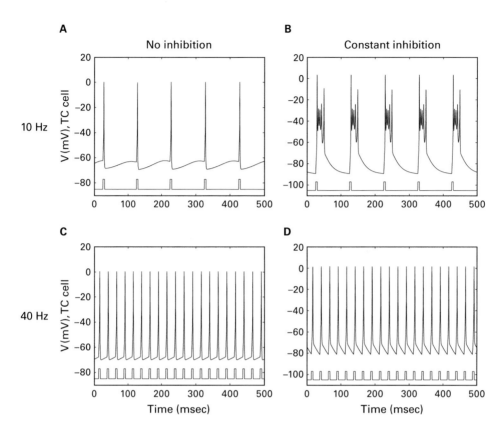

Figure 10.7
Stimulation of TC cell model at 10 Hz (A and B), and 40 Hz (C and D), at low (A and C) and high (B and D) levels of tonic inhibition. Reproduced from [RT04] with permission.

Following the schematic in figure 10.5, the Parkinsonian state is recreated by increasing the striatal input to GPe, and decreasing the amount of internal recurrent inhibition within GPe. The result is that the normal reliability of the TC cell to transmit sensorimotor information, illustrated in figure 10.8, becomes impaired in the Parkinsonian state. The key quantity here is the error rate of transmitting sensorimotor input into TC spikes. An error index can be created as

$$\text{Error Index} = \frac{\text{Missed spikes} + \text{Bad spikes}}{\text{Total inputs}}$$

$$= \frac{\text{False negative} + \text{False positive}}{\text{Total inputs}}$$

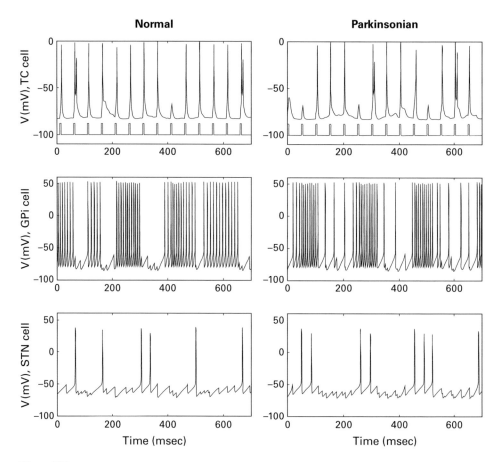

Figure 10.8
Model of the TC (upper panels), GPi (middle panels), and STN (lower panels) cells to periodic sensorimotor stimulation (top panel) in the normal (left) and Parkinsonian (right) states. Reproduced from [RT04] with permission.

Rubin and Terman [RT04] showed that the error rate was significantly elevated in the Parkinsonian[6] state compared with the normal state, and that the error rate could be normalized by simulating DBS using a constant level of high-frequency stimulation of the STN. The key to understanding these results is to know what the TC cell is receiving. In the Parkinsonian state, the amount of inhibition is fluctuating more than normal. This induces sequential excess suppression and rebound bursting in the TC cell, which destroys reliability. Applying DBS decreases the fluctuations that the TC cell receives, despite an overall increase in GPi cell firing.

---

6. It is semantically sensible to designate dynamics similar to those seen in Parkinson's disease as "Parkinsonian," and to similarly term analogous models as Parkinsonian.

A key insight of Rubin and Terman [RT04] was that the qualitative features of the above could be preserved in the TC cell *without* fast dynamics.[7] Taking their TC membrane model, they removed the fast Hodgkin-Huxley currents for sodium, I_{Na}, and potassium, I_K, and created a two-variable description of TC cell membrane dynamics:

$$\frac{dv}{dt} = -(I_L + I_T)/C_{Th} - I_{Gi \to Th} + I_{SM}$$

(10.7)

$$\frac{dw}{dt} = \varphi \cdot (w_\infty(v) - w)/\tau_w(v)$$

In equations (10.7), v represents membrane voltage, φ serves as a relative rate constant between the two differential equations, w represents T-current inactivation, and the availability of T-current I_T will be the key to reliability.

The current $I_{Gi \to Th}$ will be modeled as a true synaptic inhibitory input onto the TC cell as

$$I_{Gi \to Th} = g_{Gi \to Th} \cdot s_{Gi} \cdot (v - E_{Gi \to Th})$$

(10.8)

where s_{Gi} will be a small positive constant in the normal, a periodic square wave in Parkinsonian, and a larger positive constant under conditions of DBS. But because a reversal potential is attached to all synaptic ionic channels, $E_{Gi \to Th}$, the current $I_{Gi \to Th}$ will fluctuate with voltage v even when s_{Gi} is a constant. The equations for the T-current, I_T, are as given in equations (10.3) and (10.4).[8]

The beauty of the two-variable reduction of the TC cell is that the nullclines can be plotted in two dimensions and visualized,[9] providing further insight into the dynamics. It is also a simpler system to fit, if you are interested in observing actual data or developing control laws, than using the full model.

The nullclines for v and w are shown in figure 10.9A. In excitatory cells and their models, there is almost always a cubic or N-shaped nullcline for the fast excitatory variable, the voltage v in this case (see chapter 4). Keep in mind that in this reduced model there are no true action potential spikes (no I_{Na} or I_K currents). These phase space plots show us the slow dance between voltage changes, v, and I_T inactivation, w. In the middle panel, we see the effect of DBS. Increasing the synaptic current from GPi, s_{Gi}, as the DBS parameter

---

7. This is only one example of a case where slow excitability dynamics may be more important than the faster spiking dynamics in determining pathological activities in dynamical disease. In recent work, we have shown that in models of epileptic seizures, relatively slow fluctuations in potassium dynamics may underlie the more complex spike dynamics seen in individual cells [CUZ+09, UCJBS09].

8. The square on p_∞ can be dropped in the reduced model. See [RT04] appendix for all equation details.

9. Nullclines are the curves where the rate of change of the variables are equal to zero. The intersection of more than one nullcline gives us the solutions, stable or unstable, of the system variables.

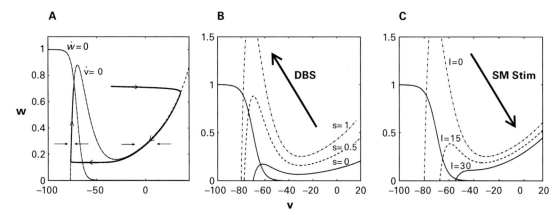

Figure 10.9
Nullclines for $\dot{v} = 0$ and $\dot{w} = 0$. In A, the heavy line is a trajectory initiated by depolarizing the rest state at the intersection of the nullclines from -65 mV to -25 mV, and the subsequent trajectory in the phase plane closely tracks the faster equilibrating v nullcline, and slowly works its way up along the direction of increasing w as the T-current deinactivates (that is, the availability of T-current, w, *increases*). B shows the effect of DBS, elevating the v nullcline. C shows the effect of sensorimotor stimulation, SM, which lowers the v nullcline. Reproduced from [RT04] with permission.

in (10.8) literally adds the $I_{Gi\to Th}$ current term in (10.7) to the solution of the v nullcline.[10] This has a qualitatively opposite effect to increasing excitatory sensorimotor stimulation I_{SM} in (10.7), which decreases the height of the v nullcline. At the point where these nullclines intersect is the resting steady state for v and for w. The T-current inactivation, w, is a key factor in whether this system will respond with a rebound burst, respond reliably to a sensorimotor input, or be inactive and unreliable by not responding at all.

When the intersection is brought lower on this phase plot, the system requires enough T-current, by having high enough w, to be able to push over the peak in the $\dot{v}$ nullcline, the *knee* of this curve (the upgoing bump). As sensorimotor stimulus steadily increases, the value of w gradually decreases setting the system up for trouble.

What this means is that if the position of the intersection of the nullclines is such that a stimulus (I_{SM}) comes when the value of w is above the knee of the $\dot{v}$ nullcline, then the cell can fire a spike. If the system were subjected to *episodic* bursts of inhibition through a Parkinsonian GPi input to TC, then when the intense inhibition is released the cell would rebound. But if you hit the cell with an intense enough *steady* DBS input, the cell will

10. The reason is that when $dv/dt = 0$, the solution of v is

$$v = (I_L - I_{Gi\to Th} + I_{SM})/(g_{Ca}p_\infty(v - E_{Ca}))$$

where the very high positive reversal potential for calcium, E_{Ca}, always makes $(v - E_{Ca})$ negative. Because of the negative reversal potential $E_{Gi\to Th}$ for the inhibitory synapses, $(v - E_{Gi\to Th})$ is almost always positive in equation (10.7). So the net effect is that increasing $I_{Gi\to Th}$ leads to an *increase* in the v nullcline.

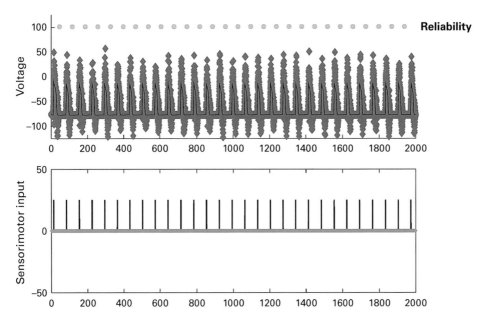

Figure 10.10 (plate 23)
Normal response of TC cell to periodic sensorimotor stimulation. Voltage in reduced TC cell model (thin black line, upper panel), with substantial added noise to serve as a noisy observable (blue markers, upper panel), and sensorimotor input (red pulses, lower panel). Each time a sensorimotor pulse is reliably transmitted, a marker (green) is placed above the successfully transmitted spike. This cell is 100% reliable. Reproduced from [Sch10] with permission.

maintain sufficient T-current and has no opportunity to rebound (there is no abrupt turn-off of inhibition). Because in the Parkinsonian state the T-current availability fluctuates on and off, the result is tremor at this fluctuation frequency. DBS can level this out and simultaneously keep the cell in a state ready to fire reliably to excitatory inputs by keeping w moderately high.

Now let's look deeper into these reduced dynamics in normal, DBS, and Parkinsonian states. In figure 10.10 (plate 23) we see a normal state. The input to the reduced TC cell from the GPi is a modest constant, which represents a state where there is asynchrony among the GPi cells, and there results a modest steady level of inhibition to the TC cells. Periodic sensorimotor stimulation is represented by brief square wave excitatory inputs (second panel). A substantial amount of random measurement noise is added to the actual TC voltage. These noisy measurements are what we record from in these models as our *observable* variable. Note that for each sensorimotor stimulus, a "spike" is transmitted from the TC cell. This is reliable transmission (100% reliable in this case).

The DBS state is modeled as increased tonic inhibition on the TC cell. The Parkinsonian case can be idealized as a periodic fluctuation in the GPi input onto the TC cells. In [RT04], such an effect can be shown as a slow on and off square wave modulation of GPi output. Using the reduced model, when excess inhibition turns on, w gradually increases, T-current becomes more available, and spiking becomes more reliable after an initial suppression by the inhibition. When inhibition suddenly stops, w remains too high, and an excessive response, with a period of rebound excitation, occurs. During the trough of the Parkinsonian inhibition fluctuations, spikes are reliably transmitted but w gradually inactivates, and when the next inhibitory pulse hits, there is spike failure. This sequence of recurrent spike failure and rebound creates unreliability. Recall that in the reduced model there are no fast sodium spikes—the slow "spikes" are periods of increased *excitability*; if we had I_{Na} and I_K, we would get a burst of fast sodium spikes riding on the excitability rebound event, and this would also reflect unreliable information transmission. Steady DBS, literally taking out the troughs in the Parkinsonian fluctuations, can restore reliability, as we will shortly examine in more detail.

10.8   A Cost Function for Deep Brain Stimulation

Xiao-Jiang Feng and colleagues sought to further explore the work of Rubin and Terman by introducing *optimization* principles [FSBG+07, FGR+07]. They, like Rubin and Terman [RT04], employed the sparse structured network of Terman and colleagues [TRYW02].

Why not just use the high-frequency stimulation that presently demonstrates efficacy in clinical use? For one thing, the more energy we apply per day, the sooner the batteries of the stimulators wear out and the devices need to be surgically replaced.

But beyond batteries, as stimulation intensities increase, so will negative cognitive side effects [AVRH+08].[11] I would pose a general principle that in the treatment of any dynamical disease of the brain, one needs to optimize the beneficial effects of symptom relief (e.g., tremor), with the inherent effects of increasing cognitive dysfunction with increasing DBS energy.

In addition, it is not trivial at all to make clinical adjustments to a patient's stimulation parameters. Once set, the parameter space of stimulation intensity and frequency, along with duty cycle, is enormous. Making changes, waiting several days or weeks to see the steady-state effect, and trying to optimize settings is difficult in patients. So if a stimulator is working and showing benefit, optimizing in this ad hoc fashion is generally avoided. Nevertheless, there is evidence that some patients' symptoms can be improved at frequencies different from the standard starting frequency of 130 Hz [MDD+08]. And who is to say that a more

---

11. There are recent reports that steady low-amplitude DC stimulation to some parts of the brain might improve cognitive capabilities [MHMB06]. But this is a polarization effect on certain parts of the cortex. To the author's knowledge, all pulse stimulation of deep brain structures will have negative effects on cognitive information processing as stimulation is increased. This will likely apply to DBS treatment of cognitive disorders, such as depression and obsessive compulsive disorder, as well as to Parkinson's disease.

complex nonperiodic stimulus delivered in open-loop might not be more beneficial than the standard periodic ones?

Last, the beneficial effects of DBS over the long term in a patient with a neurodegenerative disease (and perhaps those without degeneration) is a moving target. Patients with DBS treatment of Parkinson's disease continue to deteriorate within the tempo of progression of Parkinson's disease [KBVB+03]. Neural networks learn and change in response to stimulation. A means to automatically adapt and optimize such stimulation over time will be a very valuable advance in our stimulation technology.

One requires metrics for optimization. Feng and colleagues introduced a *reliability* measure similar to Rubin and Terman [RT04], as well as a GPi cell *correlation* metric. Neither metric has a clear route to practical application in human subjects using our present optimization done *by hand*. But these tools can be immediately applied if we construct a customized computational model for an average or specific patient. And this is the model-based approach that can be employed in future automated closed-loop optimization.

The structure of their cost function was [FSBG+07]

$$J^i = x^i + wR$$

$$x^i = \text{criterion } i \text{ (reliability or correlation)}$$

$$R = \text{energy required}$$

where w is a weighting parameter. They employed a genetic algorithm to search their parameter space and accomplish optimization.

One way of implementing such a cost function is to integrate the current coming out of the stimulator, I_{DBS}, and subtract this weighted integral from the reliability, Rel, as

$$J = Rel - w \int_{t=0}^{T} I_{DBS} dt$$

So starting from a sparse structured network, with two TC cells, Feng and colleagues replicate the fundamental findings of Rubin and Terman [RT04] in figure 10.11A–F.

They then use their genetic algorithm to explore the parameter space of frequency and amplitude as shown in figure 10.11G. In the far rear corner of figure 10.11G lies the high-frequency and high-amplitude parameter regime typically employed in clinical use, and the reliability there is high. But notice the other peaks. The most prominent secondary peak is between 40 and 50 Hz, and indeed, there is some clinical evidence that this frequency range might be therapeutic [MDD+08]. Figure 10.11H shows the results of a minor peak at a period of 80 ms (12.5 Hz); there is actually some clinical study of stimulation in the range of 5 to 20 Hz, which was (unfortunately) singularly unimpressive [ECL+08]. We note that another exploration of frequency in an extended model of Rubin and Terman [RT04] demonstrated that the frequencies greater than 100 Hz were most effective [PRSC08].

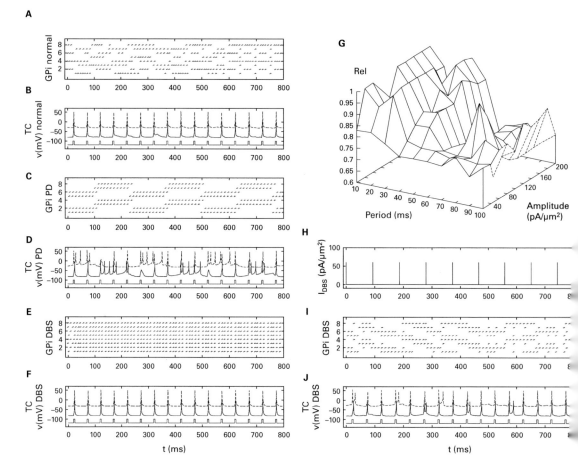

Figure 10.11
Exploration of the parameter space of period and amplitude focusing on reliability: Normal (A,B), Parkinson's disease (PD) (C,D), and DBS in PD (E,F). Optimization of reliability as a function of period and amplitude (G). Slower DBS frequency at 80 ms reliability (Rel) peak in H. Reproduced from [FSBG+07] with permission.

Reliability is an interesting measure. There seems no way to directly measure in real life the quantities needed to calculate reliability: sensorimotor inputs and thalamic relay cell output in response to this input. But such inputs can be readily modeled using a data assimilation framework. Feng and colleagues [FSBG+07] also employ a correlation measure, which was a combination of autocorrelation and crosscorrelation among GPi cells. This would require a microelectrode array to be placed within GPi. Such microarray technology is now becoming available on the shafts of clinical DBS electrodes. A simpler way to infer such correlation in practice would be to estimate spectral concentration in GPi local field potentials, and this is directly observable from an ordinary DBS electrode placed within GPi.

An important contribution of Feng and colleagues [FSBG+07] is that they demonstrated in this computational model that stochastic stimulation as well as complex waveforms can demonstrate potential efficacy.

10.9   Fusing Experimental GPi Recordings with DBS Models

A creative next step was performed by Guo and colleagues [GRM+08]. They took recordings of GPi spike timings from normal and MPTP primates, with and without DBS. They applied a structure function to these spike timings to recreate an estimate of the synaptic currents produced in the thalamic relay cells from such spiking. They used the model from Rubin and Terman [RT04] to estimate the reliability of the transmission of spikes through the thalamus. Their results demonstrate that the error index (described earlier) is higher for the MPTP monkeys without DBS or with subtherapeutic DBS, and consistently lowest for normal monkeys and those MPTP monkeys with therapeutic levels of DBS.

This is another example of how we can apply models in real life applications to Parkinson's disease treatment. Recording of an array of GPi cells would be required. One could then follow the prescription of Guo and colleagues [GRM+08] to estimate ongoing reliability, and potentially perform closed-loop optimization.

A very interesting insight from Guo and colleagues [GRM+08] was to *reverse correlate* what the inhibitory current drive from GPi was *prior* to a spike. Spikes can be faithful in following a sensorimotor input, they can miss their chance to follow, or multiple bad spikes can ensue. Guo and colleagues found that missed spikes correspond to a relatively rapid rise in inhibition, and bad spikes correspond to a relatively rapid decrease in inhibition. This is exactly what one might have expected from the reduced model effects illustrated in figure 10.9.

10.10   Toward a Control Framework for Parkinson's Disease

Parkinson's disease will be the first *dynamical disease* [MG77] amenable to management through model-based feedback control systems.

First, examine the schematic in figure 10.3. We can now relatively safely place electrodes in the STN, GPi, or thalamus.[12]

What about the small amount of damage that we get from just passing a 1 to 2 mm-thick depth electrode? Recall that in Parkinson's disease, making a lesion in these small structures is clinically beneficial. Amazingly, we observe that in the course of placing DBS electrodes in Parkinson's disease patients, there is an immediate clinical improvement in about *half* of patients at the conclusion of electrode placement, *before* the electrode is connected to the stimulator [Tas98]. This has been termed the *microthalamotomy effect* [Tas98]. A nearly identical experience has been reported in the placement of DBS electrodes in small thalamic targets for epileptic seizure suppression [HWDL02]. *Our DBS treatments are likely a combination of small lesions overlain with chronic electrical stimulations.*

Despite the beneficial effect of some of the microlesions created by electrode insertion, the best clinical strategy is almost certainly to keep the number of electrodes inserted to the absolute minimum.[13] But the network that is important in Parkinson's disease control is spread out over at least four separate structures: STN, GPe, GPi, and thalamus.

From a clinical perspective, the GPi would likely be an easier and safer target than the smaller STN or thalamic targets. A small hemorrhage or lesion effect in GPi has a reasonable chance of actually benefitting some of the Parkinson's disease symptoms. But with a superb animal model in the MPTP primates, exploring which nuclei are optimal to record data from, and work out the control algorithms, offers a valuable way to learn how to build such systems prior to their implantation in human patients. Nevertheless, given that our clinical opportunities with routine implantation in patients is ongoing, the chance to take advantage of existing STN or GPi human placements in algorithm development is highly attractive.

The models we have just discussed permit us to sample from a single nucleus within this network and reconstruct what the remainder of the network is doing. Our present state-of-the-art models appear sophisticated enough to consider using these models in a data assimilation framework. And the technology to perform chronic real-time sampling from these structures is presently available and in clinical use.

So let's begin to outline a control theoretical framework from what we have discussed.

---

12. The best current data, from A. Benabid (2007, personal communication), suggests an 8% risk of a complication per depth electrode pass. Half of these events (such as small hemorrhages) are asymptomatic. Half of the remaining give transient effects. A reasonable estimate is that 2% of electrode passes into the brain using modern stereotactic techniques will result in a significant neurological complication [LGT09].

13. This is a conjecture by the author, and one can argue in the abstract that we are accumulating a tremendous body of data showing that microlesions plus electrical stimulation through single electrodes is efficacious. But we have no such body of data suggesting that multiple microlesions in separate unilateral structures will be beneficial. And from a surgical standpoint, every electrode insertion increases operative time, increases the risk of hemorrhage and neurological deficit, and increases the risk of infection. Surgical minimalism should be a guiding principle in the absence of strong evidence supporting the contrary.

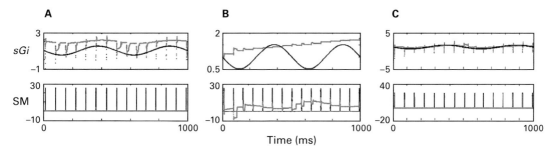

Figure 10.12 (plate 24)
Tracking and estimating parameters as a function of process noises Q. Estimates of synaptic current from GPi (sGi, upper), and sensorimotor input (SM input, lower). Process noise parameters are (A) $sGi\ Q = 30$, $SM\ Q = 0.01$, (B) $sGi\ Q = 0.01$, $SM\ Q = 30$, and (C) $sGi\ Q = 10.0$, $SM\ Q = 0.01$. The algorithmic incorporation of such process noise can be explored in the code archive with [SS08]. Reproduced from [Sch10] with permission.

The first issue with parameter estimation for equation (10.7) is to notice that there is a symmetry with the terms $(-I_{Gi \to Th} + I_{SM})$. The tracking algorithm will simply apportion such current equally between these two sources. One option is to combine the currents into one sum, which physically is what is happening to the TC cells. But this defeats our purpose here, where we would like to estimate quantities such as $(-I_{Gi \to Th})$ in isolation. A more extensive discussion of symmetries in such equations can be found in Sauer and Schiff [SS09] (see chapter 4).

When faced with such symmetries, it is best to get rid of them. If that is not feasible, then an empirical rule of thumb seems to be to set process noise in rough proportions to the average magnitudes of the corresponding variables. One could adaptively tune these process noises over time by tracking innovation error. Process noise, Q, is uncertainty commonly added to the model of the process (the *plant* in control jargon) in analogous applications (chapter 2). In this particular instantiation of ensemble Kalman filtering, we will use several Q values as the assumed variance in the respective parameters to be tracked. In addition to apportioning variance to the respective parameters, this also has the benefit of preventing a Kalman filter from driving the parameter covariance to zero.[14] In figure 10.12A, the Q for $(-I_{Gi \to Th})$ is 30, while the Q for I_{SM} is 0.01. Note that the rhythmicity of $(-I_{Gi \to Th})$ is resolved, while none of the features of I_{SM} are picked up (figure 10.12A). Now reverse the situation, setting Q for $(-I_{Gi \to Th})$ to 0.01 and the Q for I_{SM} to 30; the $(-I_{Gi \to Th})$ will be poorly tracked, but the sensorimotor inputs I_{SM} are better tracked (figure 10.12B). A more balanced set of process noises, where Q for $(-I_{Gi \to Th})$ is 10, and Q for I_{SM} is 0.01, yields a more optimal tracking of $(-I_{Gi \to Th})$ (figure 10.12C) (see plate 24.).

14. A detailed discussion of process noise can be found in [Sim06a].

We can calculate a running *reliability* index over the past 10 msec (ignoring for now having to objectively define "bad" spikes unrelated to sensorimotor inputs) as

$$\text{Reliability} = \frac{\text{Number of spikes that got through (within 10 msec)}}{\text{Total number of impulses}}$$

Let's now focus on the heart of the dynamics critical for the Parkinsonian state—the T-current inactivation w. This is a rather independent variable, in time scale and with respect to symmetry, and lies at the heart of the issues of unreliability.

In figure 10.13A (plate 25), I employ the reduced TC cell model from Rubin and Terman [RT04] in a data assimilation framework. The observable will be noisy voltage from the TC cell in the upper panel of the figure. Below that is shown the reconstructed estimated T-current inactivation, w, and the reconstruction of estimated GPi and sensorimotor activity (panels 2 through 4, respectively). The techniques used to accomplish this reconstruction are detailed in [SS08, SS09, US09], and a code archive for the basics of such data assimilation can be found in supplementary data with [SS08].

One could similarly record the average GPi output from a DBS electrode and reconstruct the estimated TC cell activity.

The sensorimotor input, in the fourth panel of figure 10.13, causes the model TC cell to generate or fail to generate reliable spike activity in the upper panel. I build up a reliability index by averaging the last ten sensorimotor spike responses, and plot this at the very top of the upper panel.

It is tempting to examine and consider control from the nullclines. There is an extensive literature in control theory on what is generally termed *variable structure control*. First described by Utkin in the 1970s [Utk77], this control law strategy is now more commonly referred to as *sliding mode* control [DZM88, YUO99]. We could use such nullclines to generate control functions toward which we seek to target the system. In figure 10.14A, we see the actual nullcline intersection of our simulation of the reduced TC cell as the GPi current switches from off to on (refer to figure 10.9). The w nullcline does not change here as GPi input fluctuates. In figure 10.14B, I show an estimation of these nullclines from a reconstruction of these curves using an unscented Kalman filter. There is a great deal of uncertainty in the v nullcline. So estimating a control surface in this phase space is not a trivial problem that we know how to solve at present.

But recall that, as discussed Rubin and Terman [RT04], it is not the nullcline intersection so much as the value of w that is the most valuable single feature in explaining the dynamical response of the TC cell. So let's estimate w by itself.

In the second panel of figure 10.13A is the estimate of the reconstructed T-current avail-ability w (red line). Note that w increases when the GPi current increases in the third panel (solid black line). I have deliberately adjusted the ratios of the process noises so that the sensorimotor input is *not* tracked well (fourth panel), in order to minimize the symmetry

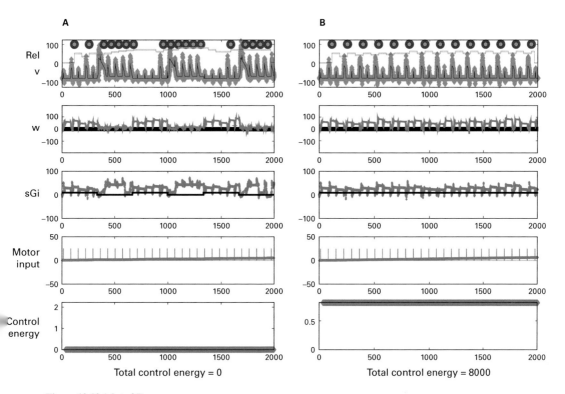

Figure 10.13 (plate 25)
(A) Uncontrolled reduced TC cell dynamics in the Parkinsonian state with fluctuating current from GPi (sGi). (B) Perfect DBS stimulation filling in the troughs in the fluctuating current from GPi. Top panels show noisy observable voltage (blue symbols), reliability as piecewise continuous plots without (blue) and with (red) control on, the green circles are the timing of SM spikes, and the smaller red circles are transmitted spikes. The second panels show estimated w (red). The third panels show real (black) and estimated (magenta) synaptic current from GPi (sGi, estimated values multiplied by 10 for discriminability from the true values). The fourth panel shows real (red) and estimated (magenta) motor input (we are deliberately not trying to reconstruct motor input in the reconstruction through Q ratio adjustment). The bottom panel shows the running control energy, the squared value of the control signal at each time point, and the total sum of squares given as total control energy. Reproduced from [Sch10] with permission.

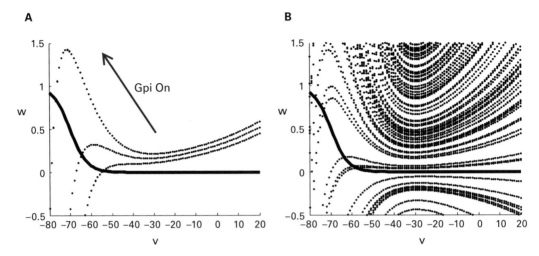

Figure 10.14
Nulclines for w (thick line) and v (dotted line). In A, with increased GPi inhibition onto the reduced TC cell, the v nullcline is elevated with respect to the w nullcline. B shows reconstruction estimates of the nullclines from noisy measurements. Reproduced from [Sch10] with permission.

of this current in the voltage equation. The bottom panel shows the running plot of control energy, which is zero for this example without feedback.

Now let's use *perfect* deep brain stimulation. We will, in the sense of Rubin and Terman [RT04], fill in the gaps of the fluctuating synaptic current from GPi, and show these results in figure 10.13B. Providing a steady level of DBS increases the amount of inhibition arriving on the TC cell. This has the effect of removing the fluctuations in inhibition creating the gaps in reliability shown in figure 10.13A, but it also dampens down the responsiveness of the TC cell. Only about half of the incident sensorimotor spikes get through reliably. And the cost, in terms of control energy, is high (8,000 on an arbitrary scale). On the other hand, the fluctuations in w are reduced, and we maintain an overall high level of w (second panel figure 10.13B).

Now let's use adaptive feedback based on the estimated w. In figure 10.15A (plate 26), I show the effect of an optimal amount of proportional feedback gain based on a moving average (35 msec) of the estimate of w. This feedback is very effective in restoring most of the unreliable (missing) spikes. The total cost in control energy is about half of the perfect DBS case shown in figure 10.13B. Now let's present a more realistic DBS scenario than in figure 10.13B, one in which a constant stimulation (open loop) will be *added* to the fluctuating Parkinsonian GPi signal. Figure 10.15B shows the largest (and most effective) additive current that is stable in this model. Here we are limited by the relatively large peak fluctuating GPi currents being applied already in the Parkinsonian state, as the dynamics of the TC cell become unstable if the impinging currents become excessive. The figure shows

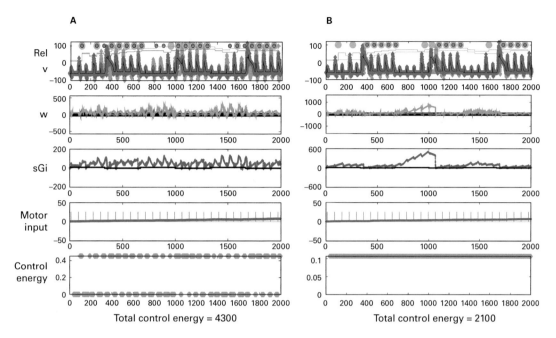

Figure 10.15 (plate 26)
Feedback control scenario based on turning on and off deep brain stimulator based on a running average of the estimated T-current availability shown in A. B shows a scenario where a constant amount of deep brain stimulation is simply added to the fluctuating Parkinsonian GPi output. No adjustment of GPi constant stimulation comes close to the reliability achieved with the closed loop feedback scenario shown in A. Symbols as in figure 10.13. Reproduced from [Sch10] with permission.

that such constant DBS does not appear capable of achieving the reliability possible with feedback control.[15]

It is important to note that in this simple scenario, a range of additional adjustable parameters are important. First, there is the ever-present issue of covariance inflation [AA99].[16] In figure 10.16A, we see that adjusting the small covariance inflation parameter has a substantial effect on the reliability of the adaptive system. Similarly, the gain on the feedback control is important. In Figure 10.16B, we see that optimizing gain readily reveals a region

15. In real life, such model instability would not arise. So long as we stay within the safety limit of the electrical stimulation, we might be able to deliver more GPi current to the TC cell. But the reliability steadily decreases in this scenario as this type of additive DBS increases in the model, since too much inhibition shuts off the TC cells. We find no inference that such open loop stimulation is going to be as effective as an intelligent adaptive approach as illustrated in figure 10.15A.

16. Recall that even Julier and Uhlmann [JU97a] had an adjustable inflation parameter that they required to tune their original unscented filter. As in so many of these recursive filter designs, there is a trade-off between tracking accuracy versus stability with greater uncertainty. One also needs to counter the tendency of these filters to recursively drive the uncertainty to zero, which ruins tracking ability.

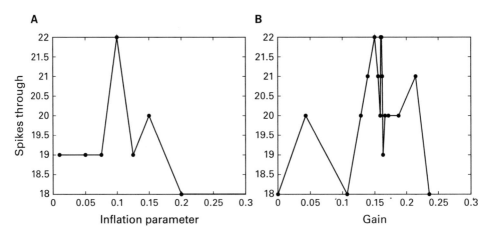

Figure 10.16
Optimizing reliability as spikes that are transmitted reliably through the TC cell as a function of the covariance
inflation parameter (A) and proportional control gain parameter (B). Reproduced from [Sch10] with permission.

where spike throughput is best. Both of these functions are not smooth, and implementing
such an algorithm should be done with continual adaptation of such parameters based upon
the monitoring of the system performance.

Another alternative is to generate a control signal based on the estimated GPi output,
shown in figure 10.17 (plate 27). Since we know the control signal added, we can subtract
this to follow just the underlying estimated GPi input to the thalamus. Since the estimates
of GPi input to thalamus are noisy (dotted blue line in figure 10.17), it is helpful to create a
moving average filter of this estimate (solid blue line) to prevent the controller from turning
on and off too often. Since at the heart of Parkinson's disease physiology are the large-scale
slower fluctuations, we can create a long-term running average of the GPi output, much
longer than the noise-reducing short-term moving average, and let this serve as an adapting
threshold (magenta line). Control is turned on whenever the short-term moving average
(blue solid line) falls below the long-term moving average (magenta line). The control is
applied in this case by turning on the stimulator with the same constant amplitude, shown
as the red lines. The underlying true GPi output fluctuations are shown as a black line for
comparison. Note that a *control reliability* can be calculated as the fraction of time that the
control signal (red) is on or off in correct reflection of the peaks and valleys in the true GPi
signal (black). In this example, the control reliability is 67%. But the effect on the neuronal
reliability in the TC cell, our goal, is not very impressive (a few additional transmitted
spikes in the controlled case, but also some missed spikes).

Since reliability is our goal, and since we are estimating it, why not use it as the
control parameter? Recognize that the way I have been calculating TC cell reliability in
figures 10.14 through 10.17 employed a moving average of relatively infrequent events.

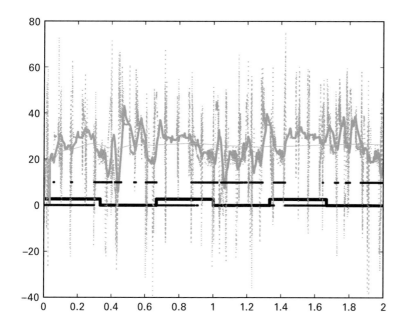

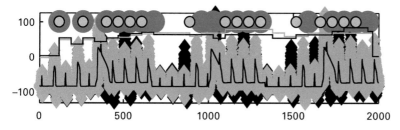

Figure 10.17 (plate 27)
Estimated GPi input to TC cell (blue dotted line), and smoothed short-term moving average of this GPi input (blue solid line). We take a long-term moving average of this current (magenta line) as an adapting threshold to tell when the more instantaneous GPi input is fluctuating up or down. Crossing below the threshold determines when to turn the control on (red). The actual GPi fluctuations are shown (black lines). In the lower panel are the results with control off (blue markers), and on (red markers), and the uncontrolled (green markers) and controlled (magenta markers) spikes transmitted. The running reliability of the TC cell is plotted as a piecewise continuous line for uncontrolled (blue line) and controlled (red line) scenarios. Reproduced from [Sch10] with permission.

(The incoming spikes are on a time scale significantly slower than membrane dynamics such as w.) So this formulation of reliability is substantially delayed with respect to the dynamics of the system. This delay creates the type of results seen in figure 10.18A (plate 28) for control based on turning the stimulator on when estimated reliability is too low.

On the other hand, what is delayed for the past is predictive of the future. If you use low reliability as a control variable, by the time the moving average of reliability decreases to an arbitrary threshold, turning on stimulation will likely come too late. But for periodic fluctuations, for which Parkinson's disease abounds, you can invert your control variable. This uses *high* reliability as a control threshold. By the time reliability becomes high enough, the lack of inhibition will be drawing to a close, and the incipient upturn in inhibition will be imminent. Such results can be interesting, as shown in the improved control in figure 10.18B. In these examples attempting to use estimated reliability as a control variable, the control reliability with respect to GPi output never rose above 46 to 50%. On the other hand, our goal is to treat thalamic reliability, not the waveform of GPi output, and calculating control through running estimates of thalamic reliability is something we are presently exploring in more complete models.

Given these insights, we have begun to explore the full sparse structured Rubin and Terman model [RT04] with feedback control. In figure 10.19A, we show an individual TC cell response in the Parkinsonian state without feedback control. We now apply periodic high-frequency DBS to the STN nucleus (166 Hz), and calculate that the reliability of the TC cells improves from 45% to 100% in figure 10.19B. Now we simulate recording from GPi with a single macroscopic electrode, produce a composite summed measurement as if we were using a typical DBS electrode contact to record. We have found that abetting proportional control with a bias term, basically a proportional-integral control law strategy, we can achieve high reliability with a reduction of over 2 orders of magnitude in control energy applied, as shown in figure 10.19C, by delivering stimulation back to GPi.

To summarize this section, we have explored a variety of dynamics for controlling the thalamic cells. These examples envision direct stimulation of GPi, and our results are consistent with those of Pirini and colleagues [PRSC08], demonstrating that open loop stimulation of GPi can be a strong suppressor of thalamic activity (figure 10.13). Control through estimation of nullclines is not a trivial problem (figure 10.14), but we have not approached this with the addition of nullcline constraints and inertia that would further improve this strategy. It is interesting that direct estimation of the synaptic currents bombarding the thalamus (figure 10.17) was not as effective as estimating the T-current activation (figure 10.15) in these proportional control algorithms. Similarly, control from reliability estimates (figure 10.18) was not as effective as estimating T-current activation. One reason why T-current may have been so effective is that the time constant on the T-current activation equation is intermediate between the rapid estimation of GPi input currents and the relatively slow reliability estimates. This intermediate time constant may help damp out the relatively high

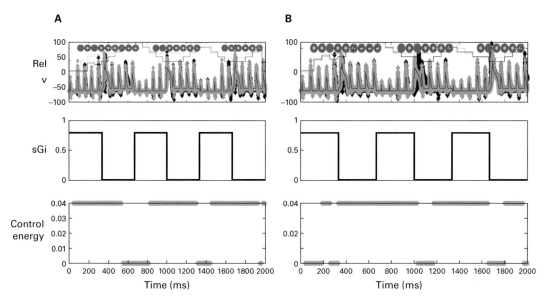

Figure 10.18 (plate 28)
Control of a TC reduced-cell model using reliability as a control parameter. A shows a threshold of turning on GPi stimulation when reliability is <0.9. B shows a different strategy, using an inverse approach. In B, control is turned on when reliability is >0.5. Note that the relevant reliability in both examples is the controlled (red) piecewise continuous line in the upper panel (the blue reliability line is the uncontrolled state shown for comparison). The inherent delays in employing the moving average of reliability can be exploited so that inverse reliability control can be more reliable than when using a more intuitive strategy based on turning on stimulation when reliability falls. Reproduced from [Sch10] with permission.

levels of noise that we have imposed on this system, but still offer suitable responsiveness for an essential component of the pathological dynamics. Whether more instantaneous estimates of reliability would serve to improve this situation is unclear (the reliability uncertainty would increase with less averaging). Nevertheless, in principle, and in full model simulations, reliability is a compelling control variable at the heart of the control problem in Parkinson's disease. Last, regardless of which method is chosen, adaptively adjusting covariance inflation and control gain (figure 10.16) is a fundamental necessity to optimize such algorithms over time.

10.11 Looking Foward

I have just touched on the potential strategies for the use of computational models in Parkinson's disease control in this chapter. Sketching out the unstudied issues is worthwhile.

The calculations in figures 10.14 through 10.18 assume that two electrodes are inserted—a recording electrode in the thalamus and a stimulating electrode in the GPi or STN. The ideal Parkinson's controller would work off of a single electrode, albeit one with multiple

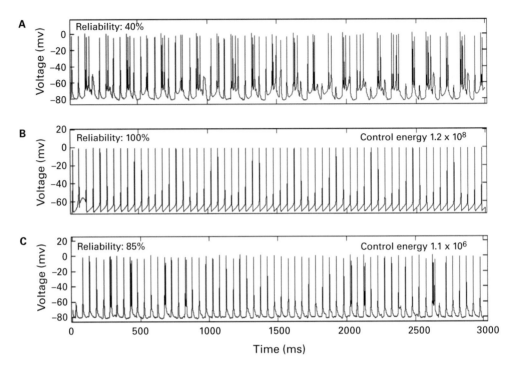

Figure 10.19
Feedback control using full model of Rubin and Terman [RT04]. A shows a TC cell in the Parkinsonian state. The reliability is 45%. B shows high-frequency DBS (166 Hz) applied to STN. The reliability of the TC cells is 100%. C shows feedback control using a proportional-integral scheme (see text for details). The reliability is 85%, in the setting of a 100-fold reduction in control energy. Figure and results are courtesy of collaboration with Patrick Gorzelic and Alok Sinha.

contacts, inserted into just one nucleus (an example is modeled in figure 10.19C). We can employ separate contacts for recording and stimulation along the electrode shaft. Perhaps picking up the oscillatory rhythms from GPi or STN would be sufficient for feedback control of those same nuclei. On the other hand, the models presented here give us the freedom to take such oscillatory dynamics from GPi or STN and estimate the reliability of the thalamus. We might never be able to record a good estimate of sensorimotor input to the actual thalamus, but we can provide such signals, or a range of such signals, to a model thalamus that is functioning as an observer model system when only GPi or STN serves as the recording site.

    The reduced model of the TC cell is valuable for gaining intuition. Adding back the fast Hodgkin-Huxley sodium and potassium currents may be helpful with such models in tracking the thalamic dynamics, as might using an ensemble of such cells. But one of the appeals of such reduced models is that they can represent ensemble activity. Using a scaled-up version of this reduced TC cell, *renormalized* [Fis98] to represent an ensemble of such cells, might well be an effective way to track thalamic dynamics in real brains.

As shown in figure 10.19, we are presently using the full Rubin-Terman model [RT04] to explore set-up of a control system by working with the more complete computational Parkinsonian brain. All of the same principles mentioned previously for the reduced model tracking are applicable in the full model. But one of the interesting facets of contrasting full and reduced models is that, beyond the obvious computational overhead of computing additional variables and parameters, the need to fit these additional variables and parameters may render more complete models *less accurate* in tracking complex systems than more reduced models. This seems counterintuitive, but the use of reduced models is commonplace in meteorological forecasting in place of the full atmospheric dynamical equations [Kal03], and in recent fluid dynamics experiments, reduced models have also proven useful in such data assimilation frameworks [CHO+09] (chapter 8). An important feature of such reduced models for tracking scenarios is that the optimal parameters in these inadequate models are often nonphysical [CHO+09, SS09]. But what matters for Parkinson's disease patients is that we optimize our controllers and improve symptoms better than with open loop approaches, not that we reverse engineer their failing basal ganglia. Model validation is not our primary goal in neural system control.

Recent work has extended the model of Rubin and Terman [RT04] to take into account more biologically relevant connections, the *direct* versus *indirect* pathways, from striatum to the structures of the basal ganglia (figure 10.20). The further incorporation of relevant model components may well give greater fidelity to the realistic basal ganglia dynamics in Parkinson's disease. But with control as goal, such fidelity through complexity will need to be balanced against accuracy of data assimilation and control metrics. One of the important issues raised by Pirini and Colleagues [PRSC08] is that a more complex model of the thalamic cell's function, beyond the simple relay, is likely important. One example of this is *action selection* theory, which envisions that the basal ganglia serve to select from competing neuronal efforts for access to the final common path of motor movement

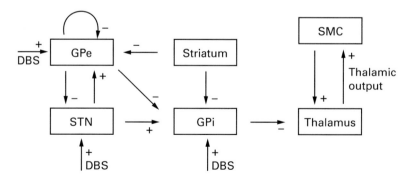

Figure 10.20
The extended Rubin-Terman model as proposed by Pirini and colleagues. Reproduced from [PRSC08] with permission.

[HSG06]. Furthermore, as in human patient experience, the DBS target sites are not equivalent. Indeed, maintaining the flexibility to perform model-based control off of STN, GPi, or VIM, depending on a patient's symptom complex, is a challenge for future work.

Parkinson's disease will likely be the first dynamical human disease in which model-based control principles will show efficacy. The bar is not really that high. Were we merely to demonstrate that we can significantly extend the battery life of implanted stimulators, the control algorithms would be valuable. But as I hope the preceding calculations indicate, such device longevity may well be accompanied by improved performance. The parameter space that we presently face clinically in adjusting a patient's stimulator is, for all practical purposes, infinite. Adaptive algorithms that can continually optimize stimulator performance will give our present empirical trial-and-error efforts very good competition. As long as our devices operate within the electrical safety limits, and incorporate present DBS protocols as alternatives if the adaptive protocals are not better, our patients have nothing to lose and improved lives to gain.

11 Control Systems with Electrical Fields

Despite the ubiquitous use of electrical stimulation there have been few basic studies of the biophysics of the process.
—Tranchina and Nicholson, 1986 [TN86]

11.1 Introduction

Most neurophysiologists have an affair with electrical fields. For some, such as the author, the infatuation starts early and never abates. My guess is that reading Hodgkin and Huxley's papers at an impressionable age may be responsible. It can lead to obsession.

If we assimilate data from the nervous system, what can be done with such information has been the subject of considerable discussion. Some suggest ringing little alarm bells from a device to tell a person that something bad is about to happen in their nervous system; this could allow you to park your car at the side of the highway, for example, before you have an epileptic seizure—the electronic analog of a service dog. Another alternative is to inject drugs into part of a nervous system in response to signals from a data assimilation system— not the worst idea, especially if you want those drugs to target a particular place in the brain to modify a disease state such as epilepsy or Parkinsons disease, as opposed to bathing the whole brain or body in the medication. Most of us gravitate toward electrical stimulation. Such stimulation can be delivered by induction through changing magnetic fields, or applied directly as quasi-static electrical fields (ignoring induced magnetic fields) [NS06].

11.2 A Brief History of the Science of Electrical Fields and Neurons

Digging into the history of electrical fields fairly rapidly drags the reader back to the eighteenth-century experiments of Luigi Galvani and his wife Lucia Galeazzi [Bar89, Pic97]. In experiments in their backyard, they connected an exposed frog spinal cord to a wire leading to their house's lightning rod, and grounded the frog's leg to their drinking well. When lightning struck nearby, the electrical field developed sufficient potential in their circuit to cause the frog leg to twitch (figure 11.1).

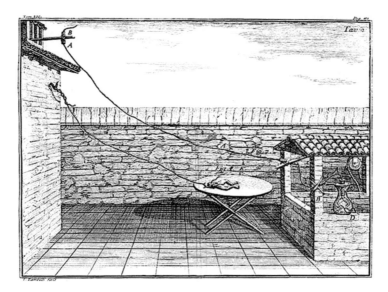

Figure 11.1
Galvani's plate from his 1791 text [Gal91]. From the facsimile generously made available from the efforts of the
International Centre for the History of Universities and Science, Department of Philosophy, University of Bologna.

By the early twentieth century, William Rushton performed the seminal electrical field
experiment that lays out most of what the researcher would need to replicate the rest of this
chapter [Rus27]. He knew of Galvani's work, and fashioned an ingenious experiment. He
suspended a frog sciatic nerve between two insulating glass receptacles. The nerve could be
stimulated within one insulator, and the action potentials then traveled along the length of
the nerve suspended in a bath of Ringer's solution. The other end of the nerve was attached
to the calf muscle, which would twitch if the impulse arrived. The nerve was suspended
between two silver-silver chloride plates, which generated a uniform electrical field in the
region of the nerve. The apparatus was placed on a turntable, graduated in degrees, so that
he could change the angle of the nerve with respect to the electrical field (figure 11.2). He
found that the threshold of current required, using a short square wave stimulus created
through the swinging of a pendulum, was related to the cosine of the angle of the electric
field, with respect to the long axis of the nerve.

Rushton's work was known to Bernard Katz and Otto Schmitt [KS40], who demonstrated
that action potentials in one nerve fiber could influence the transmission of action potentials
in a neighboring fiber. The physics of such influence would serve to synchronize nearby
action potentials so that they traveled together. These effects were termed *ephaptic*.[1]

---

1. It goes without saying that Katz, who would later receive the Nobel Prize in medicine and physiology, was
fortunate in collaborating with the young Schmitt. Schmitt would later become one of the most important electronics
inventors of the twentieth century. His *Schmitt Trigger* was directly inspired by the knowledge he gained from
experiments on all-or-none action potentials in nerve fibers.

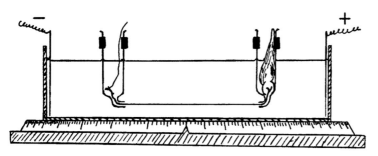

Figure 11.2
Rushton's experiment exploring the relationship between the threshold of a nerve fiber and the angle with respect to electrical field. Reproduced from [Rus27] with permission.

By 1956, Terzuolo and Bullock [TB56] demonstrated that the electrical field required to modulate an active neuron was far lower than that required to stimulate one to fire from rest. Remarkably, they estimated that the strength of the electrical field required to perform such modulation was of the order of 1 mV/mm, an estimate not to be broken by theory [WVAA98] or experiment [FGS03] for over four decades. It was clear to Terzuolo and Bullock [TB56] that these magnitudes of fields were "available" to neurons within active nervous systems, and therefore might play a role in synchronization. They further recognized that the morphology of such cells would play a role in their sensitivity to electrical fields, that their dendrite fields would act as *antennae* when oriented parallel to the field direction, and that the arrangement of neurons in mammalian cortex were especially set up with such alignments.[2]

By the late 1970s, we developed the ability to cut thin slices of brain and keep such slices alive for many hours in perfusion chambers [And77]. The living and physiologically active brain slice was soon viewed as an ideal new window to begin exploring the effects of electrical fields on neuronal and network activity [Jef81].

In a bizarre set of experiments in the early 1980s, researchers noted that neurons could form organized activity and synchronize in the absence of chemical transmission in brain slices [JH82, TD82, YKH83, KHY84]. In a series of computational modeling studies, Traub and colleagues [TDTK85, TDSK85] explored the origin of such pathological synchronization as well as the implications for more physiological synchronization of individual neurons to populations.

But the remaining piece of this puzzle was to be worked out by Charles Nicholson and his colleagues, in a remarkable set of experiments in the mid-1980s. Using isolated turtle cerebellum in an elegant experimental setup, they demonstrated in neuronal populations that the asymmetry of the soma-dendritic axis correlated with the responsiveness of such cellular populations to modulate their firing in response to relatively slow sinusoidal stimulation

---

2. Many of us will sadly mark the recent loss of Ted Bullock. I was deeply impressed by the interest that this preeminent scientist took in inspiring school students.

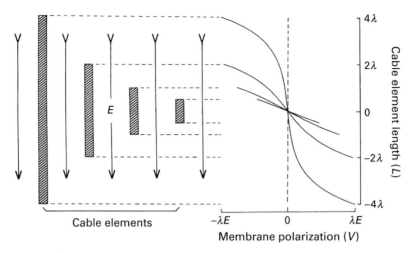

Figure 11.3
Semileaky cables exposed to a constant electrical field will have sigmoid-shaped (hyperbolic tangents) voltage profiles. Reproduced from [CN86] with permission.

[CN86]. They proposed that from passive cable calculations[3] [Ran75] the potential profile, $V(x)$, would vary with distance x as a sigmoid (hyperbolic tangent):

$$V = \lambda E \sinh(x/\lambda - L/\lambda)/\cosh(L/\lambda)$$

where λ is a space constant, E the strength of the constant polarizing electrical field, $2L$ the length of the cable, and x the distance along the cable (figure 11.3). Two years later, Chan and colleagues [CHN88] demonstrated from intracellular measurements that such field effects on neuronal spiking were consistent with these cable equations.[4]

But embodying all of this experimental science was the seminal computational work by David Tranchina and Nicholson [TN86]. In this work, they considered that, since the dendritic branches in a neuron could be electrically equivalent to a simpler structure, they created an equivalent cylinder to account for the soma and dendrite fields, as shown in figure 11.4.

It is important to realize that whereas the gradient (first derivative) of the potential (the electrical field) is the crucial element in polarizing the equivalent cylinder in figure 11.4, it

---

3. Such cable equations are actually derivative of the cable equations for the nineteenth-century transatlantic cables.

4. It is worth noting that there was an older literature demonstrating opposite effects of DC polarization of neurons embedded within the intact brain, cotex [PM65] and hippocampus [PM66], which these new findings now explained. For large asymmetric principal (pyramidal) cells, the orientation of the large apical dendrite with respect to the soma was the key to predicting whether polarization would be excitatory or inhibitory.

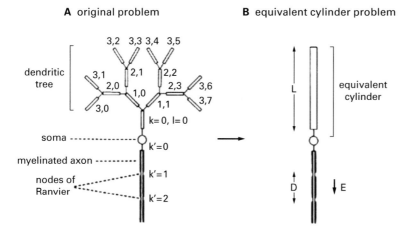

A original problem **B** equivalent cylinder problem

Figure 11.4
Converting a neuron with a complex dendritic tree in A, to the equivalent cylinder model of dendrite, soma, and axon in B. Reproduced from [TN86] with permission.

is the second derivative of the potential[5] that is important if you wish to activate the middle of a cylindrical structure [WGD92] (see also related discussion in chapter 10).

A few more facts need to be mentioned in this too-brief summary. The time constant to polarize a neuron in the way we have been discussing is about 25 msec [BIA+04]. This corresponds to maximal frequencies of about 20–40 Hz.[6] Interestingly, higher-frequency sinusoidal stimulation will induce a high potassium outflow that can block neuronal activity (depolarization block), and is orientation independent [BLH+01].

So we are done with the basic science of this chapter. Much of it can be summarized in the extremely simple diagram in figure 11.5.

11.3 Applications of Electrical Fields in Vitro

Our studies also indicate the utility of this relatively simple stimulation method that enables populations of cells to be stimulated without penetration of individual cells. Moreover the current density is quite low at any point in the tissue since the current is not delivered by a local punctate electrode ...
Such a stimulus is potentially valuable in prosthetic devices.
—Chan and Nicholson, 1986 [CN86]

This book is concerned with pushing the limits of neural control engineering. The vast majority of applications of stimulation in both experimental and clinical nervous system

---

5. For better or worse, this second derivative is often termed an *activating function* in the literature.
6. There is evidence that synaptic afferents might analogously polarize [BIA+04].

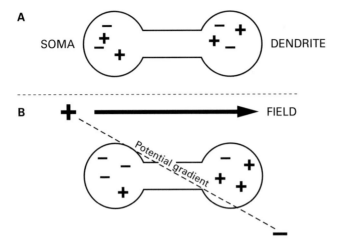

Figure 11.5
The completely reduced view of low-frequency electrical field polarization of neurons along their soma-dendritic axes. A shows the situation without an applied field. Electroneutrality resides along the axial direction of the neuron. In B, with steady polarization along the soma-dendritic axis, there is a small separation of charges, which actually build up along the membranes at the ends of the neuron as a Debye layer. This produces a counterfield (opposite in direction) to the applied field. Since the transmembrane (inside versus outside) potential difference near the soma determines the threshold of the neuron to fire, the polarization of such a neuron will make it either more or less excitable, depending on whether the potential near the soma is hyperpolarized (less excitable) or depolarized (more excitable).

applications are from pulse stimulation. In such scenarios, one delivers a *square wave* pulse of either voltage or current, depending on the source of energy. There are advantages and disadvantages of each. If one delivers constant voltage, the junction of the electrodes with neural tissue can vary widely, and thus the current delivered can similarly be unreliable. So in many experimental situations, constant current produces more reliable results. On the other hand, if one exceeds certain voltage limits, one can readily form irreversible reduction-oxidation reactions at the electrode surface with neural tissue [MBJ05], so constant voltage is safer.

But the problem with square wave–type pulse stimulation is that the rapidly changing electrical field associated with such stimuli, which turn on and off with a theoretically "infinite" rate of change in voltage or current, is that the ability to simultaneously measure the response of the neural activity is lost. One can, of course, record simultaneously, but we have no way of accurately separating the stimulation artifact from neural activity during such rapid shifts in stimulation energy.

The Chan and Nicholson quotation from 1986 at the start of this section was one of the least noticed paragraphs in neuroscience for about a decade. After nailing down the advantages of low-frequency electrical fields for neural prosthetic applications using extracellular, intracellular, and computational models, their work lay in wait for the rest of us to study. So

I am going to pick up on this thread for the rest of this chapter. This is not to say in any way that the usefulness of pulsed stimulation is not well established in many applications. Certainly we think it is required in areas of the brain, such as stimulation of the basal ganglia for Parkinson's disease (chapter 10), where neuronal orientation may not permit low-frequency field stimulation. Nor do I intend to suggest that we understand pulsed stimulation within the brain—the situation is anything but clear [MGST04]. But low-frequency field stimulation has serious advantages for a control engineer, and the literature on applications is a growing area of focus. As we will show, it offers a powerful way to separate the stimulation from neural response in continuous feedback scenarios.

Bruce Gluckman built a miniature version of the Rushton chamber (without the turntable), and mapped the electrical field throughout (figure 11.6). We placed a reference electrode in the chamber outside of the slice as well as a differential electrode within the slice. By turning on a test field such as a sinusoid, and moving the reference electrode with respect to the field isopotentials, the isopotential aligned with the slice electrode is found when the test signal disappears. Once this is set, one can apply any arbitrary field and it is removed from the recording. If only this could be readily accomplished in the intact brain.

It was relatively straightforward to demonstrate that suppression of seizure-like events could be achieved with DC electrical fields aligned with the neurons firing seizures [GNN+96a]. But these steady polarizations were largely temporary. After about 30 or more seconds of steady polarization, seizures would return. An interesting effect of nonorientation-specific higher-frequency sinusoidal stimulation was that the poststimulation effect on activity suppression was longer lasting—likely due to potassium extrusion from cells [BIA+04].

From a control point of view, examine the results in figure 11.7. In panel A, as the applied electrical field polarity is flipped positive and negative, and with graded amplitude, the burst rate of the ensemble of neurons responds in a graded manner [GNN+96a]. What about individual ensemble bursts? In figure 11.7B, we demonstrate that by randomly selecting the field polarity, amplitude, and duration, we could parametrically control the individual burst intervals (unpublished data, courtesy B. J. Gluckman).

11.4 A Brief Affair with Chaos

In 1994, we published a demonstration of chaos control in such brain networks [SJD+94]. Using pulse stimulation, we could shorten the interval reliably from a previous burst until the next burst. This gave us some ability to manipulate the ratio of the interval from one interval to the next, as shown in figure 11.8 (plate 29). The reason for these findings is that in chaotic systems, although all periodicities can be unstable, all are not equal. Indeed, such systems contain a hierarchy of unstable cycles called *unstable periodic orbits* [Cvi88]. In 1990 Edward Ott, Celso Grebogi, and James Yorke proposed that such systems could be controlled through manipulating about such low-period unstable periodic orbits

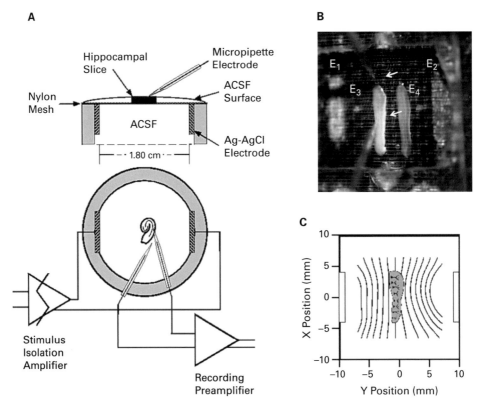

Figure 11.6
Electrical field control chamber for brain slice experiments. A shows the overall schematic; the photograph in B shows an example of two brain slices within such a chamber; and C is an actual mapping of field within the chamber. Note the presence of four electrodes in B: E_1 and E_2 generate the field through a low-source impedance current source, while the smaller E_3 and E_4 are connected to high-impedance field effect transistors for sensing the field with minimal current flow. This configuarion is a large-scale voltage clamp, with a four-electrode technique, as discussed in chapter 3. The arrows in B point to the position of the tips of the electrode in the slice and the electrode in the chamber. A and C are reproduced from [GNN⁺96a] with permission.

[OGY90], and William Ditto and colleagues experimentally verified this theory shortly thereafter in a physical [DRS90] and then a cardiac [GSDW92] system. But what we never could do in such control experiments was to lengthen, as well as shorten, such intervals. The primary motivation for figure 11.7 was to be able to find a true control parameter that could produce more refined control than the rather crude approach shown in figure 11.8.

The underlying idea for control of chaotic systems is actually simple. Chaos in its pure sense is an infinite number of frequencies that are *all* unstable. But there is a hierarchy of these cycles, and the low-period ones are experimentally accessible. You see them in

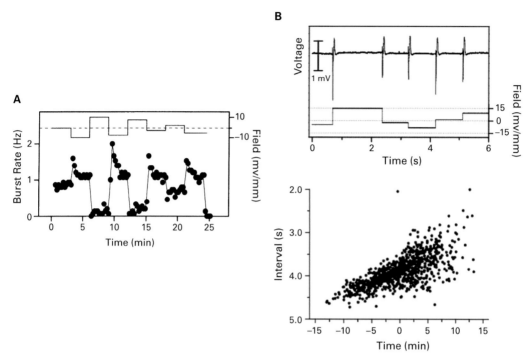

Figure 11.7
A shows how the neuronal ensemble burst firing rate was dependent on the amplitude and polarity of the applied electric field. B shows that even individual burst intervals could be parametrically manipulated using electric field control. A is reproduced from [GNN+96a] with permission. B is unpublished data courtesy of B. J. Gluckman.

the data as multiple near-miss cycles. Look at the time series from chaotic mathematical models—most of them hover near certain values. This is a reflection of such cycles or orbits. Since these orbits have both attracting and repelling regions of phase space where the system will naturally gravitate to or away from such cycles, a simple control scheme can be applied to exploit these inherent dynamics of the system. In figure 11.8, we show examples of both control (C) and anticontrol (D).

We spent years seeking to test the integrity of the orbit structures shown in figure 11.8. We developed a much more rigorous means of extracting such period orbits from experimental data [SOS+96, SOS+97]—only further confirming their presence in neuronal systems. Expanding our search for such orbits in single-cell, small networks, and large-scale networks, we proposed unstable periodic orbits as a novel language for neuronal dynamics [SFN+98].[7]

---

7. A Chinese translation of this paper appeared in [Qia99].

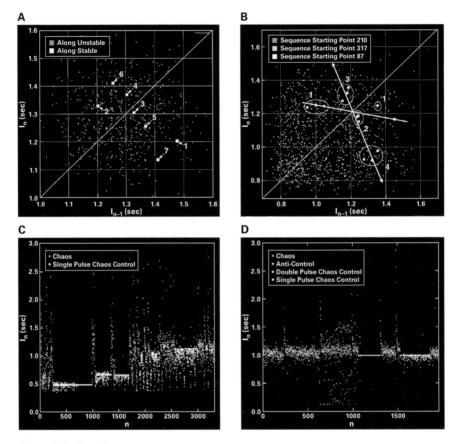

Figure 11.8 (plate 29)
Controlling chaos in the brain. A shows the identification of a motif that shows an approach to the line of identity where sequential intervals, I_n and I_{n+1}, are equal. B shows multiple repetitions of this motif. C demonstrates that by directing the system to fall near the contracting direction (manifold), that the chaotic intervals could be regularized. D shows that by pushing the system away from the contracting direction, that we could prevent regularization. Reproduced from [SJD+94] with permission.

Such chaos control techniques are now known to be a subset [YCX+01] of the more general class of control laws known as variable-state or sliding-mode control [Utk77, DZM88, YUO99]. Armed with electrical field control systems, and with our growing sophistication of sliding mode control strategies, the prospects to fuse these techniques into a much more effective control strategy for neuronal systems is certainly at hand.

A powerful variant of such control ideas for nonlinear systems, where the parameters create a strongly attracting limit cycle, was proposed by Danzl and colleagues [DHM09]. Most neuron models will have such strongly periodic solutions, and they are *not* chaotic. Such solutions often arise concurrently with the creation of an unstable periodic orbit, contained in the phase space *within* the stable limit cycle, the result of a "Hopf" bifurcation as you change a parameter of the neuron model. If you want to desynchronize such a population of neurons, you push it with stimulation to a region where the sets of different phases (termed *Isochrons*) are concentrated, such as an unstable fixed point. In this context, different phases imply different times to the next event or spike. Push a system to a little ball near this unstable point, and noise will randomize the phases of the neurons for the next spike. This control technique can also be used to optimally increase the firing rate by pushing the neuron toward the point on the stable limit cycle corresponding to the spike. The time-optimal nature of the control results in the maximal spike rate increase for a given stimulus magnitude. When applied to a population of neurons, this may result in synchronization with maximal firing rates. My sense is that this type of adaptive control scheme can also be developed from empirical strategies as shown in figure 11.8—an essential step if the rigor and assumptions of [DHM09] cannot be met in practice. Reading this book, you would also wish to consider placing this scheme within a data assimilation framework. The implications for modifying dynamics of conditions such as epilepsy or Parkinson's disease are intriguing.

11.5   And a Fling with Ice Ages

A related experiment possible with such an experimental apparatus was to test the brain for *stochastic resonance*. In stochastic resonance, the response of a nonlinear system to an otherwise subthreshold signal is optimized with the addition of noise [BSV81]. Since its proposal as a mechanism for amplifying the effects of the Earth's small periodic orbital variations by random meteorological fluctuations leading to ice age periodicity [BPSV82], stochastic resonance has been observed in a diverse range of physical systems [WM95]. There was an interesting prediction that in two-dimensional excitable media, spatiotemporal stochastic resonance should be seen in the creation of coherent structures such as spiral waves [JMK95]. Since electrical fields linearly superimpose, we blended sinusoidal signals with random noise, and were able to cleanly demonstrate that stochastic resonance could be observed in neuronal networks [GNN$^+$96b] (figure 11.9). In figure 11.9A, we show the comparison of raw data where random noise is applied through an electrical field to a neuronal network, and below this the combination of random noise and a sinusoidal electrical field. Note that in the presence of the sinusoidal signal, the bursts of the field components tend to occur on the peaks of the periodic signal. In figure 11.9B a family of curves are shown, each at different amplitudes of sinusoid A_{sin}, as a function of noise amplitude A_{noise}. Note that for each amplitude of signal, there is an optimal noise for which the signal is transmitted to the neuronal network. To my knowledge, the two-dimensional spatiotemporal neuronal version of this experiment remains to be done.

    Why is stochastic resonance important? Certainly we do not worry about ice age glacia-tion within the brain. The subtle point is that the *noise* in these examples is a signal that

A   **B**

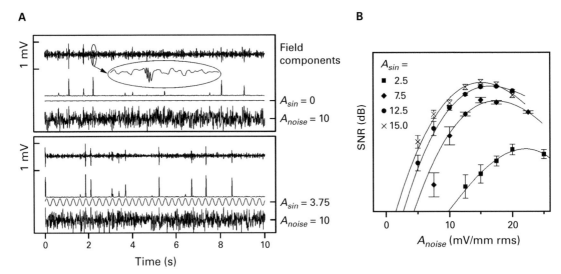

Figure 11.9
A compares between adding noise to a neuronal network through an electrical field (upper) and combining noise
with a sinusoidal electrical field signal. Note that, in the combination, the neuronal responses tend to occur at the
peaks of the sinusoidal signal. B shows families of curves at a range of sinusoidal signal amplitude A_{sin}. As a
function of noise amplitude, A_{noise}, the signal-to-noise ratio (SNR) in decibels (dB) peaks at an intermediate level
of applied noise. Reproduced from [GNN$^+$96b] with permission.

is *uncorrelated* with the simple sinusoidal signal at hand. In the brain, the functional
result is that *stochastic resonance can occur whenever two uncorrelated signals impinge
on neuronal networks*. Now the range of implications explodes: the activity of ongoing
inputs to the auditory cortex in the presence of external sounds that are uncorrelated, the
multimodal confluence of sensory information processing the lateral interparietal cortex,
the list is endless.

Stochastic resonance in a speculative sense is the added energy in ensembles of interacting elements with thresholds that likely fuels neuronal interaction throughout the nervous
system. Perhaps the only thing surprising about stochastic resonance would be if it were not
found in the nervous system when searched for. One major unsolved question is whether
the levels of activity of different parts of the brain are tuned in any way to optimize information flow. Another question that naturally arises is whether the resonance from independent
signals promotes enhancement of multiple information streams simultaneously in the same
network. Needless to say, the range of implications of this inherent property of nonlinear
systems with thresholds is enormous for the nervous system.

11.6 Feedback Control with Electrical Fields

We spent years searching for an effective way to incorporate the electrical field effects from
Gluckman and colleagues [GNN$^+$96a, GNN$^+$96b] into a feedback control strategy. There

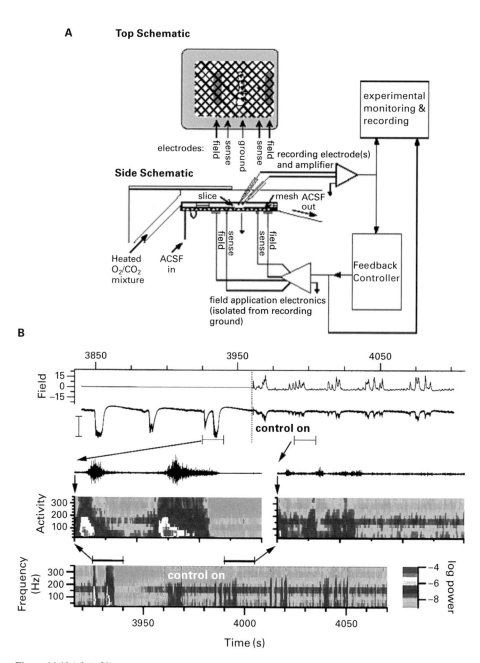

Figure 11.10 (plate 30)
Adaptive feedback control of neuronal activity. Schematic of experimental chamber and electronics shown in A.
In B are results demonstrating that a proportional control law could effectively suppress seizure activity in the
brain slice. Reproduced from [GNWS01] with permission.

were some obvious electrical engineering issues that needed to be addressed. We needed to isolate the electronics of the recording system, with an adjustable reference electrode on a field isopotential, from the electronics of the stimulation system, as shown in figure 11.10A (plate 30).

One of our goals was to find a way to suppress epileptic seizures. An advantage of using brain slices is that they preserve the native neuronal architecture from which they are harvested. Interestingly, it was found that exposure of such brain slices to elevated levels of potassium in the perfusate, raising the physiological level of about 3.5 to 8.5 mM [RLJ85], would produce burst firing similar to epileptic spikes.[8] Under certain conditions, such slices exposed to elevated potassium would generate much more prolonged patterns of activity that had a remarkable resemblance to human epileptic seizure patterns [TD88].[9]

It is important to pay attention to the current-generating electrodes that supply the controling electric field. In the experiments in figure 11.10, we used silver-silver chloride electrodes. These electrodes form a very efficient chemical reaction with saline fluids, exchanging chloride ions with solution as charge is passed to the electrodes. But they are toxic if implanted into the brain. For in vitro experiments, where they are used without direct contact with the brain networks, this issue is avoided. But the author must confess that earlier experiments attemped to use stainless steel electrodes, which polarize and do not pass low-frequency currents well. An obvious hallmark of such mistakes is the frequent appearance of gas bubbles around such electrodes—water hydrolyzes at electrode polarizations a bit larger than ± 1 V. Platinum-based electrodes have better electrical characteristics. A very complete review of the often neglected aspects of the materials in such electrodes can be found in Merrill and colleagues [MBJ05].

In figure 11.10B, we show the effect of a simple proportional control law. The control parameter was taken to be the root-mean-square of the high-frequency neuronal activity, from 100 to 500 Hz. The power in such a frequency band was relatively low except during seizures, and was especially prominent at the onset of seizures. With negative gain multiplying this control parameter, and delivering low-frequency (<30 Hz) feedback electrical fields, seizures were reliably suppressed.

Our first efforts employed the rational but unfortunately poor decision to half-wave rectify the control signal. We presumed that we wanted to suppress seizures, and that there was

8. Epileptic spikes are brief intense electrical events seen in the EEG of patients with epilepsy, and only rarely in the EEG of people without a history of epileptic seizures. In EEG terminology, these spikes are less than 70 msec in duration. The burst firing events seen in high potassium are about 100 msec, and have often been presumed (hoped) to be the analogs of epileptic spikes.

9. Producing seizure-like events in elevated potassium is a tricky business. Often Traynelis and Dingledine [TD88] used a period of transient hypoxia to bring out these events. Such posthypoxic hyperexcitability was well described [SS85], but it complicates the relationship of the seizures to the ion manipulations. One can remove the magnesium or calcium from the perfusate, but this seems to the author a drastic physiological approach. A purer "ionic" formulation of reliable seizure generation is to use 4-aminopyridine with a partial reduction in magnesium levels [ZCBS06]. Which is best? No one knows. Such unknowns remind us of the limits to the use of in vitro models for disease analogs.

only one polarity, with reference to the schematic in figure 11.5, that would accomplish such consistent suppression. This was very successful at suppression so long as the control was maintained. As soon as the controller was turned off, there were intense withdrawal seizures. Our mistake was that this half-wave rectification would have a net polarization effect on the brain region in bulk. We suspect that building up potassium asymmetrically within the slice or layers was likely a culprit [GMN83].

By employing a balanced charge (nonrectified) proportional-integral control law, these withdrawal seizures were eliminated. We calculated the net current output over the most recent 0.5 to 1.5 sec of controller operation, and then compensated for this recursively in our controller output. The full analysis of proportional versus proportional-integral control can be found in Gluckman and colleagues [GNWS01].

Such difficulties in applying control engineering principles to control of neuronal circuits emphasizes the need for animal experiments to test even the most logical of ideas. In chapter 12, we will discuss ways of fusing models of the dynamics and diffusion of potassium ions in brain tissue with compartmental neuronal models expressing Hodgkin-Huxley dynamics. The more refined our models become, the more we hope to minimize the need for animal experiments in what should be the rational design of neuronal controllers. But in the end, testing such controllers on something as complex as the actual brain is, for the foreseeable future, a necessity. Unlike Boeing and their 777 airframe, we still need to test fly the designs of even the simplest of brain control systems to see if they crash.

11.7   Controlling Propagation—Speed Bumps for the Brain

If one pharmacologically blocks the fast (GABAa) inhibition in brain, then small stimuli can produce traveling waves across the cortex. They tend to travel (or perhaps are led by) the middle cortical layers such as layer 5 [TC98].

As mentioned in chapter 6, the Wilson-Cowan equations can produce traveling waves. Such nonlinear wave phenomena are common in both physics and biology, and it was noted some time ago that inhomogenous "bump" type solutions of even stationary waves would characteristically have a spatially larger stable and a spatially smaller unstable solution [AMa77]. Intuitively, one needs, as with nonlinear waves in general [Sco03], enough regenerative energy supplied to offset the tendency of such waves to dissipate. In Wilson-Cowan equations, this requires a sufficient amount of "input" from other regions of the network to keep the core spatial region sufficiently activated. In figure 11.11A, we show a traveling wave solution $u(x, t)$ for Wilson-Cowan equations, with threshold θ and propagation speed C. Figure 11.11B shows the characteristic plots of threshold θ versus propagation speed C. There are two solutions along each branch—the upper one is stable, and the lower one (within the gray region) is unstable.

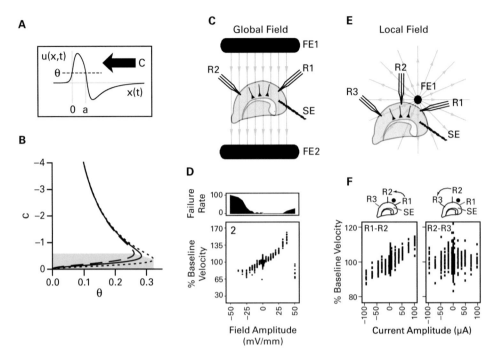

Figure 11.11
Traveling waves in cortex, theoretically and experimentally. A shows a schematic of a traveling wave solution, where we plot a traveling wave solution $u(x, t)$ for Wilson-Cowan equations, with threshold θ and propagation speed C over distance x. B shows the characteristic relationships for velocity C versus threshold θ. C shows a schematic of an experiment where global electrical fields are applied, between field electrodes FE1 and FE2, to waves initiated by stimulating electrode SE, and recorded with recording electrodes R1 and R2. D shows the modulation of velocity with applied field amplitude, and the failure rates of initiated waves. E shows the application of radial electrical fields, and F shows the modulation of velocity with only close proximity to the punctate field source. Reproduced from [RSG05] with permission.

Figure 11.11C shows the experimental setup whereby a global electrical field is imposed on a cortical brain slice. A traveling wave is initiated by the stimulating electrode (SE) and travels first by recording electrode 1 (R1) and then reaches recording electrode 2 (R2). The wave velocity, C, can be increased or decreased by alternating the field polarity about zero, as shown in figure 11.11D. As theory predicts in figure 11.11B, when the waves are slowed to a sufficiently small value of C, they become unstable, which is shown as an increasing failure rate in figure 11.11D.[10]

10. Note that at high field strengths where velocity is peaked, there are also some increasing propagation failures. Modeling suggests (unpublished data) that such failures may be related to the increasing hyperpolarizing blockade that will develop in the dendrite fields of asymmetric neurons when the somata are depolarized. Pursuit of this conjecture requires further study.

Now note the effect of a radial electrical field produced by a punctate stimulating electrode made from a pellet of silver-silver chloride, shown in figure 11.11E. When the waves pass by close to the source of these punctate field sources, the propagation velocity can be increased or decreased by alternating the field polarity about zero, as shown in figure 11.11F. When the wave passes between electrodes a short distance away, its velocity is unaffected by the field. Blockade of these waves is difficult to produce with such punctate field sources, perhaps reflecting inhomogeneity in the field, or transmitting fibers coursing *en passage* through the region of the most intense field polarization. For traveling waves, such punctate fields act less as brick walls than as speed bumps.

In terms of controller design for seizure activity, one would ideally predict and prevent a seizure from initiating. Seizure prediction at the time of this writing remains more a goal than reality [LL05, MKR+05]. The next goal would be to rapidly interact with a seizure to stop it, as shown in figure 11.10. If a seizure escaped being crushed early on, one could implement a third strategy—blocking propagation, as shown in figure 11.11. One can, of course, implement all three goals in a tiered device design. We envision that placing both a set of electrodes at a seizure focus, in an effort to predict and disrupt, and then a more distant set of electrodes designed to cut off propagation if the first strategies fail would be a sensible design for such a controller and electrode deployment. For seizure-prone regions of the brain, such as the hippocampus, this strategy can be employed with a single electrode shaft, with contacts at the tip of the shaft targeting the epileptic focus, and electrodes more distal from the tip available for propagation blockade. Such axial electrodes are in common use in clinical practice and can be targeted precisely from an entrance site at the occiput using image-guided stereotactic surgical placement. Such minimally invasive approaches to control electrode deployment drive our configuration for the translational in vivo experiments discussed in more detail later.

11.8 Neurons in the Resistive Brain

Computational neuroscience has, not surprisingly, tended to focus on neurons [KS98, Dal01]. But as the previous material has been leading us to conclude, one needs to consider embedding our model neurons within a resistive lattice that represents the electrical characteristics of the brain. As should also be obvious, we also need to take into account the physics of lower-level ionic dynamics representing the transmembrane pumping, cellular buffering (especially by glial cells), and diffusion of ions that are intimately linked to determining excitability. We will address metabolic ion dynamics in considerably more detail in chapter 12, but first let's examine the resistive brain.

The pioneering computational work on the electrical field interactions between neurons through their endogenous fields was performed by Roger Traub and colleagues [TDTK85, TDSK85]. It is essential to use multicompartmental neuronal models to represent polarization effects. Whereas Traub and colleagues focused on multicompartmental representations of dendrites, soma, and axons [TM91], we have already seen how Tranchina and colleagues [TN86] reduced such models to equivalent cylinders because of the symmetries inherent in polarizing symmetrical dendrite fields. Pinsky and Rinzel [PR94] demonstrated how much of the qualitative properties of multicompartment pyramidal cell firing could be represented in a simpler two-compartment model where the properties of dendrites and soma could be lumped into single composite compartments. The Pinsky-Rinzel model is the minimally reduced model that can represent polarization effects from an applied electrical field.

In Park and colleagues [PBG$^+$05], we embedded Pinsky-Rinzel two-compartment neurons (figure 11.12A) within an electrically resistive matrix (figure 11.12B). We created equivalent terminal resistances (z) to represent the equivalent networks of a resistive matrix of (for all the neuron cares) infinite extent. We coupled these neurons, only excitatory here, with two types of excitatory synapses.

We were keen to study the effect of applied electrical fields on neuronal syncyhrony. By varying the nature of the cells (the internal conductances), and the natural frequency of their spiking, we were able to study the degree of synchronization as a function of applied electrical field and neuronal disparity. The results demonstrated a complex landscape between synchrony and asynchrony, as shown in figure 11.12C. Especially interesting were the phase relationships that emerged from a study of the influence of the relative phase difference between cellular firing and the response of the neighboring cells. In contrast with the phase response curves dependent on synaptic connectivity (figure 11.12Da, Dd), the phase response curves as a function of the electrical ephaptic effects alone were of considerably smaller magnitude (figure 11.12Db, Dc). Nevertheless, at particularly sensitive phases, the ephaptic effects were important. In this modeling, we incorporated a simple model relating potassium extrusion from activity in cells to a change in the proportionality of intracellular to extracellular electrical resistance.

11.9   How Small an Electrical Field Will Modulate Neuronal Activity?

As discussed previously [TB56], the electrical field strength required to modulate neurons that are actively firing is much lower than that required to fire neurons from rest. How small an electrical field will do this?

Theoretical calculations for elongated cells exposed to an electrical field predicted that fields as small as 100 μV/mm would be a lower bound on such effects [WVAA98]. The key here was that an external electrical field needed to overcome the *molecular shot noise* produced by biochemical processes within cells [WVAA98]. It was clear from experiments that the limit of interaction was lower than 1 mV/mm [GBD00], but the discrepancy between theory and experiment remained.

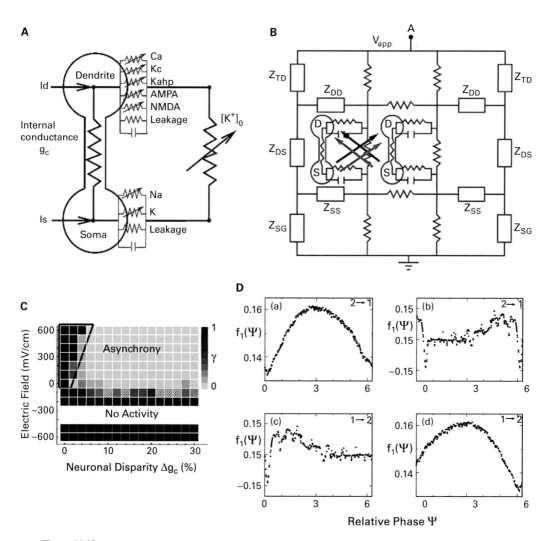

Figure 11.12
A shows the Pinsky-Rinzel lumped two-compartment model of a pyramidal cell, and the relevant conductances and membrane currents. In B, these neurons were embedded within an electrically resistive array with terminal resistances to represent an array of infinite extent in comparison with the dimensions of the neurons. C shows the complex interplay between applied electrical field, neuronal disparity, and 1:1 synchrony (solid black). Gray scale γ reflects intensity of synchronization. The hatched boxes represent higher-order phase locking beyond 1:1. D shows numerical simulations of the phase response curves of each neuron influencing the neighboring neuron from synaptic (a,d) and ephaptic electrical field (b,c) effects. The analytical derivation of these curves is outlined in detail in [PBG+05].

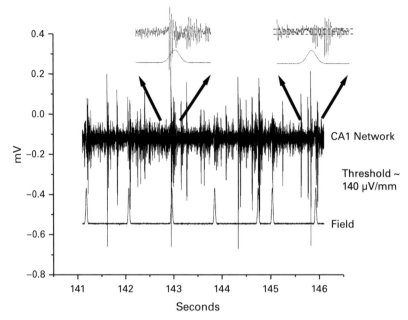

Figure 11.13
Experimental search for the electrical field threshold to modulate neuronal firing. We employed small field pulses with Gaussian waveform, to emulate naturally occurring burst firing activities. On the CA1 network, the insets show the raw data illustrating modulation of ensemble firing in response to Gaussians with a 140 μV/mm root-mean-square amplitude shown above. Careful statistics, including perturbations in the timing of the Gaussian pulses, demonstrate that modulation at this amplitude is statistically significant. Reproduced from [FGS03] with permission.

We sought to explore this lower limit [FGS03]. In figure 11.13, we used an electrical field similar to that of burst firing within a nervous system. Such pulses of intense neuronal activity can be seen within neuronal networks during active exploration, for instance [BUZ86], where local field intensities of over 10 mV/mm can be measured within the brain. We demonstrated that the smallest such pulsed field that can modulate neuronal firing was 140 μV/mm root-mean-square, 295 μV/mm peak amplitude; this validated the theoretical predictions. We further showed that these networks are more sensitive than the average single neuron threshold (185 μV/mm root-mean-square, 394 μV/mm peak). The origin of the enhanced network sensitivity might well reflect array enhancement, perhaps through a stochastic resonance effect [LMD+95, Gai00, KSK00].

These findings have several implications of importance. First, the strength of the electrical field required to modulate the firing frequencies of elongated and asymmetrical neurons is very small. It is about 2 orders of magnitude less than the strength of the electrical fields generated in the course of normal functioning of the brain [BUZ86]. So there is little question that the brain synchronizes its neurons in part through electrical field effects of its own neurons [AMB+10]. The functional implications of this remain unclear. Certainly during population events, such as navigational theta (4 to 12 Hz in rodents; see chapter 7) rhythms, the phase relations of individual neuronal firing to the population may well be modulated at least in part through endogenous electrical field interactions. But these findings also raise the spectre of the possible public health effects of exogenous environmental electrical fields, which has been a subject of intense controversy [JDBF03, SJ07].

Environmental electrical fields vary widely in frequencies. Our cell phones produce megahertz-gigahertz frequencies—far faster than the polarization time constants of neurons. Such frequencies may affect macromolecular bonds and nucleic acids, and the scientific understanding for such radiofrequency radiation remains a subject of controversy [FAK05, Woo06]. Cell phone frequencies may cook molecules within your cells, but such interactions are useless for our focus here on neural system control design. One of the problems in the environmental assessment of electrical fields is that the actual measurements of field attenuation by scalp, skull, and the coverings of the brain has often been lacking in our assessments [Ada91]. More exacting science, whether at radiofrequencies, or at the extremely low frequencies in the ranges where modulation of neuronal firing is expected, seems warranted.

11.10   Transcranial Low-Frequency Fields

The thresholds for electrical field interaction with brain seem so low [FGS03] that the bold might attempt some truly strange experiments. The application of slow transcranial potentials (0.75 Hz) across the scalp at small current densities during sleep as performed by Marshall and colleagues, [MHMB06] could enhance the retention of declaritive memories in human volunteers. In a comparable experiment during wakefullness [KWS+09], the encoding of memories was enhanced during learning. Interpretions of such experiments leaves many unanswered questions at present. Yet recent experiments suggest that both slow oscillatory and constant transcranial DC stimulation can alter excitability of motor cortex [BGS+09]. Recent experiments have demonstrated that similar transcranial electrical fields in animals can indeed alter the firing frequency of neurons [OSB+10]. Whether such slow or steady fields work through polarization, as discussed in this chapter, or also involve ionic shifts, as we will discuss in more detail in the next chapter, remains an exciting area for future research.

11.11   Electrical Fields for Control Within the Intact Brain

The goal of this chapter was to get to this section. Can we apply the principles of closed loop control to directly interact with the dynamics of neurons within the brain? And can we attempt to modulate such neurons to specifically target pathological dynamics such as epileptic seizures?

From the preceding discussion, we see several parts of the mammalian brain where the neuronal architecture is such that the large outflow excitatory neurons, often termed *pyramidal cells* for their shape and more generally labeled as *principal cells*, are aligned in parallel formation. The hippocampus has the principal cells within the *cornu ammonis* aligned as such, although the curvature of this structure presents some engineering issues in terms of optimal electrode configuration, as shown in figure 11.14.

We spent considerable time, following Gluckman and colleagues [GNWS01], attempting to implant small platelike electrodes to replicate the arrangement of polarization electrodes, as shown in in vitro experiments illustrated in figure 11.6. There was sufficient damage to the structures connecting to the hippocampus to render such implants useless, regardless of perceived surgical skill. We then explored a series of experiments using an axially placed wire electrode that coursed through the long central axis of the hippocampus. Such experiments were motivated by our clinical experience with such axial electrodes, where their use in the localization of epileptic seizure foci is a common approach [HSW+91]. An example of such a posterior placement of an axial depth electrode is illustrated in figure 11.15A (plate 31). In our experiments, we prepared steel electrodes with surfaces of electrodeposited thin

A   **B**

   Hippocampus   Cortex

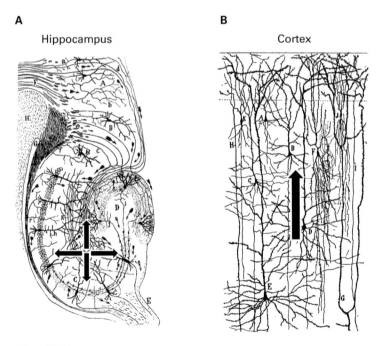

Figure 11.14
A shows image of the hippocampus. An added radial electrical field source is illustrated as arrows within the curve of the *cornu ammonis*. Reproduced from [Caj09]. B illustrates cortex, again demonstrating field alignment as an arrow. Reproduced from [Caj09].

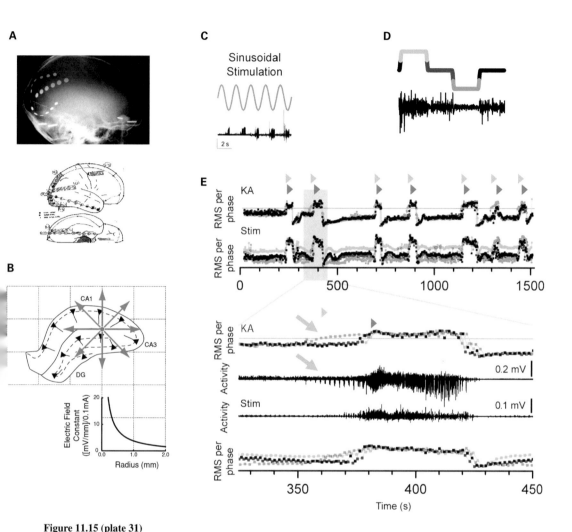

Figure 11.15 (plate 31)
A shows an example of occipitally placed recording depth electrode in human epilepsy patient. Reproduced from [SSKW05] with permission. B is schematic of an axially placed stimulating electrode for epilepsy control in experimental hippocampal seizures. C shows experimental verification in animal models of sinusoidal entrainment of hippocampal neurons with low-frequency sinusoidal electrical fields. Using DC electrical fields of alternating polarity, in D, we demonstrated in E that by stimulating the contralateral side to the kainic acid (KA)–injected side with subthreshold alternating pulses, that the network demonstrated progressively increasing responses to these small stimuli in the seconds leading up to the start of the seizure (red arrows). B through E are reproduced from [RGW+03] with permission.

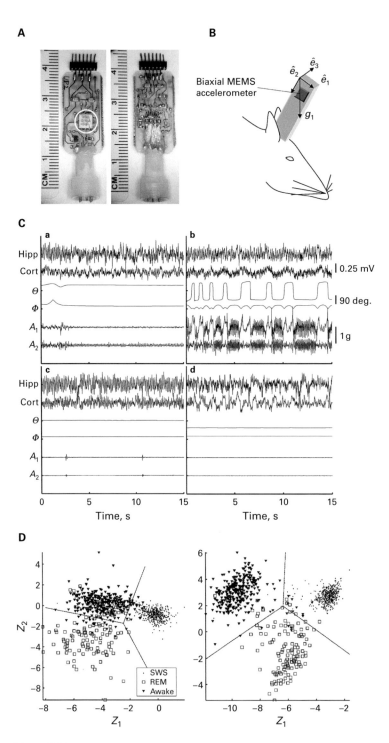

Figure 11.16
(Caption on facing page)

films of iridium oxide [MCNR01]. These surfaces were designed to minimize the polarization and maximize the charge-carrying capacity of such stimulation electrodes. This is critical to prevent the polarization of the electrode surfaces to the extent that irreversible chemical reactions take place at the electrode surface [MBJ05], and to maximize the amount of charge that can be safely passed while constraining the potential developed to within strict safely limits [MBJ05].

We could thus generate a radial electrical field, as illustrated in figure 11.15B [RGW$^+$03]. It was apparent, in these experiments, that we could sinusoidally entrain hippocampal neurons as shown in figure 11.15C. We then performed a series of alternating DC pulse stimuli (figure 11.15D) opposite to the side of the kainic acid–induced acute epileptic focus, and recorded from the side ipsilateral to the focus. We used pulses that were subthreshold to stimulate the neuronal circuitry in the interictal (between seizure) state. One of our findings was that, in the tens of seconds prior to a seizure, we could detect an increasing amplitude of fluctuations from the neuronal network, as shown in figure 11.15E. Whether such weak dynamical *probes* of epileptic networks could be employed to increase the accuracy of seizure prediction is a question that warrants further exploration.[11]

11.12   To Sleep Perchance to Dream

Absent from this discussion, and from most of the literature on EEG dynamics and seizure prediction, is the issue of sleep. Nevertheless, the *state of vigilance* of an animal or human has a dramatic effect on both behavior and EEG dynamics. Our present quantification of state of vigilance tends to be an amalgam of EEG, electromyography, and video recording of a person's behavior. Electromyographic electrodes are uncomfortable—they impale your muscles. It is infeasible for an automated control device to rely on video interpretation of wakefulness.

We have accordingly explored the use of microelectromechanical (MEMS) accelerometers as an adjunct to EEG recordings [SCP$^+$07]. An example of the integration of MEMS accelerometers into an electronic head stage is shown in figure 11.16. Panel B is a schematic of

11. One would certainly like to speculate that seizures might be analogous to phase transitions in physical systems [SRSRSL99]. In such systems, near the critical points of phase transitions, the fluctuations within ensemble systems can increase [Sta87]. Attractive as this notion might be, it is too speculative at present to warrant more than a footnote in this text.

Figure 11.16
Incorporation of MEMS accelerometer into head stage EEG apparatus in A, and schematic of use in B. C shows examples of various behaviors, reflected in hippocampal electrode (Hipp), cortical electrode (Cort), steady (DC) head tilt angles θ and ϕ, and dynamic (>1 Hz) biaxial acceleration plotted in A_1 and A_2. D shows linear discriminant analysis without (a) and with (b) acceleration data included in the discriminator. Z_1 and Z_2 represent discrimination variable (optimal linear combinations of original variables). Reproduced from [SCP$^+$07] with permission.

the head acceleration data measurements. The types of signature of EEG and accelerometer data in drinking, grooming, rapid eye movement sleep, and slow wave sleep, are shown in panel Ca–d, respectively. By creating a linear discrimination classifier for these combined accelerometer and EEG data [Flu97, SSKW05], we were able to show that the addition of acceleration data improves the state of vigilance classification of many of the presently used sleep discrimination algorithms [SCP+07].

No epilepsy control device, and perhaps no sophisticated Parkinson's disease controller, can be designed without taking state of vigilance discrimination into account in the algorithms.

11.13   Toward an Implantable Field Controller

The next stage in the development of an electrical field controller is to design it to function in awake behaving animals. Can we interact with seizures in open loop in this setting?

In [SCP+09] we used an axial hippocampal trajectory, which in people would be placed from an occipital approach, but in rodents a vertical approach is easier, as shown in figure 11.17. Making certain that the response of the neuronal responses in such settings is not caused by stimulation artifact is a complex task. We calculated a transfer function, α, relating the applied current to the electrical potential as

$$V(t) = \alpha \otimes *I(t)$$

where $\otimes$ denotes convolution. Ideally, one compensates for frequency-dependent amplitude and phase of a transfer function such as this [CSG08]. For now, we will assume that with periodic stimuli, α is constant with zero phase lag. We assume that the recorded signal, $X(t)$, is a summation of the neuronal signal, $N(t)$, with the voltage created by the applied current as

$$X(t) = N(t) + \alpha I(t)$$

If $N(t)$ and $I(t)$ are uncorrelated, then

$$\alpha = \langle X(t)I(t) \rangle / \langle I^2(t) \rangle$$

where $\langle \cdot \rangle$ indicates time average. Artifact is then removed from the signal, leaving an estimated neuronal signal $\hat{N}(t)$ as

$$\hat{N}(t) = X(t) - \alpha I(t)$$

Using such an artifact-reduced signal, we show spectrograms using multitaper spectral estimates [Tho82, PW93] in figure 11.17Ba and Be for an unstimulated and stimulated seizure, respectively. In this particular type of tetanus toxin seizure [FJ00], the waveforms

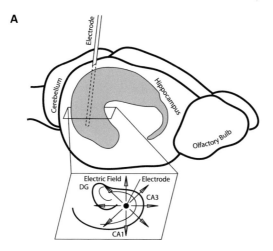

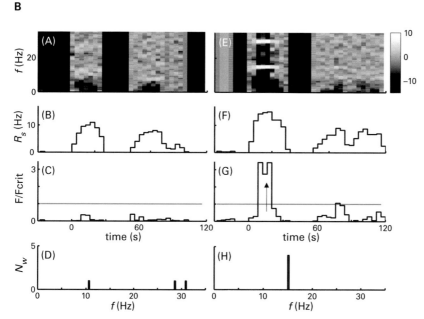

Figure 11.17
A shows a schematic of axial placement of hippocampal stimulating electrode within the ventral axial hippocampus.
B shows examples of seizures with and without low-frequency electric field stimulation. Both seizures were taken
from the same animal, recorded from a depth electrode in the same hippocampus as the stimulation. Analysis was
done for 150 sec starting 30 sec prior to seizure onset. The seizure on the left (Ba–c) was unstimulated, and the
one on the right (Be–g) was stimulated at 15 Hz. Analysis was done in 4-sec nonoverlapping windows. Shown
are (Ba, Be) spectrograms computed from the spike-time series, (Bb, Bf) average spike rate R_s per 4-sec window,
(Bc, Bg) Thomsons F-statistic normalized to a critical value (0.1%) at 15 Hz. No obvious persistent spectral lines
are observed from the unstimulated seizure, and the observed values are consistent with chance. The stimulated
seizure has a significant line at 15 Hz that persists for four windows, which is highly unlikely ($p \sim 10^{-7}$) from
chance. Reproduced from [SCP^{+}09] with permission.

can be decomposed into spikes that are readily assigned times as point processes. These spike rates are plotted in figure 11.17Bb and Bf. Because we are stimulating here periodically, the spectral analysis would show a single spectral line resulting from the stimulation, and we can employ Thomson's F statistic [Tho82] to test whether the amount of power observed at that line of interest is more than we expect by chance after the artifact has been removed. The confidence line for this F statistic is shown in figure 11.17Bc and Bg, demonstrating that there is a very significant entrainment of this seizure at the stimulation frequency.

So we can entrain seizures. The next step of creating feedback control laws to intelligently interact with such seizures is presently underway.

The use of low-frequency electrical fields to modulate neuronal firing is not well suited to different parts of the brain. What we have described here is best suited to regions of the brain where neurons—perhaps only the large outflow neurons such as pyramidal cells—are all oriented in a fashion that allows us to devise electrical fields to align with the long axis of the neurons. Such neurons need to be asymmetrical, so that more radially configured neurons, like many of the inhibitory neurons (or at least their dendrite fields), are less affected. The hippocampus is a region of the brain where the pyramidal cells are generally aligned in a curved sheet so that a radial electric field as we have presented here will effectively modulate the neurons. In cortex, neurons are again oriented vertically so that their somatodendritic axes point normal to the cortical surface. In primates, the curvature of the gyri and sulci of the folded cortex complicate this alignment, and cortical stimulation in the manner of Marshall and colleagues [MHMB06] will tend to preferentially polarize only a subset of the surface area (experiments such as [OSB+10] were performed in lissencephalic rodents, where there are no sulci to interfere with field effects). At present, it is not at all clear (although in fairness, it has not been explored either) whether stimulation of deep nuclei in Parkinson's disease can take advantage of such polarizing fields. Although chapter 10, on Parkinson's disease, focused on pulse stimulation, a separate line of research explores the electrical field interaction from such deep brain stimulation pulses [MGST04].

Further work on the interaction of electrical fields, whether low-frequency continuous or pulsed, is very much warranted in order to understand how to improve controller design.

12 Assimilating Seizures

As meteorologists have long known, the atmosphere exhibits no periodicities of the kind that enable one to predict the weather in the same way one predicts the tides. No simple set of causal relationships can be found which relate the state of the atmosphere at one instant of time to its state at another. It was this realization that led V. Bjerknes in 1904 to define the problem of prognosis as nothing less than the integration of the equations of motion of the atmosphere. But it remained for Richardson to suggest in 1922 the practical means for the solution of this problem. He proposed to integrate the equations of motion numerically and showed exactly how this might be done. That the actual forecast used to test his method was unsuccessful was in no sense a measure of the value of his work. In retrospect, it becomes obvious that the inadequacies of observation alone would have doomed any attempt however well conceived, a circumstance of which Richardson was aware. The real value of his work lay in the fact that it crystallized once and for all the essential problems that would have to be faced by future workers in the field and that it laid down a thorough ground-work for their solution.
—Jule Charney, 1951 [Cha51]

12.1 Introduction

By the turn of the twentieth century, Vilhelm Bjerknes had proposed, in 1904 [Bje04], that weather prediction required the use of the fundamental equations of motion of the atmosphere. The full equations of such motion [Kal03] were intractable to integrating by hand, there was no regional or global network of weather sensors, and digital computers would not be invented for decades.

When these issues began to be addressed, the first results were worthless. The introduction of simplified dynamical models that retained the most important of the physical dynamics was a critical development [Kal03]. The turning point came when integrating simplified models gave "first approximations that bore a recognizable resemblance to the actual motions" (Jule Charney, 1951 [Cha51] discussing his historic 1950 demonstration with John von Neumann that integrating a fundamental model made a meaningful 1-day weather forecast feasible [CFVN50]).

It is the author's conjecture that at the time of the writing of this book, neuroscience shares similarities with Charney's predicament nearly six decades before. This chapter will be the first description of our efforts at integrating the equations of motion of biophysically realistic models of neurons, using data assimilated from experiments. Our findings will indeed bare a recognizable resemblance to the actual neuronal dynamics. I have placed this chapter at the end of this book, but it is just the beginning.

12.2   Hodgkin-Huxley Revisited

Neuronal activity is almost always measured by observing just a single variable such as voltage (or calcium using optical techniques). Such measurements are always uncertain, due to a combination of noise in neurons and amplifiers, as well as uncertainties in recording equipment such as electrode access resistance and capacitance. And such measurements only reflect one of the many variables that are important in the dynamics of an electrically active neuron.

In chapter 3 we introduced the foundational Hodgkin-Huxley equations [HH52a, HH52b, HH52c, HH52d]. In subsequent chapters, we demonstrated that reduced models of Hodgkin-Huxley type dynamics could be employed in a control framework (chapters 5 and 10). But what about using the original equations? The distinction is important—the full Hodgkin-Huxley formalism includes the actual biophysical variables such as voltage and ionic currents.[1]

Of course, Hodgkin and Huxley had already placed their system in a control framework. Their goal was to use feedback to control the transmembrane voltage to a given constant command potential. The currents required to maintain the voltage in response to transient voltage changes are the currents that entered into their original equations. More recently, *dynamic clamp* techniques have introduced the manipulation of a conductance by adjusting the amount of current injected into a neuron while simultaneously measuring the membrane potential [GM06, PAM04]. This has typically been done by using the conductance equation of a particular current, measuring voltage across the membrane, and reconstructing the current required to add or subtract the particular current of interest. The reader can likely see where this line of thought is going to lead, but we need to address the original Hodgkin-Huxley equations first.

The transmembrane potential V of a single neuron, normalized to zero in the resting state, is modeled with the Hodgkin-Huxley equations (3.9). We can write these equations in compact form as

$$V = e + \frac{1}{C_m} \int_0^T (I_K + I_{Na} + I_L + I_{stim}) dt,$$

$$I_K = -g_K n^4 (V - V_K), \quad I_{Na} = -g_{Na} m^3 h (V - V_{Na}),$$

$$I_L = -g_L (V - V_L), \, dq/dt = \alpha_q (1 - q) - \beta_q q, \, q = m, n, h, \tag{12.1}$$

---

1. The author prematurely slipped into the chapter on Parkinson's disease (chapter 10) a Hodgkin-Huxley model in figure 10.19, without offering an adequate discussion for such modeling in control frameworks.

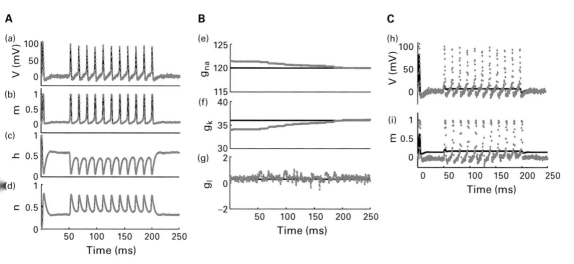

Figure 12.1
Reconstruction of full Hodgkin-Huxley dynamics through noisy voltage measurements alone. A(a–g) tracks the dynamics of the HH neuron (true values from model are shown in the black line, and uncontrolled estimates are gray symbols and lines). Fixed reversals of -12 mV and 10.6 mV (relative to 0 mV resting potential), respectively, are used for K^+ and leak currents (external potassium concentration $[K]_o$ fixed), and a constant stimulus $I_{stim}=10\,\mu A/cm^2$ is applied to the neuron for $50 \leq t \leq 200$ msec. In B (h, i) inaccurate model gives accurate estimates—same as a, b except constant α_m is used ($\alpha_m = 0.5$ black, α_m estimated gray). Reproduced from [US09] with permission.

where e is the V measurement noise;[2] n^4 and m^3h represent gating variables for potassium, I_K; and sodium, I_{Na}, currents; and I_L is the leak current. The rate equations for the gating variables, written with the modern convention for voltage sign, are $\alpha_n = (0.1 - 0.01V)/(\exp(1 - 0.1V) - 1)$, $\beta_n = 0.125 \exp(-V/80)$, $\alpha_m = (2.5 - 0.1V)/(\exp(2.5 - 0.1V) - 1)$, $\beta_m = 4 \exp(-V/18)$, $\alpha_h = 0.07 \exp(-V/20)$, and $\beta_h = 1/(\exp(3 - 0.1V) + 1)$.

In contrast with our previous employment of reduced models [VTK04, SS08], we here seek to reconstruct the full Hodgkin-Huxley ionic dynamics, including gating variables of the channels and various conductance parameters, using noisy voltage measurements. Although there are more sophisticated methods to explicitly account for model inadequacy (chapter 8), here we minimize model inadequacy by carefully choosing the process noise added to the voltage estimate.

Figure 12.1 shows the voltage measurements, with the estimated gating variables m, h, and n in panels a–d, respectively. We also tracked the three current conductances—g_{Na}, $g_k K$, and g_L—shown in panels e–g, respectively.

---

2. I am going to substitute e for r in this chapter, in order to reduce the chance that a reader will missread r for "resistance."

The reconstructions shown in figure 12.1 of the full Hodgkin-Huxley equations are far better than what seems reasonable. Granted, this is the assimilation of data, contaminated with noise, where the model producing the data and the model doing the assimilation are the same. But this is far better than what one would see using, for instance, the Wilson-Cowan equations [SS09] (chapter 6). It is very likely that symmetries are important.[3]

In the first of Hodgkin and Huxley's four 1952 papers [HH52a], they assumed "only that the chance that any individual ion will cross the membrane in a specified interval of time is independent of the other ions which are present." This *independence principal* was borne out by experiments (chapter 3), and it is reflected in the Hodgkin-Huxley equations. Each major ion, Na and K, was indeed shown to travel through separate channels in work that followed Hodgkin and Huxley's original curve fits. Of considerable help to us is the fact that these two major currents are also operating on different time scales. But this does not completely account for the absurdly high-quality fits in figure 12.1. Even the parameters of maximal conductance—g_{Na}, g_K, and g_L—converge rapidly to their true values (the process noises, Q, used were 1.5, 0.5, and 0.5 for g_{Na}, g_K, and g_L, respectively).

Let's formally make this an inadequate model. By replacing the dynamics of I_{Na} turn-on with a constant, we destroy the ability of the equations to spontaneously spike. We do this by substituting the actual $\alpha_m(V)$

$$\alpha_m(V) = (2.5 - 0.1V)/(\exp(2.5 - 0.1V) - 1)$$

with the trivial

$$\alpha_m(V) = a \text{ (a constant)}$$

This is not the first time that we have employed a constant to track a dynamical parameter. A similar example of such use was the trivial dynamics in the reconstruction of the unknown signal input in the Fitzhugh-Nagumo equations in chapter 4. Recall that the Kalman filter will continually compare the value of such a floating constant with the reconstruction, and fit a compromise value using Bayes's rule. Each iteration propagates this new value forward as the next value of a. Trivial dynamics such as these are not predictive. They assume that the parametric values are just not very different from the previous value. For parameters that are relatively slowly changing compared with faster-moving variables, such trivial dynamical assumptions can be very good. In this particular case, we have done this with one of the fastest changing variables in the set of equations, α_m. Intriguingly, although the reconstruction is off by a modest amount (20 to 30% root-mean-square, time-step dependent, of the full functional form of α_m), the reconstruction is sufficiently good to track the spiking as shown in figure 12.1B. Kalman filters are robust for this type of tracking even in the face of potentially crippling model inadequacy. Kalman filtering constantly reestimates the

---

3. An important discussion of symbolic observability can be found in Letellier and Aguirre [LA09] and references therein.

trade-off between model accuracy and measurements, expressed in the filter gain function [Kal60, Sim06a].

Hodgkin and Huxley performed their experiments on the giant axon of the squid in a perfusion chamber where they assumed that the ion concentrations outside of their nerve membrane were constant. This is not unreasonable given that they stripped out a nearly bare axon, and immersed it in an ionic bath of, for all practical purposes, infinite extent. Hodgkin and colleagues would later revisit the important issue of how ions change in the extracellular space, especially potassium, when "No attempt was made to remove the layer of connective tissue (about 20 μm in thickness) which clings tightly to the axon" [FH56]. Even a single action potential leaves a measurable and physiologically important accumulation of $[K^+]_o$ outside of the neuronal membrane in its wake. In [FH56] the increase with each action potential was about 1.6 mM, and it dissipated with a time constant of 30 to 100 msec. There is a rich subsequent history of the consequences of such potassium changes in the extracellular space of brains [Som04].

So let's force $[K^+]_o$ to vary substantially. The potassium and leak reversal potentials will be updated based on the instantaneous $[K]_o$ using the Nernst equation

$$V_k = 70 + 26.64\ln([K]_o/[K]_i) \tag{12.2}$$

and the Goldman-Hodgkin-Katz equation [Gol43, HK49]

$$V_l = 70 + 26.64\ln\left(\frac{[K]_o + 0.085[Na]_o + 0.1[Cl]_i}{[K]_i + 0.085[Na]_i + 0.1[Cl]_o}\right) \tag{12.3}$$

In figure 12.2, we show the results of forcing the Hodgkin-Huxley equations through cyclically modulating the $[K^+]_o$ between 5 and 10 mM. Note that the neuron spikes on the

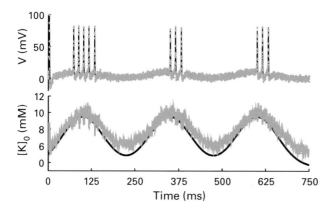

Figure 12.2
Driving Hodgkin-Huxley equations through variable external potassium concentration, $[K^+]_o$. Only noisy voltage, V, is measured. Reconstructed voltage and $[K^+]_o$ are in gray, with underlying true values in black. Reproduced from [US09] with permission.

peaks of this variation in $[K^+]_o$. We now measure noisy voltage. Again, reconstruction of voltage and the external potassium $[K^+]_o$, as well as the remaining variables and parameters from figure 12.1 (not shown), are well estimated.

12.3   The Dynamics of Potassium

We have learned much about the dynamics of potassium flow in the brain in the years following the study of isolated squid axons. In general, the intrinsic excitability of neuronal networks depends on the reversal potentials for ion species. During neuronal activity, both the extracellular potassium and intracellular sodium concentrations ($[Na^+]_i$) increase [MFP74, HLG77, RRS00, AMM02]. Glia help to reestablish the normal ion concentrations, but require time to do so. Consequently, neuronal excitability is transiently modulated in a competing fashion: the local increase in $[K^+]_o$ raises the potassium reversal potential, increasing excitability, while the increase in $[Na^+]_i$ leads to a lower sodium reversal potential and thus less ability to drive sodium into the cell. The relatively small extracellular space and weak sodium conductances at normal resting potential can cause the transient changes in $[K^+]_o$ to have a greater effect over neuronal behavior than the changes in $[Na^+]_i$, and the overall increase in excitability can cause spontaneous neuronal activity [McB94, RLJ85, TD88]. Nevertheless, the competing influences of these two ions on excitability also sets up neurons and their networks to oscillate in their excitability, as we will shortly demonstrate.

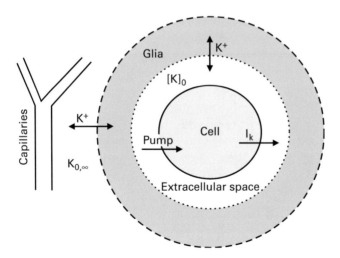

Figure 12.3
A schematic of the model dynamics: the cell releases K^+ through I_k, which is either pumped back into the cell through K^+-Na^+ exchange pumps or buffered by the glia from the extracellular region. K^+ can also diffuse to the nearby reservoir (capillaries in tissue or bath solution in vitro). Reproduced from [US09] with permission.

A rich literature on the interrelated dynamics of ion concentration and neurons is evolving [BTSS04, KWS00, KWS02, Som04, PD06, KWS07, FTSB08, SKW08, CUZ+09, UCJBS09]. We will employ a formulation of these dynamics suitable for our purposes, as more fully discussed in [CUZ+09, UCJBS09].

A schematic of the spatial relationship between the spaces involved in our ion dynamics is illustrated in figure 12.3.

J. Robert Cressman and colleagues [CUZ+09] let the extracellular potassium dynamics, $[K]_o$, be represented in a model based on I_K, activity of the pump exchanging K^+ and Na^+, I_{pump}, diffusion of potassium to the microenvironment, I_{diff}, and glial buffering, I_{glia}.

$$[K]_o = e_k + \int_0^T (0.165 I_k - 2\beta I_{pump} - I_{diff} - I_{glia}) dt, \tag{12.4}$$

where e_k is the $[K]_o$ measurement noise.

The equations for I_{pump}, I_{diff}, and I_{glia} are, from [CUZ+09]

$$I_{pump} = I_{max} \left(\frac{1}{1 + exp((25 - [Na]_i)/3)} \right) \left(\frac{1}{1 + exp(8 - [K]_o)} \right) \tag{12.5}$$

$$I_{diff} = \kappa([K]_o - k_{o,\infty}) \tag{12.6}$$

$$I_{glia} = \frac{G_{glia}}{1 + exp((18 - [K]_o)/2.5)} \tag{12.7}$$

where the Na^+-K^+ pump is modeled as a product of sigmoidal functions, I_{max} is the pump strength under normal conditions, and $[Na]_i$ is the intracellular sodium concentration. Additional parameters are given in the footnote.[4] Each sigmoidal term saturates for high values of internal sodium and external potassium, respectively. More biophysically realistic models of pumps, such as those in Lauger [Lau91], produced substantially similar results. $k_{o,\infty}$ in the diffusion equation is the potassium concentration in the nearby reservoir. Physiologically, this would correspond to either the bath solution in a slice preparation or the vasculature in the intact brain (noting that $[K]_o$ is kept below the plasma level by transendothelial transport). Both active and passive K^+ uptake into glia is incorporated into a simplified single sigmoidal response function that depends on extracellular K^+ concentration, with G_{glia} representing the maximum buffering strength. A similar but more physiological approach was used in Kager and colleagues [KWS07]. Two factors allow the

---

4. Membrane capacitance $C_m = 1 \mu F/cm^2$, maximum conductances (mS/cm^2) are $g_{na} = 120$, $g_k = 36$, $g_l = 0.3$, sodium reversal $V_{na} = 115$ mV, ion concentrations (mM) are intracellular $[K]_i = 130$, extracellular $[Na]_o = 130$, intracellular $[Na]_i = 20$, extracellular $[Cl]_o = 130$, intracellular $[Cl]_i = 8$ [BTSS04], maximum pump flux $I_{max} = 0.1$mM/sec, and glial buffer strength $G_{glia} = 100$ mM/sec. The diffusion coefficient of $[K]_o$ to the nearby reservoir κ, from Fick's law, is $\kappa = 2D/\Delta x^2 = 4.0$ sec^{-1}, where $D = 400 \times 10^{-6}$cm^2/sec is K^+ diffusion constant in neocortex [FPP76] and $\Delta x \approx 15 \mu m$ for brain reflecting the average distance between capillaries [Sch44a]. The factor 0.165 mM·cm^2/μcoul in eq.(12.7) converts ionic current to concentration rate of change and is calculated using $\beta A/FV$ [CUZ+09], where A, V and F represent cell area, volume and Faraday constant, respectively.

glia to provide a nearly insatiable buffer for the extracellular space. The first is the large size of the glial network. Second, the glial end-feet surround the pericapillary space, which, through interaction with arteriole walls, can effect blood flow; this in turn increases the buffering capability of the glia [KWBT71, MEW82, PN87].

To complete the description of the potassium concentration dynamics, we assume that the flow of Na^+ into the cell is compensated by flow of K^+ out of the cell. Then $[K^+]_i$ can be approximated by

$$[K]_i = 140.0\,mM + (18.0\,mM - [Na]_i)$$

where 140.0 mM and 18.0 mM reflect the normal resting $[K^+]_i$ and $[Na^+]_i$ respectively. The limitations of this approximation are addressed in Cressman and colleagues [CUZ$^+$09].

The intra- and extracellular sodium concentration dynamics are modeled by

$$\frac{d[Na]_i}{dt} = 0.33\frac{I_{Na}}{\beta} - 3I_{pump}$$

and

$$[Na]_o = 144.0mM - \beta([Na]_i - 18.0mM)$$

where we assume that the total amount of sodium is conserved. Here, 144.0 mM is the sodium concentration outside the cell under normal resting conditions for a mammalian neuron.

We will let I_K flow from a spherical cell with a radius of 13 μm for now. We set K$^+$ in the nearby infinite reservoir $[K^+]_\infty = 4.0$ mM, and intra- to extracellular volume ratio $\beta = 7$ [MTD90]. *Note that this reservoir is not the extracellular space, where potassium dynamics are quickly changing.*

As with Parkinson's disease dynamics (chapter 10), it is instructive to reduce this complex ion model so that we can examine the essence of the excitability changes and study the bifurcation structure of qualitative changes in the dynamics. We formulated this reduction by eliminating the fast-time-scale spiking behavior in favor of the slower ion concentration dynamics. This was accomplished by replacing the entire Hodgkin-Huxley mechanism with empirical fits to time-averaged ion currents. Using the membrane conductances from the full model, we fixed the internal and external sodium and potassium concentration ratios and allowed the model cell to attain its asymptotic dynamical state, which was either a resting state or a spiking state. Then, the sodium and potassium membrane currents were time averaged over 1 sec. These data were fit to products of sigmoidal functions of the sodium and potassium concentration ratios, resulting in the (infinite-time) functions

$$I_{Na\infty}\left([Na]_i/[Na]_o, [K]_o/[K]_i\right)$$

and

$$I_{K\infty}\left([Na]_i/[Na]_o, [K]_o/[K]_i\right)$$

The interested reader can find full details in the appendix of Cressman and colleagues [CUZ$^+$09]. $I_{Na\infty}$ is nearly identical to $I_{K\infty}$, differing significantly only near normal resting concentration ratios, due to differences in the sodium and potassium leak currents. These currents replace the original I_{Na} and I_K in the full model described earlier. We examine the estimated $[K^+]_o$ and $[Na^+]_o$ produced by the full and reduced model during episodic seizure like activity when the $[K^+]_\infty$ of the distant reservoir is increased to two times normal levels, as shown in figure 12.4.

I have dragged the poor reader through this morass of ionic notation to be able to show the following. Using the reduced model, we can analytically examine the interrelationships of varying external potassium concentration in the distant reservoir, $[K^+]_\infty$, ion pump strength, I_{pump}, diffusion strength, I_{diff}, and glial buffering strength, I_{glia}, as shown in figure 12.5. We see that as is well known experimentally [RLJ85], increasing $[K^+]_\infty$ to near twice normal results in oscillations in extracellular potassium (figure 12.5A), which will cause the excitability of the neurons to oscillate as illustrated in figure 12.2. At higher concentrations of reservoir potassium, a stable level of extracelluar potassium reemerges, which corresponds to a cessation of spiking during a depolarization block (the far right

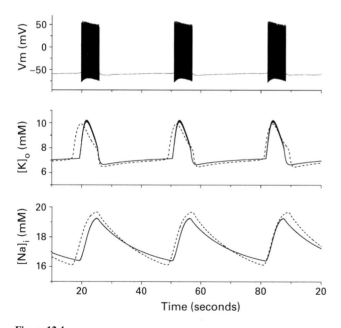

Figure 12.4
Comparison of full (solid line) versus reduced (dashed line) model of ion dynamics when potassium in the distant reservoir $[K^+]_\infty$ is doubled over normal. Episodic periods of high firing activity, shown in the upper tracing, last many seconds, similarly to intermittent seizures. These periods of high activity are reflected in the substantial ion shifts shown. The reduced model is a good qualitative approximation to the full ion dynamics occurring during these events. Reproduced from [CUZ$^+$09] with permission.

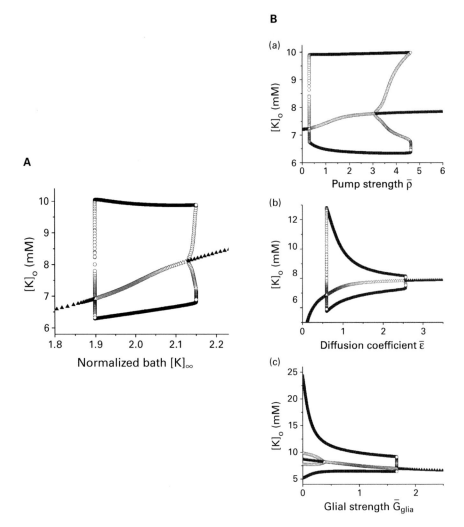

Figure 12.5
A shows the bifurcation structure of extracellular potassium, $[K^+]_o$ as a function of distant reservoir (bath or vasculature) concentrations of potassium $[K^+]_\infty$ normalized to the normal physiological bath concentration. Thus, 2.0 indicates twice normal bath concentration where oscillations typically begin in experiments [RLJ85]. B shows comparable bifurcation diagrams for $[K^+]_o$ as a function of ion pump strength, I_{pump}, diffusion strength, I_{diff}, and glial buffering strength, I_{glia}. The overbars, ¯, again indicate normalized with respect to normal values. Reproduced from [CUZ⁺09] with permission.

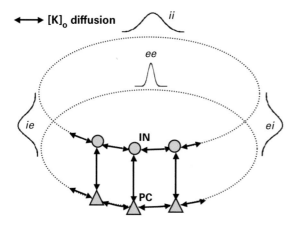

Figure 12.6
Topology of the network. The network consists of one pyramidal cell layer and one interneuron layer, both arranged in a ring (dotted lines). The two neuronal types make synaptic connections in the same layer as well as across the layers, with Gaussian synaptic footprints (represented by Gaussian curves, where *ee*, *ei*, *ie*, and *ii* stand for excitatory-excitatory, excitatory-inhibitory, inhibitory-excitatory, and inhibitory-inhibitory synaptic connections, respectively). Note the narrower *ee* Gaussian curve. The $[K^+]_o$ concentration around each neuron diffuses to the nearest neighbors in the same layer and the nearest neighbor in the next layer (represented by solid lines with double arrows). Reproduced from [UCJBS09] with permission.

side of figure 12.5A). In figure 12.5B, we show comparable bifurcation diagrams at a fixed value of $[K^+]_\infty$ of twice normal as a function of pump strength, diffusion, and glial buffering strength. Note that, as glial buffering strength approaches zero, the dynamics remain unstable. Oscillations abound in this quadrangle of interactions. Importantly, such oscillatory regimes are broad in parameter space. As we will shortly see, *modeling these interrelations is critical for the assimilation of data from real dynamics within the brain.*

Let's now apply our intuition from the reduced model to networks of neurons with full-model ionic dynamics. There are few parts of the brain or spinal cord where excitatory and inhibitory neurons are not tangled in an interactive web from which the dynamics of our thoughts emerge. In recent years, we have come to recognize, to our computational horror, that there are many more types of inhibitory cells than excitatory ones. Since they are small, they are hard to impale with electrodes. And they appear to orchestrate the more complex dynamics of the larger excitatory neurons [SK05]. These inhibitory neurons have unique spatial locations and particular electrical characteristics [GWM00], and they are chemically and transcriptionally unique [MTRW+04].

We start with a simple framework of interacting excitatory and inhibitory neurons, as illustrated in figure 12.6.

Gutkin and colleagues [GLC+01] counterintuitively demonstrated that such networks would form states of *persistent activity* by *asynchronous* interactions between its neurons (as shown in figure 12.7A). Such persistent states are a model of brief working memory [MBRW03, Wan03] that we feel is held electrically in the activity of neurons [FBGR89, Fus95, GR95, MED96, RAM98, RBHL99]. Recently, the phenomenon of *up* states have been similarly linked to persistent states [SVM00, SHM03]. The brain requires a certain degree of instability to form such an activity balance, often likened to a Turing instability [Mur03]. We asked, What are the network properties that render such physiological persistent states stable to perturbations and, especially, in what circumstances do such perturbations lead to seizure-like activity? In figure 12.7A, we replicate the fundamental findings of [GLC+01]. Stimulating such a network can produce persistent activity that is asynchronous, and a second stimulus can synchronize and stop the activity. In order to take into account the findings of Jokubas Ziburkus and colleagues [ZCBS06] of excitatory-inhibitory cell interplay during seizure-like events, Ghanim Ullah modified the synaptic current entering the pyramidal cell and interneuron to account for the lack of spiking during depolarization blockade.[5] In figure 12.7B, we show a broad regime for given levels of $[K^+]_\infty$ within which stable persistent activity is a function of the strength of excitatory synaptic activity (α_{ee}). We then studied how such persistent activity would respond to perturbations (figure 12.7C) as a function of $[K^+]_\infty$ (figure 12.7D). We see that, as a function of $[K^+]_\infty$, the persistent state can be stable, or lose stability and involve the entire network in a seizure-like state. *Network ion dynamics may be an important aspect of the determination between normal and pathological dynamics.*

In recent years, the role of glial cells in such ion homeostasis has become increasingly recognized. Astrocytes in particular have been shown to affect virtually every aspect of neuronal function through glia-neuron crosstalk. They have the ability to buffer excitatory ions such as K^+ from the interstitial volume [AMM02], preventing an undue increase in excitation of the neuronal network. They can also enhance the excitability of the network

---

5. We use

$$I_{syn}^e = -\frac{(V_j^e - V_{ee})}{N} \sum_{k=1}^{N} g_{jk}^{ee} s_k^e \chi_{jk}^e - \frac{(V_j^e - V_{ie})}{N} \sum_{k=1}^{N} g_{jk}^{ie} s_k^i \chi_{jk}^i$$

and

$$I_{syn}^i = -\frac{(V_j^i - V_{ei})}{N} \sum_{k=1}^{N} g_{jk}^{ei} s_k^e \chi_{jk}^e - \frac{(V_j^i - V_{ii})}{N} \sum_{k=1}^{N} g_{jk}^{ii} s_k^i \chi_{jk}^i$$

where $V_j^{e/i}$ is the voltage of the jth excitatory/inhibitory neuron, $s_k^{e/i}$ is the variable giving the temporal evolution of the synaptic efficacy emanating from the kth excitatory/inhibitory neuron, and $\chi_{jk}^{e/i}$ takes into account the interplay between pyramidal cells and interneurons described earlier; that is, if a cell goes into depolarization block, then the synaptic inputs from that cell to others are reduced toward zero by the factor $\chi_{jk}^{e/i}$. V_{ee}, V_{ei}, V_{ie}, and V_{ii} are the reversal potentials for the excitatory-excitatory, excitatory-inhibitory, inhibitory-excitatory, and inhibitory-inhibitory synaptic inputs, respectively. This is necessary in order to be able to handle the voltage region of steady depolarization block, especially in the presynaptic cells (generally about -30 to -10 mV with respect to outside the cell). See [US09] for further details. Such a formulation leaves the synaptic effects of fast-spiking neurons (when potential overshoots zero) unchanged.

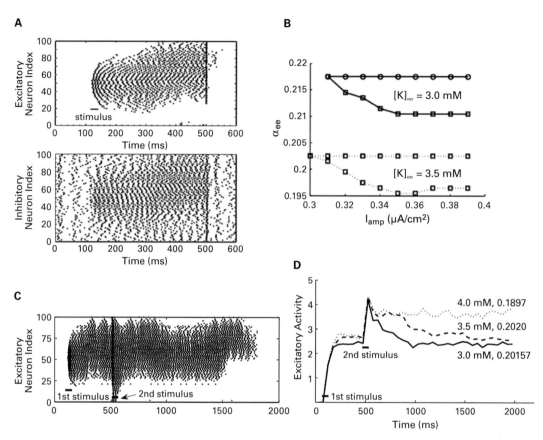

Figure 12.7
The existence and stability of spatially restricted foci of activity in neuronal network of excitatory and inhibitory neurons. In A, a 20-msec excitatory stimulus with amplitude $I_{stim} = 1.5\mu A/cm^2$ applied to pyramidal cells 21–79 causes a persistent and spatially restricted activity packet in the network. This activity is turned off by synchronizing neuronal activity using a stronger 1-msec stimulus of $100\mu A/cm^2$ at $t = 500$ msec. The top (bottom) panel shows a raster plot for the pyramidal cell (interneuron) network where each dot represents a single-cell spike in the network. Here, $\alpha_{ee} = 0.215$, and $[K^+]_\infty = 3.0$ mM. B shows regions of persistent activity stability for different values of $[K^+]_\infty$. Beyond a certain threshold of the stimulus ($I_{stim} = 0.31$) the weakness of the stimulus is compensated by the synaptic strength and external potassium level. In C we show the results of a first stimulus at $t = 112$ msec that causes a stable focus of activity in the network. A second stimulus $t = 500$ msec causes the focus to take over the entire population. After the stimulus is removed, the network decays to zero-activity state, returns to the stable focus, or expands to the state where the activity is spread throughout the entire population. D shows the effect of different $[K^+]_\infty$ values on the first and second stimuli. For lower extracellular potassium, the network activity decays back to the stable focus after the second stimulus is turned off. However, if extracellular potassium is higher, the network maintains its enhanced activity even after the second stimulus vanishes. Reproduced from [UCJBS09] with permission.

by releasing glutamate and ATP in a Ca^{2+}-dependent manner [PBL+94, PH00]. In spite of a direct role of glia in modulating the activity of neuronal networks, detailed models of glia-neuron interaction are few [NJ03]. In our modeling results presented earlier, we demonstrated how dysfunction of glial cells can cause the *occasional* transition from normal stable neuronal activity to seizure-like uncontrolled activity. Several experimental studies support glial dysfunction during epileptic seizures [HGJ+00, HSH+00]. In Ullah and colleagues [UCJBS09], we further modeled the effect of glutamate released from glia in a Ca^{2+}-dependent manner [TAT+05, KXX+05]. Our simulations showed that the ability of glial cells to induce seizure-like activity depends on the initial baseline state of the network. If the network is in a relatively low excitatory state, then a transient glial glutamate perturbation could not cause epileptic activity no matter how strong the perturbations were. However, even a small perturbation by glial cells is enough to cause seizure-like activity in the network if its excitability setpoint is sufficiently high.

An extensive literature describes the stability of ensemble systems [For90], and indeed the fundamental hallmark of a stable (equilibrium) system in physics is reflected in how such systems dissipate fluctuations. Might an analogous principle regarding the response to perturbations help us characterize the stability of biological systems? The earlier models show that *how neuronal networks respond to perturbations, and how this response decays (or grows), are likely critical determinants of the state of the brain.* How neuronal networks respond to perturbation determines how transient patterns of activity emerge from a background activity in response to internal fluctuations of activity or external stimuli. We propose that the response of a balanced, or mildly imbalanced, network to such perturbations is a fundamental feature that may underlie the initiation of a variety of disparate transient phenomena such as seizures, working memory, up-states, cortical oscillations [HTY+04], and spinal cord [CMO06] and hippocampal neonatal burst firing [LKC+02].

There is gathering evidence characterizing the interplay between excitatory and inhibitory neurons during seizures [FTIKT04, PVC99]. Ziburkus and colleagues [ZCBS06] observed interplay between pyramidal cells and interneurons during in vitro seizure-like events, as shown in figure 12.8A. In particular, pyramidal cells were seen to go into a silent state when interneurons were burst firing, followed by burst firing in pyramidal cells when interneurons went to depolarization block. The interneurons appeared to be firing in a much more erratic and high-frequency manner than the pyramidal cells, and a particularly large fluctuation might have tipped the interneuron network into depolarization blockade. What might be the origin of the depolarization block in such interneurons? Figure 12.8B shows that such blockade can be emulated in this model network by the effect of potassium dynamics. In particular, $[K^+]_o$ builds up in the extracellular space, sculpting the activity of the interneurons in a fast-slow-fast temporal pattern, with depolarization blockade in the middle of the episode. It is our view that such ion dynamics likely are a factor in these events, but it is hard for someone who has experience with oxygen physiology to be able to observe such depolarization blockade and not consider the contribution of oxygen limitation during high-frequency firing to this phenomenon as well [Som04, SS87].

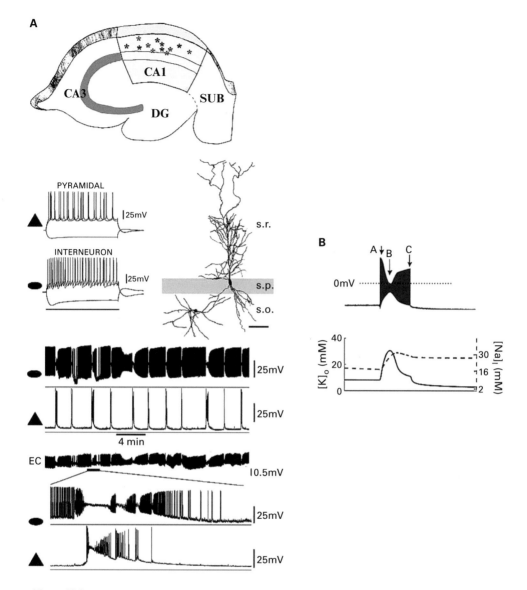

Figure 12.8
Excitatory-inhibitory interplay during seizurelike events in vitro. In A, at top, is a schematic with gray stars indicating interneurons that underwent depolarization block in seizures. Below this are examples of pyramidal (triangle) and oriens interneuron (ellipse) membrane properties following 500-msec (scale bar) negative and positive current injections. Camera lucida microscope drawings of the dendrites of the cells. s., stratum; o., oriens; p, pyramidale; r, radiatum. Scale bar: 50 μm. The lower electrical recordings show repetitive seizures and the relationship of excitatory and inhibitory cell interplay. The interneuron (ellipse) is more active than the pyramidal (triangle) cell. The extracellular (EC) electrode was placed in the oriens, about 75 μm from the interneuron soma. Note the cessation of axonal spikes in EC during interneuron depolarization block. Reproduced from [ZCBS06] with permission. B shows model-based network results for excitatory and inhibitory cell dynamics in the presence of active potassium dynamics. The sculpting of the interneuron activity (top panel) shows a period of depolarization blockade flanked by high-frequency activity. Reproduced from [UCJBS09] with permission.

Keep in mind the pattern of excitatory and inhibitory neuron interplay shown in figure 12.8. We will return to this phenomenon shortly, and show how ion dynamics are as critical as neuronal interactions in a control framework for data assimilation.

12.4 Control of Single Cells with Hodgkin-Huxley and Potassium Dynamics

Ghanim Ullah [US09] followed the strategy of [SS08] to estimate the control vector to apply to the Hodgkin-Huxley equations. In figure 12.9A we show, on the left, the schematic of direct proportional control, where the control signal is generated from the noisy measurement y, which in turn is generated by adding noise to the observation variable. In the right panel, we show Kalman observer control where the model observer system is used to generate the control vector from the estimated observable $\hat{y}$. In both schemes, the control vector is applied to both the system and observer (contrast with figure 6.22 in chapter 6). In figure 12.9B, we demonstrate use of voltage measurements (top) to control the dynamics of the cell. The extracted Na^+-channel gating variable, m, is shown in the middle panel. As is clear from the bottom panel, the energy required for the control vector estimation through Kalman observer control is smaller than that used in direct proportional control. In fact, the energy requirements for direct proportional control relative to the energy requirements for Kalman observer control increases exponentially with increased noise in the measurements (inset right bottom), although beyond a certain limit the direct proportional controller becomes unstable (not shown).

Since we can estimate the rest of the gating variables and parameters along with $[K]_o$, it is possible to construct control signals based on the estimated gating variables. Such a control framework offers a novel means to modulate neuronal dynamics through various conductances using dynamic clamp methods [GM06, PAM04]. In figure 12.10A, we show a dynamic clamp simulation where seizure-like activity is controlled through g_k. In typical dynamic clamp methods, the current corresponding to a given ion species, I_i, is calculated in isolation from the relation $I_i = g_i(V - V_i)$, where g_i is the voltage-dependent dynamical conductance, and V_i is the reversal potential for a given ion species i [GM06, PAM04]. In the bottom panel we show $[K]_o$ estimated by the filter. Our state estimation framework has the advantage over this previous methodology in that the current to be injected into the cell to modulate a conductance can be calculated from more complete dynamical descriptions of available conductance and gating parameters, combined with an optimal strategy for handling model and measurement uncertainty, which, heightened by computational constraints, are major issues in applying dynamic clamp [BLSW08]. Wrapping dynamic clamp algorithms within our framework will increase the computational load for such applications, and a trade-off between increasing algorithmic complexity for accuracy versus lengthening the discrete time interval for computation will be required. Experimental testing of the predicted advantages of such a strategy given presently available hardware is under investigation.

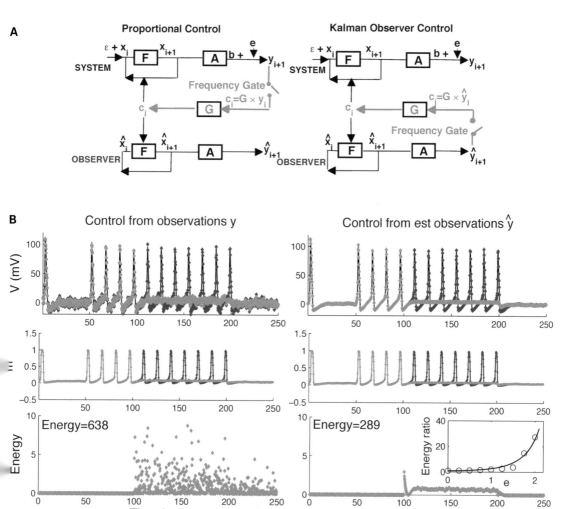

Figure 12.9
In A are the schematics of Kalman filter observers: The system function F acts on state variable x, and the observation function A produces output vector b. Output observable y is generated by adding measurement noise e to b. In the left panel of A, the control signal is constructed directly from actual noisy data by multiplying it with gain G; in the right panel, the filter uses the estimated state from the observer. B shows the results of controlling the dynamics of a Hodgkin-Huxley neuron (controlled estimates are light gray; uncontrolled estimates and true values are black) using direct proportional (left) and Kalman observer control (right). The panels from the top to bottom show V, m, and energy used by the controller (sum of squares of control vector). The control vector c is turned on at $t = 100$ msec and is only injected into V. Inset shows the ratio of energy required in direct proportion to the energy required in Kalman observer control, versus measurement noise (circles), fitted by function $\exp(e^{1.7})$. Reproduced from [US09] with permission.

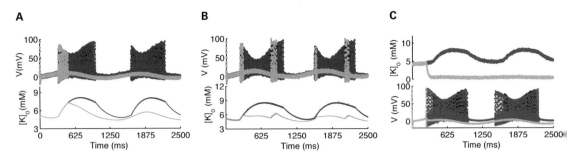

Figure 12.10
Single-cell seizures controlled through g_K (A), V (B), and $[K]_o$ (C) In A, the top panel shows V measurements with (light gray) and without (black) control. At $t = 500$ msec, g_K is doubled using dynamic conductance clamp, which suppresses the seizure-like events. Bottom panel shows estimated $[K]_o$. In B we repeat simulation from A, but the control signal is applied to V when the spiking frequency rises above 50 Hz. $[K]_o$ is lower after seizures are blocked (bottom panel). In C, neuronal dynamics including V are generated from, and control signals are applied to, $[K]_o$ measurements. The control vector is injected only into $[K]_o$ (top panel). After the injection of control signal V stays close to the resting value (bottom panel). Reproduced from [US09] with permission.

In figure 12.10B, we use V measurements to control seizure-like events by applying the control signal to V (top). Here we place a frequency gate on the filter (figure 12.9A), tuning the filter such that it applies the control signal only when the frequency of measured or estimated neuronal spiking goes above a certain value (50 Hz). The frequency gate in the filter allows the controller to modulate only the high-frequency spiking attributed to seizure-like behavior, while allowing the neurons to perform their more normal (low-frequency) functions. This not only minimizes the energy requirements by shutting the controller off when it is not needed, but it can also minimize the long-term risk of damage to brain tissue due to continuous electrical stimulation.

An interesting variant using our framework is to measure $[K^+]_o$ instead of voltage. This is a slower signal, but one that sets and reflects the excitability of the network. The relatively small extracellular space causes a small I_K to generate moderate changes in $[K]_o$, which enhances the membrane excitability significantly by changing the reversal potentials for K^+ channels. We therefore reconstruct neuronal dynamics including V based on $[K]_o$ measurements. In the top panel of figure 12.10C, we show $[K]_o$ measurements, and apply K^+ control at $t = 250$ msec. Driving $[K]_o$ to lower levels causes the seizure-like burst firing in V to go to silence (bottom). Controlling seizures with $[K]_o$ has two advantages: (1) the time scale for $[K]_o$ is slower than the fast dynamics of V and gating variables, and hence can be more easily implemented in real time, and (2) the $[K]_o$ controller allows us to control the neuronal network rather than a single cell. It has been shown, using ion-selective microelectrodes to measure K^+ activity changes in extracellular space beneath the surface of the brain, that electrical current could be used to decrease or increase $[K]_o$ [GMN83]. A recent theoretical study [BMS08] predicts a functional block along axons due to K^+ accumulation in the extracellular (submyelinated) space during deep brain stimulation for Parkinson's disease. Recent optical techniques offer the prospect of increasing the speed and decreasing the scale of $[K]_o$ measurements [MFL89].

12.5   Assimilating Seizures

The purpose of this long chapter was to get to the point of utilizing all of the preceding in an experimental situation. In the experiments shown in figure 12.8A, we recorded from two or three neurons simultaneously, including at least one inhibitory and one excitatory neuron. Such data offers us a way to employ our network models and *validate* our reconstruction through data assimilation. We can use the data from one excitatory (inhibitory) cell and reconstruct the other inhibitory (excitatory) cell. Can we achieve Jule Charney's [Cha51] description of "first approximations that bore a recognizable resemblance to the actual motions" [Cha51]?

In Ullah and Schiff [US10], we created a network model of two cells, one inhibitory and one excitatory, as shown in figure 12.11. Using this model, we can take recordings from either pyramidal inhibitory cells, and reconstruct the remaining unobserved variables in the system as shown in figure 12.12.

Including the dynamics of intra- and extracellular ion concentrations in the network model is *necessary* for accurate tracking of seizures. Using Hodgkin-Huxley–type ionic currents with fixed intra- and extracellular ion concentration of K^+ and Na^+ ions fails to track seizure dynamics in pyramidal cells, as shown in figure 12.12Bc. We used physiologically normal concentrations of 4 mM and 18 mM for extracellular K^+ and intracellular Na^+, respectively, for these simulations. The conclusion remains the same when higher $[K]_o$ and $[Na]_i$ are used. A similar tracking failure is found while tracking the dynamics of inhibitory interneurons during seizures (data not shown).

Reconstructing Network Interaction

Since the interaction of neurons determines network patterns of activity, it is within such interactions that we seek unifying principles for epilepsy. To demonstrate that an unscented Kalman filter framework can be used to study cellular interactions, we reconstructed the dynamics of one cell type by assimilating the measured data from another cell type in the network in [US10]. In figure 12.13, we show the estimated membrane potentials, but we also reconstructed the remaining variables and parameters of both cells (not shown). We first assimilated the membrane potential of the pyramidal cell to estimate the dynamics of the same cell and also the dynamics of a coupled interneuon in figure 12.13A–D. Conversely, we estimate the dynamics of pyramidal cell from the simultaneously measured membrane potential measurements of the interneuron (figure 12.13E–F). As is evident from

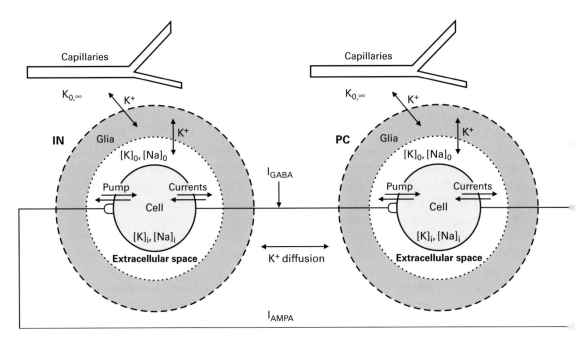

Figure 12.11
A schematic of the model dynamics: Potassium is released to the extracellular space and is pumped back to the cell by the ATP-dependent K^+-Na^+ exchange pump, buffered by glia, and diffuses to the microenvironment where it interacts with capillaries. Sodium entering the cell through Na^+ channels is pumped out of the cell by the ATP-dependent pump. Pyramidal cell (PC) and interneuron (IN) from the CA1 region of the hippocampus are coupled both synaptically and through lateral K^+ diffusion. Figure courtesy of G. Ullah.

figure 12.13, the Kalman filter framework is successful at reciprocally reconstructing and tracking the dynamics of these different cells within this network.

It appears that we have met Charney's challenge [Cha51].

12.6   Assimilation in the Intact Brain

I view the reconstruction of individual cellular activity from multiple intracellular impalements in brain slice experiments as proof of concept. A bit of the neuroscientist's equivalent of a party trick. We are presently working with an even harder trick: the assimilation of three types of neuron from such experiments—oriens interneurons, basket cells, and pyramidal cells. These appear to be the minimum three canonical cell types to even begin to consider that we may be able to capture essential dynamics in small bits of the brain. Much of the nervous system has small inhibitory cells that lay near the principal excitatory output cells. These cells receive excitation from branches of the excitatory cells, and deliver inhibition back near the cell bodies of those excitatory cells. Such inhibition near the perisomatic sites

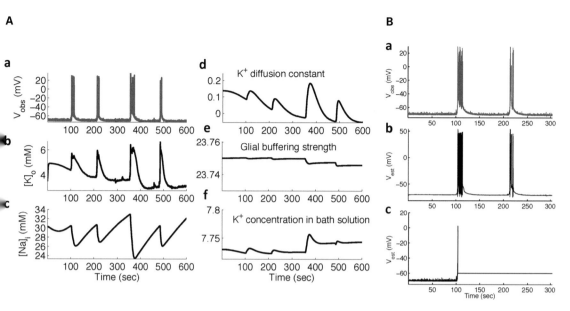

Figure 12.12
Assimilation of spontaneous seizure data by whole-cell recording from pyramidal neurons. Aa shows measured
V (gray) from a single pyramidal cell during spontaneous seizures. Estimated (black) $[K]_o$ (Ab), $[Na]_i$ (Ac),
K^+ diffusion constant (Ad), glial buffering strength (Ae), and K^+ concentration in bath solution (Af). Data
generously provided by Jokubas Ziburkus; panel Aa modified from [ZCBS06] with permission. B shows an
attempt to assimilate such data without including microenvironmental K^+ and Na^+ dynamics in the model.
Observed (Ba) and estimated (Bb) membrane potential using the model with ion concentrations dynamics. In Bc,
we show estimated membrane potential using the model without ion concentrations dynamics. Reproduced from
[US10] with permission.

where action potentials are initiated are well positioned to cut off the outputs of those cells.
In the hippocampus, these cells have a high level of calcium-binding protein parvalbumin
within them, which helps in their identification [SK05]. In cortex we tend to call these "rapid
spiking" cells. In the spinal cord, the Renshaw cell takes this role. Our nervous systems
are full of cells that act as speed governors of our large neurons—they keep them in line.
Perhaps given recent data, they also play an important role in forming patterns of activity.

There is also another type of inhibitory cell with wide distribution. We focused on these
in the preceding work. In the hippocampus, about 80% of the neurons in the *stratum
oriens* layer of the hippocampus are known as oriens-lacunosum-moleculare interneurons—
a dreadful name that derives from the course of their axons, from the oriens layer to the
border of the lacunosum-moleculare layer. In the cortex, these are known as low-threshold
spiking interneurons [GBC99]. The importance of this is that the output of these inhibitory
cells, after being activated by the pyramidal cells, bypasses the pyramidal cell bodies and
courses all the way to the tips of the principal (apical) dendrites of these pyramidal cells.
This places the inhibition of these neurons in a position to cut off the inputs to these cells.

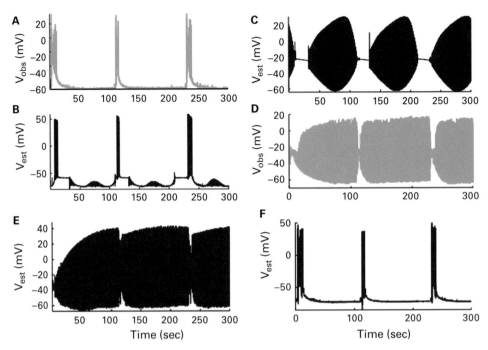

Figure 12.13
Validation of reconstruction in simultaneously recorded cells. Measured (shown in A, gray) and estimated (B, black) V for pyramidal cell. C shows estimated V for interneuron. We used the membrane potential recorded from the pyramidal cell (A, gray) to not only reconstruct the full dynamics of the same pyramidal cell (only membrane potential is shown in B) but also reconstructed the dynamics of the interneuron (only membrane potential shown in C). Simultaneously recorded V from the interneuron is shown in (D, gray) for comparison. Estimated V for interneuron in E and pyramidal cell in F by assimilating measured V from interneuon shown in D. D through F are converses of A through C. That is, in D through F we used membrane potential recorded from the interneuron (shown in D) to not only reconstruct the full dynamics of the same interneuron (only membrane potential shown in E) but also the coupled pyramidal cell (only membrane potential shown in F; compare with actual values shown in A). Simultaneous membrane potential measurements, shown in (A and D), were from a pyramidal cell and oriens interneuron in the hippocampus using simultaneous dual whole-cell patch clamp recordings demonstrating the firing interplay between these cells during in vitro seizures. Data provided by Jokubas Ziburkus. Panels A and D are modified from [ZCBS06] with permission. Overall figure reproduced from [US10] with permission.

Gibson and colleagues [GBC99] discovered that these different types of inhibitory interneurons in cortex tend to electrically couple with their own cell types and act as a larger subnetwork in their synchronized activity. It was further discovered by Gloveli and colleagues [GDR+05] that these subnetworks can, at least in the hippocampus, be spatially orthogonal to each other. Tort and colleagues proposed [TRD+07] a way to link the computational models of Gloveli and colleagues [GDR+05] together to reflect the binding of our hippocampal slices back together into their original functional arrangement.

We have taken the model from Gloveli and colleagues [GDR+05] and simplified it somewhat for data assimilation, as shown schematically in figure 12.14. We employ our three canonical neurons—pyramidal, basket, and oriens—and arrange their synaptic connections in the native way they interconnect. We then add the potassium dynamics as before. The schematic shows an extracellular recording, since this is the customary arrangement that we use in the awake behaving brain [SCP+07, SCP+09]. The modeling must emulate an appropriate observing function reflecting the resistive and capacitive properties of the brain lattice within which these neurons are embedded, and helping to best process the extracellular signal.

We use our intracellular experiments, as before, to set many of the unknown parameters of this observer model. Ideally, we will track only the parameters that are most essential to track in such scenarios. The schematic then shows the assimilation of the estimated potential of the lumped dendrite compartment or equivalent cylinder from such data recording. We then reconstruct the potential of the pyramidal soma (E), basket cell (I), and oriens cell (O). We are presently employing such model-based data assimilation in several in vivo paradigms: spontaneous chronic seizure activity, and navigational oscillatory rhythms in maze settings. The remainder of this figure, showing the results of this effort, will gradually materialize in the years to come.

12.7 Perspective

To our knowledge, these are the first demonstrations of data assimilation from neuronal networks using fundamental model-based techniques. It is important to frame out this statement. There has been intense interest in the neuroscience communities in bringing control-theoretical tools to bear on neuronal encoding and decoding problems as discussed in chapter 9 [BNF+01, BWFB05, SB03, SB07, SEMB07, SWSB07, WGB+06, EBE+07, LOH+09]. In all of this work, statistical models (continuous or point process) were fit to data recorded from neurons, and these empirical models incorporated into applications. Our use of control theoretic tools is very different. We built computational models from the physiological properties of neurons and their networks, as well as the properties of ion metabolism, *without data fitting*. Using these fundamental models of the physics of neuronal systems, we fuse these models with data—*data assimilation*—in a manner commonly applied in

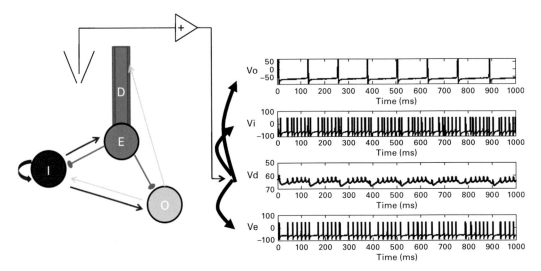

Figure 12.14
Model of three canonical neurons interacting within a transverse "slice" of the hippocampus: pyramidal cells
(soma, E, and dendrite, D), basket cells (I), and oriens interneurons (O). We envision the recording of extracellular
voltage with the electrode and amplifier shown, and through appropriate filtering assimilate this potential into the
dendrite voltage, V_d. One could similarly record and assimilate the pyramidal soma potential. In the schematic,
we use arrows to represent the reconstruction through the model of the potential within the oriens neurons, V_o,
basket cells, V_i, and pyramidal soma potential, V_e. Figure courtesy of Yina Wei.

meteorology [Kal03, HKS07, OHS⁺04, BHK⁺06, SBIJ08, SKJI06]. Other authors have
also recently discussed the importance and power of going beyond statistical empirical
models in neuronal systems, and simulations have begun to explore the feasibility of carrying this out [PAF⁺, HP09]. To our knowledge, our findings are the first experimental
validation that a fundamental biophysical model of part of the brain can be employed to
assimilate incomplete data and reconstruct its network dynamics.

Brain dynamics emerge from within a system of apparently unique complexity among the
natural systems we observe. Even as multivariable sensing technology steadily improves,
the near infinite dimensionality of the complex spatial extent of brain networks will require
reconstruction through modeling. Since, at present, our technical capabilities restrict us
to only one or two variables at a small number of sites (such as voltage or calcium),
computational models become the *lens* through which we must consider viewing all brain
measurements [MB08]. We need reconstruction to account for unmeasured parts of the
neuronal system, relating microdomain metabolic processes to cellular excitability. As we
learn how to do this, we can validate our cellular dynamical reconstruction against actual
measurements.

At the end of this book, we have come to the beginning. We have constructed a framework
of model-based data assimilation and control for spiking neuronal systems based on the

foundational Hodgkin-Huxley ionic model. This framework has a wide range of potential applications to various systems in biology, where one can use a single measured variable to estimate the variables and parameters that are experimentally inaccessible, and allows us to predict the future trajectory of the system. These methods are applicable to a variety of model-based scenarios not discussed in this book. Recent developments in the modeling of blood pressure [ZRC07] make this a fruitful area for study. Other systems with sophisticated models and sensing available include the spiking dynamics of heart cells, or the insulin-secreting cells in the pancreas and blood glucose regulation.

Nevertheless, in terms of brain dynamics, I would wish to suggest to the reader that as our models of the nervous system improve, we should strive to never again settle for inspecting raw data. Our microscope of understanding will consist of filtering through our biophysical models what we measure and observe from the brain.

13 Assimilating Minds

State estimation is intimately connected with the meaning of life.
—Dan Simon, 2006 [Sim06a]

13.1 We Are All State Estimation Machines

It is a quiet secret of most neuroscientists and neurophysicians that they were drawn to the study of the brain in the hope of understanding the mind. Understand the mind and you understand life.

Coming to grips with the reality that, in our lifetimes at least, we will never understand the mind was a blow to most of us. The profusion of information about the nervous system has become overwhelming in an information age, where even high-quality publications about the brain and mind expand beyond the reach of any one of us to keep track of.

So it comes with some initial surprise to find, at the end of one of the best texts on state estimation yet written, a chapter on the relationship between state estimation and the meaning of life [Sim06a].

Similarly, in reading deeply within the geophysical literature, one stumbles on a quite serious discussion: "Like a data assimilation system, the human mind forms a model of reality that functions well, despite limited sensory input, and one would like to impart such an ability to the computational model" [DTW06].

The opportunity to speculate on these broader issues, prior to confirmed senility, is an opportunity that the author cannot pass on. I hope that these thoughts are welcomed by any reader who has had the patience and long-suffering temperament to have come this far in this book.

13.2 Of Mind and Matter

[The mind] can never produce any new quality in the object, which can be the model of that idea, yet the observation of this resemblance produces a new impression in the mind, which is its real model.
—David Hume, 1748 [Hum07]

You walk outside and need to decide whether your perceptions are delivering to your mind sensations of a banana or an airplane.[1] The empirical Scottish philosopher David Hume, in his 1748 book, *An Enquiry Concerning Human Understanding* [Hum07], eloquently considered this problem (albeit without the airplanes):

[N]othing is ever really present with the mind but its perceptions or impressions and ideas, and that external objects become known to us only by those perceptions they occasion. To hate, to love, to think, to feel, to see; all this is nothing but to perceive. [Hum07]

The mind is a state estimation machine, with regard to external objects as Hume speculated on, and internally, with regard to introspection. Just as in control engineering, there are many ways to accomplish state estimation. Kalman's original filter was optimal for linear systems with Gaussian noise. There is no rationale to suspect that our minds use Kalman's filter in its linear or linearized guise. There is no indication that evolution might have optimized our mental "filter" for the external world. We seem to use and demand a luxurious amount of energy to keep our minds running (about 20% of our basal metabolic rate) [Len03].

But filters are essential. As discussed this book, dynamical data assimilation filters are really filters. They remove noise in an effort to extract the relevant dynamics they seek. Simon speculates on the moral philosophy and theology of such filters [Sim06a]. In the theory of the cognitive dynamics of schizophrenia, the inability to parse and filter the incoming sensory stream has been postulated to be at the core of the symptoms, and correlative physiological data support this view [NAW+91]. Optical illusions [HP05] might be viewed as reflective of our error-prone reconstructions of the external world. We think that normal minds have more effective (more optimized?) filters than disordered ones.

13.3   Robot Beliefs versus Cogito Ergo MRI

We shall adopt an approach that we call model-dependent reality. It is based on the idea that our brains interpret the input from our sensory organs by making a model of the world.
— Stephen Hawking and Leonard Mlodinow [HM10]

In control robotics, the robot has a model of its world—perhaps a map it needs to navigate to perform its tasks. It takes in sensor data in various forms: video, laser range finding, global positioning system, and so on. And then it forms an estimate of where it thinks it is—a *belief*, in control jargon [TBF05].

When our minds are intact, we also estimate where we are and navigate. When our minds are severely damaged, we may not give external indications that we are conscious.

---

1. This phrase was shamelessly lifted from Dan Simon's discussion [Sim06a].

Nevertheless, in recent years, we have developed the unprecedented ability to peer within people's minds with functional magnetic resonance imaging (fMRI). Patients who survive severe traumatic brain injury typically emerge to a state of wakefulness within a period of weeks. This state may not give evidence of conscious response to internal thought or external interaction; it may be a state of wakefulness without apparent awareness of one's self or surroundings. Such states are described with the unfortunate term *vegetative*.

A recent report documented the case of a patient who was vegetative following a severe traumatic brain injury. On fMRI imaging, when asked to imagine navigating through all of the rooms of her house, she demonstrated activity in the parts of the brain involved in spatial imagery navigation [OCB+06]. This finding has generated much discussion in our society. Is the act of mentally navigating (or a motor activity such as swinging a tennis racket) equivalent to consciousness? This evidence is perhaps not convincing, but too risky to dismiss in error.

More recently, Monti and colleagues [MVC+10] described a patient initially diagnosed as vegetative[2] who was seen to display such navigational versus motor (tennis) activity on fMRI scanning. He was then taught to respond *yes* to one form of imagery and *no* to the other form of activity. He was able to demonstrate consistent yes-no responses to complex personal questions.

One must be careful in interpreting such measurements. We cannot make the "unjustified assumption that the association between a behavior and a pattern of brain activation implies the converse" [NH07]. Brain activation is a necessary but not sufficient condition for evidence of consciousness, unless you can show that the same activation cannot occur without it [NH07, Rop10].

I have been careful not to define consciousness in this discussion. But whatever that term might represent, whatever doubts we might have regarding navigation or swinging a racket, that such imagery might represent more complex thoughts raises difficulties in interpreting these phenomena to a higher level.

These findings bring us to one of the deep issues that artificial intelligence has grappled with. In 1950, Alan Turing [Tur50] suggested that perhaps a digital state machine—a digital computer—would be able to think by following the rules of digital computation. He discussed numerous ways of opposing or supporting this argument. John Searle's [Sea80] *Chinese room* argument is the most cited refutation of Turing's *test*. Program a computer to take written questions in a language—say Chinese. It replies in an apparently intelligent fashion just as a conscious person who spoke Chinese might. Then place someone in a room with the same computer program, and perhaps a library of reference books. The person, knowing nothing of the Chinese language, takes the questions and uses the

---

2. This patient, on reexamination could display "reproducible but highly fluctuating and inconsistent signs of awareness" at the bedside [MVC+10]. Such a finding technically reclassifies the diagnosis from vegetative to the new term "minimally conscious state" [Far08].

computer program to create answers just as a computer would. Although the responses appear intelligent, we know that the person knows absolutely nothing of Chinese. Is this intelligence?

We appear to be at the stage where people with severely damaged brains are now manipulating symbols much in the fashion that Turing and Searle were considering in their philosophical writings. Perhaps the issue is one of uncertainty and risk. We risk little if we are wrong about a machine's thoughts. But we simply cannot make a mistake in the face of the evidence of human thought and intention. I'll side with the humans.

13.4 Black versus Gray Swans

The works and customs of mankind do not seem to be very suitable material to which to apply scientific induction [Tur50]

Turing [Tur50] in his consideration of whether machines might think, nicely discusses the appalling mistakes we make by applying scientific induction to scientific problems. If you want to ask whether a machine can think, the reader might examine his or her past experience with machines. You may have seen tens of thousands of machines. They were ugly and stupid. Does induction, extrapolating into the future based on past experience, tell us anything useful? Often not.

A marvelous rediscovery and expansion of the fallacy of logical induction is discussed by Nassim Taleb [Tal07]. His poignant example of the turkey as logician is not to be missed. Although Taleb's exposition is focused on the tracking and prediction of economics, the lessons for our observation of brain dynamics are very pertinent indeed.

In the author's admittedly biased perspective, Taleb's black swans boil down to the issue of model inadequacy. if we do not have a fundamental model of a system that truly characterizes its dynamics, then we need to rely on statistical models based on previous empirical data. Empirical models, such as our discussion of principal orthogonal decomposition in chapter 7, can capture only what has been previously observed. The epitome of such empiricism was expressed by:

I will say that I cannot imagine any condition which could cause a ship to founder. I cannot conceive of any vital disaster happening to this vessel. Modern shipbuilding has gone beyond that.
—E. J. Smith, captain of the Titanic (New York Times, April 16, 1912)

Taleb is rightfully concerned with natural dynamics that have power-law distributions of their amplitudes. Such events can certainly have underlying generative models that are inherently unpredictable [Man82]. Having dynamics with extreme events need not necessarily require that the underlying events be unpredictable. But it does require that one has a grasp of those dynamics. Airframes and pendula can be effectively modeled. The short-term atmospheric dynamics of weather can be modeled. Parts of the brain appear to be modelable. Can the economy be better modeled? Taleb might be severely skeptical, but his induction here might not be better than our induction about black swans.[3]

There is actually a theory that all of physics is, in effect, a balance between measurement of the natural world and uncertainty in those measurements. We are playing an information-theoretic game with nature, seeking to solve an extremization problem, in which information we gain comes at the expense of nature's disorder, as the system is closed [FC96]. In Taleb's *extremistan*, the appearance of large errors invalidates our models. The stock market crash of 1929 and the credit market meltdown of 2008 certainly support Taleb's perspective on financial markets. But airframes rarely crash these days. Whether or not your system belongs to extremistan may depend on what the physics are. Perhaps the brain, or at least portions of the brain, are now tractable enough that the conjectures presented in this book are not doomed to fall prey to the biting skepticism of Taleb.

Taleb also points to a key problem for our work on state estimation here. What are the consequences of losing track of the system? Were errors and uncertainty described by the neat Gaussian error distributions we have discussed, as Kalman's original filter assumed or our unscented version presumed, then things would be simple. Nothing in life is Gaussian. Julier and Uhlmann [JU97a] dealt with this in part by introducing a fudge factor—their kappa—that boosted the distribution width of their uncertainty to take unknowns into account. Taleb would jeer at us: If we had an uncertainty distribution without a well-defined mean, and with long "tails," then wherever we set our Gaussian distribution we would be hit from the rear by low-flying black swans. Control theory has long recognized the danger of such fowl events, sparking a tremendous amount of energy into *robust control* strategies [ZDG96, Sim06a].

Perhaps the question comes down to the consequences of losing track of a system we are observing. If an airframe controller loses track of a landing sequence, the consequences are dreadful. If a Parkinson's disease controller loses track of the oscillations within the basal ganglia, the patient will get a bit stiffer, and we can use any number of strategies commonly employed in particle filtering [Sim06a] to spread out the ensemble of state estimates to recapture the trajectory. If a seizure controller loses track of a system, an epileptic seizure might result—serious, but likely not lethal in and of itself. I hear Taleb scoffing at us. To my knowledge, no one has attempted to take an ensemble Kalman filter, such as the unscented filter, and parameterized the uncertainty with a power-law distribution. There is nothing that would prevent such a maneuver were it indicated. It would certainly decrease the accuracy of our state estimations. But it would make our tracking in the face of non-Gaussian distributions in critical applications more robust. And "a gray swan concerns modelable extreme events" [Tal07].

---

3. There actually are black swans. But they are in Australia. Our literature and science had a few thousand years of certainty that all swans were white.

The case for model-based observation and control rests on state estimation and short-term prediction. Taleb's cautions are for phenomena where our models are so inadequate, and the consequences so dire, that operating with such risk exposure is hazardous. Nevertheless, there is a good operational test to see whether or not you are in the insane danger zone: "Prediction, not narration, is the real test of our understanding of the world" [Tal07]. The author completely agrees, so long as the prediction horizon is short enough to encompass the nonlinearities of the system. Weather (as opposed to climate) prediction, beyond two weeks, remains pretty useless. But for the dynamical diseases we have discussed in this book, predictions by our models from tens of milliseconds to tens of seconds would be sufficient to accomplish much.

13.5 Mirror, Mirror, within My Mind

We may then lay it down for certain that every representation of a movement awakens in some degree the actual movement which is its object. . . .
—William James (*The Principles of Psychology*, vol. 2, 1890, p. 611 [Jam90])

In 1992, di Pellegrino and colleagues [dPFF+92] were studying neurons in primates that fired in response to specific motor movements, and discovered that those same neurons fired in response to similar movements on the part of the investigators. They sparked a period of intense study of such effects, which have now been observed in an interlocking hierarchical set of regions in the brain related to vision and movement [ID06]. We now speculate that this system forms part of our normal ability to imitate someone else's motor skills, and perhaps more important, might critically underlie our social skills of empathy and understanding of others. In the normal state, we speculate that such mirror systems solve the *problem of other minds* [Iac09]. We have also speculated how cognitive disorders such as autism [OHM+05, DDP+06, RFDC09] and schizophrenia [AA05] might be disorders of such systems.

Mirror neurons appear to be state estimation systems. Karl Friston and colleagues have proposed that the mirror neuronal system functions as a predictive coding model [KFF07]. In this scheme, each portion of this hierarchical system, from lower-level motor output through higher-level goal generation, has feedback and error correction with a built-in computational model of each interacting system. Indeed, the interaction of the mind with its motor system—the only way it can communicate with the external world—and social interactions, are compelling [WDK03]. Similarly, in the visual system, such feedback of efferent copies of predicted events seem critical [DR05].

At each level in the motor or visual system, the type of state estimation described in this book appears to be important. Kalman's brain probably was not born with Kalman's filter already hard wired. But as the evidence of multiple levels of state estimation within the brain and between brains in society become stronger, nature appears to have given us some type of model-based filter.

Friston's predictive coding is really an instantaneous prediction—a zero-time prediction horizon. Yet we live in a world where events and time move inexorably forward. Our motor interaction with the stream of events necessitates *prospective coding* [SBP07].

The juvenile sea squirt wanders through the sea searching for a suitable rock or hunk of coral to cling to and make its home for life. For this task, it has a rudimentary nervous system. When it finds a spot and takes root, it doesn't need its brain anymore, so it eats it! (It's rather like getting tenure.) The key to control is the ability to track or even anticipate the important features of the environment, so all brains are, in essence, anticipation machines. (Daniel Dennett, 1991 [Den91])

13.6 Carl Jung's Synchronicity

The psychologist Carl Jung deeply explored synchrony and the mind [JP55a]. Jung was inspired by an essay published in 1851 as part of a larger collection by Arthur Schopenhauer, *Parerga and Paralipomena* "Transcendent Speculation on the Apparent Deliberateness in the Fate of the Individual" [SP74]. Schopenhauer was a bit of an extreme determinist. He envisioned the events of the world as trajectories (meridians), whose intertwining at a particular instance of time might bring us simultaneity without causal connectedness:

Accordingly, all those causal chains, that move in the direction of time, now form a large, common, much-interwoven net which with its whole breadth likewise moves forward in the direction of time and constitutes the course of the world. [SP74]

Our estimates of our states within the stream of time and the events of the external world can be estimated only by what we have access to. Our efforts are abetted by our models of the inaccessible variables of the external world and future time through our internal models.

[L]ife in time can never properly be understood, just because no moment can acquire the complete stillness [*fuldelig Ro*][4] needed to orient oneself backwards. (Søren Kierkegaard, *Papirer, 1833–1843*, p. 441 [Kie69] Translation from [Han82])

Some of Schopenhauer's deterministic transcendent fatalism [SP74] is the same semantics that we face today with deterministic random number generators. All computer random number generators are deterministic (you would need to time radioactive decay, for instance, to see what we think is true randomness). But the numbers my computer spits out are, for all intents and purposes, as good as random. All chaotic systems similarly eat up their information and predictability through sensitivity to initial conditions, and after a certain time, the trajectories and numbers that characterize them are unpredictable. In Schopenhauer's worldview, the chance simultaneity of uncountable deterministic causes might underlie

---

4. The translation of Kierkegaard's choice of the archaic Danish *fuldelig Ro* is open to some controversy. Harder [Har96] offers "sufficient quiet." My Danish friends remind me that the other meaning for this phrase can be "completely drunken."

events in our world, but the fact of concealed determinism offers no practical advantage over a more stochastic worldview. Schopenhauer's transcendant fatalism has a nice way of supporting Taleb's transcendant fatalism.

Randomness in this book, our *uncertainty*, is *unknowledge* or incomplete information in the sense of Taleb [Tal07]. It is whatever you have not embodied in your state estimation model.

Schopenhauer envisioned each of us decoding and interpreting the web of events around us as "One man's fate is always in keeping with another's, and everyone is the hero of his own drama, but at the same time figures also in that of another ... and that all plurality of phenomena is conditioned by time and space" [SP74]. To the extent that our minds make sense of the external world, and that more than one of our minds agree on such interpretations, depends strongly on our internal models drawing similar state estimations.

Hawking and Mlodinow present the opposite viewpoint, one biased by their inquiry into the nature of physical law. "There may be different ways in which one could model the same physical situation, with each employing different fundamental elements and concepts. If two such physical theories or models accurately predict the same events, one cannot be said to be more real than the other; rather, we are free to use whichever model is most convenient" [HM10]. They will further assume the plausible existence of multiple coexisting realities, a cosmological perspective lacking in the less radical philosophical works to which we now return.

Jung envisioned synchronizing observers. He defines a term *synchronicity*, as "the simultaneous occurrence of a certain psychic state with one or more external events which appear as meaningful parallels to the momentary subjective state—and, in certain cases, vice versa" [JP55a]. In chapter 8 we discussed in detail the equivalence of state estimation with synchrony of the observer and the object observed.

Jung also discusses the Taoist view of the unconscious as a microcosm of macrocosmic events—clearly an allusion to our model-based state estimation. He also discusses the monads of Gottfried Leibnitz's monadology [LL25], wherein each monad is a *little world*: "minds are also images ... capable of knowing the system of the universe, and to some extent of imitating it" [LL25].[5] To Jung, "The perceptive faculty of the monad corresponds to the knowledge, and its appetitive faculty to the will" [JP55a].

Jung stated that "I need hardly point out that for the primitive mind, synchronicity is a self-evident fact" [JP55a]. Unfortunately, today, proving synchronization, especially for complex nonlinear functions, is a question at the forefront of dynamical physics [PCH95, NPS04, NCPS06]. The Taoist concept of meaningful coincidence might, indeed, now be

---

5. The reader might pause to reflect on how one might attempt to draw analogies from Leibniz's monadology, to the situation of envisioning how individual neurons, as irreducible elements within the brain, interact with each other only through their limited perceptions of their external world (personal communication, J. Y. Lettvin). Such considerations can draw one toward studying neuroscience, as many of us in Professor Lettvin's class were.

embodied in Hebb's principle of learning—neurons that fire together, wire together—by strengthen their connections [Heb49]. We know today that a precise timing is required for such Hebbian learning to take place. There must be a causality between the cause and the effect [BP98]. Spike timing-dependent plasticity [CD08] is an embodiment of the meaningful coincidence.

Synchrony implies that a functional relationship exists between dynamical systems (chapter 8). Insufficient coupling between systems does not guarantee synchronization. But any degree of coupling does imply that the system receiving information from the other contains information from the signaling system. At sufficient levels of coupling, a system that drives another will become synchronized with it; at that point, *each* system contains enough information to estimate the state of the other [SSC+96]. Coupled dynamical systems live in states of partial information. Synchronized systems understand each other. Our neurons are coupled to each other. The components of our brain are coupled to each other. Our minds are coupled to each other.

Jung discusses the philosophy of Plato, "which takes for granted the existence of transcendental images or models of empirical things, the $\epsilon\iota\delta\eta$ (forms, species), whose reflections ($\epsilon\iota\delta\omega\lambda\alpha$) we see in the phenomenal world" [JP55a]. Plato understood model-based control.

Wolfgang Pauli provided an essay published with Jung's 1955 text [JP55b]. He speculated that

The process of understanding nature . . . [is] based on a correspondence, a "matching" of inner images pre-existent in the human psyche with external objects and their behavior. This interpretation of scientific knowledge, of course, goes back to Plato and is, as we shall see, very clearly advocated by Kepler. W. Pauli in [JP55b].

Pauli goes on to propose an interpretation of knowledge, as "a 'matching' of external impressions with pre-existent inner images" [JP55b].

Perhaps state estimation of our minds will enable us to better conceive of our minds and society.

Bibliography

[AA05] M. M. A. A. Arbib and M. A. Arbib, *Schizophrenia and the mirror system: an essay; neuropsychologia*, Neuropsychologia **43** (2005), no. 2, 268.

[AA99] J. L. Anderson and S. L. Anderson, *A Monte Carlo implementation of the nonlinear filtering problem to produce ensemble assimilations and forecasts*, Monthly Weather Review **127** (1999), no. 12, 2741–2758.

[AB73] C. M. Armstrong and F. Bezanilla, *Currents related to movement of the gating particles of the sodium channels*, Nature **242** (1973), no. 5398, 459–461.

[ABH+05] V. C. Anderson, K. J. Burchiel, P. Hogarth, J. Favre, and J. P. Hammerstad, *Pallidal vs subthalamic nucleus deep brain stimulation in Parkinson disease*, Arch Neurol **62** (2005), no. 4, 554–560.

[Abl39] R. Ablowitz, *The theory of emergence*, Philos Sci **6** (1939), no. 1, 1–16.

[ABS94] P. Ashwin, J. Buescu, and I. Stewart, *Bubbling of attractors and synchronisation of chaotic oscillators*, Physics Letters. A **193** (1994), no. 2, 126–139.

[Ada79] D. Adams, *The hitchhiker's guide to the galaxy*, 1st American ed., Harmony Books, New York, 1979.

[Ada91] R. K. Adair, *Constraints on biological effects of weak extremely-low-frequency electromagnetic fields*, Phys Rev A **43** (1991Jan), no. 2, 1039–1048.

[AG00] B. Amirikian and A. P. Georgopulos, *Directional tuning profiles of motor cortical cells*, Neuroscience Research **36** (2000), no. 1, 73–79.

[Ala52] M. Alan, *Turing. The chemical basis of morphogenesis*, Phil Trans R Soc Lond B **237** (1952), 37–72.

[Ale98] C. Alexander, *The endurance: Shackleton's legendary antarctic expedition*, 1st ed, Knopf, New York, 1998.

[AMa77] S. Amari, *Dynamics of pattern formation in lateral-inhibition type neural fields*, Biological Cybernetics **27** (1977) no. 2, 77–87.

[AMB+10] C. A. Anastassiou, S. M. Montgomery, M. Barahona, G. Buzsáki, C. Koch, *The effect of spatially inhomogeneous extracellular electric fields on neurons*, J Neurosci **30** (2010Feb), no. 5, 1925–36.

[AML+05] L. Alvarez, R. Macias, G. Lopez, E. Alvarez, N. Pavon, M. C. Rodriguez-Oroz, et al. *Bilateral subthalamotomy in Parkinson's disease: Initial and long-term response*, Brain **128** (2005), no. Pt 3, 570–583.

[AMM02] F. Amzica, M. Massimini and A. Manfridi, *Spatial buffering during slow and paroxysmal sleep oscillations in cortical networks of glial cells in vivo*, J Neurosci **22** (2002 Feb), no. 3, 1042–1053.

[And77] P. Andersen, *Long-lasting facilitation of synaptic transmission*, Ciba Found Symp **58** (1977), 87–108.

[Asc02] G. A. Ascoli, *Computational neuroanatomy: Principles and methods*, Humana Press, Totowa, N.J., 2002.

[ASY97] K. T. Alligood, T. Sauer, and J. A. Yorke, *Chaos: An introduction to dynamical systems*, Springer, New York, 1997.

[AVRH+08] J. L. Alberts, C. Voelcker-Rehage, K. Hallahan, M. Vitek, R. Bamzai, and J. L. Vitek, *Bilateral subthalamic stimulation impairs cognitive-motor performance in Parkinson's disease patients*, Brain **131** (2008), no. 12, 3348–3360.

[Bal97] D. H. Ballard, *An introduction to natural computation*, MIT Press, Cambridge, Mass., 1997.

[Bar89] A. Baruzzi, *From Luigi Galvani to contemporary neurobiology*, Vol 22, FIDIA Research Series, Liviana Press, New York, 1989.

[BBBG96] T. Boraud, E. Bezard, B. Bioulac, and C. Gross, *High frequency stimulation of the internal globus pallidus (gpi) simultaneously improves parkinsonian symptoms and reduces the firing frequency of gpi neurons in the mptp-treated monkey*, Neurosci Lett **215** (1996), no. 1, 17–20.

[BBH98] O. Bousquet, K. Balakrishnan, and V. Honavar, *Proceedings of the pacific symposium on biocomputing*, Is the hippocampus a Kalman filter (1998), 655–666.

[BBK03] M. Bar, A. K. Bangia, and I. G. Kevrekidis, *Bifurcation and stability analysis of rotating chemical spirals in circular domains: Boundary-induced meandering and stabilization*, Phys Rev E **67** (2003), no. 5, 56126.

[BCM$^+$83] R. S. Burns, C. C. Chiueh, S. P. Markey, M. H. Ebert, D. M. Jacobowitz, and I. J. Kopin, *A primate model of parkinsonism: Selective destruction of dopaminergic neurons in the pars compacta of the substantia nigra by n-methyl-4-phenyl-1,2,3,6-tetrahydropyridine*, Proc Natl Acad Sci USA **80** (1983), no. 14, 4546–4550.

[BD06] K. L. Briggman and W. Denk, *Towards neural circuit reconstruction with volume electron microscopy techniques*, Curr Opin Neurobiol **16** (2006), no. 5, 562–570.

[Ber12] J. Bernstein, *Elektrobiologie: Die lehre von den elektrischen vorgängen im organismus auf moderner grundlage dargestellt*, Die Wissenschaft, vol. 44 Hft. F. Vieweg, Braunschweig, 1912.

[Beu56] R. L. Beurle, *Properties of a mass of cells capable of regenerating pulses*, Phil Trans R Soc Lon B **240** (1956), 55–94.

[BFT$^+$98] E. N. Brown, L. M. Frank, D. Tang, M. C. Quirk, and M. A. Wilson, *A statistical paradigm for neural spike train decoding applied to position prediction from ensemble firing patterns of rat hippocampal place cells*, J Neurosci **18** (1998), no. 18, 7411–7425.

[BGN$^+$00] A. Benazzouz, D. M. Gao, Z. G. Ni, B. Piallat, R. Bouali-Benazzouz, and A. L. Benabid, *Effect of high-frequency stimulation of the subthalamic nucleus on the neuronal activities of the substantia nigra pars reticulata and ventrolateral nucleus of the thalamus in the rat*, Neuroscience **99** (2000), no. 2, 289–295.

[BGS$^+$09] T. O. Bergmann, M. Groppa, M. Seeger, M. Mölle, L. Marshall, and H. R. Siebner, *Acute changes in motor cortical excitability during slow oscillatory and constant anodal transcranial direct current stimulation*, J Neurophysiol **102** (2009 Oct), no. 4, 2303–2311.

[BH06] P. Blomstedt and M. I. Hariz, *Are complications less common in deep brain stimulation than in ablative procedures for movement disorders?*, Stereotact Funct Neurosurg **84** (2006), no. 2–3, 72–81.

[BH90a] C. E. Beevor and V. Horsley, *An experimental investigation into the arrangement of the excitable fibres of the internal capsule of the bonnet monkey (macacus sinicus)*, Phil Trans R Soc Lond B **181** (1890a), 49–88.

[BH90b] ———, *A record of the results obtained by electrical excitation of the so-called motor cortex and internal capsule in an orang-outang (simia satyrus)*, Phil Trans R Soc Lond B **181** (1890b), 129–158.

[BHK$^+$06] S. J. Baek, B. R. Hunt, E. Kalnay, E. Ott, and I. Szunyogh, *Local ensemble Kalman filtering in the presence of model bias*, Tellus A **58** (2006), no. 3, 293–306.

[BIA$^+$04] M. Bikson, M. Inoue, H. Akiyama, J. K. Deans, J. E. Fox, H. Miyakawa, and J. G. R. Jefferys, *Effects of uniform extracellular dc electric fields on excitability in rat hippocampal slices in vitro*, J Physiol **557** (2004), no. Pt 1, 175–190.

[Bje04] V. Bjerknes, *Das problem der wettervorhersage, betrachtet vom standpunkte der mechanik und der physik*, Meteorologische Zeitschrift **21** (1904), 1–7.

[BK86] D. S. Broomhead and G. P. King, *Extracting qualitative dynamics from experimental data*, Physica D: Nonlinear Phenomena **20** (1986), no. 2–3, 217–236.

[BKS07] A. E. Brockwell, R. E. Kass, and A. B. Schwartz, *Statistical signal processing and the motor cortex*, Proceedings of the IEEE **95** (2007), no. 5, 881–898.

[BLH$^+$01] M. Bikson, J. Lian, P. J. Hahn, W. C. Stacey, C. Sciortino, and D. M. Durand, *Suppression of epileptiform activity by high frequency sinusoidal fields in rat hippocampal slices*, J Physiol **531** (2001 Feb) no. Pt 1, 181–191.

[BLSW08] J. C. Bettencourt, K. P. Lillis, L. R. Stupin, and J. A. White, *Effects of imperfect dynamic clamp: Computational and experimental results*, J Neurosci Methods **169** (2008 Apr), no. 2, 282–289.

[BMS08] S. C. Bellinger, G. Miyazawa, and P. N. Steinmetz, *Submyelin potassium accumulation may functionally block subsets of local axons during deep brain stimulation: A modeling study*, J Neural Eng **5** (2008 Sep), no. 3, 263–274.

[BMT+02] M. D. Bevan, P. J. Magill, D. Terman, J. P. Bolam, and C. J. Wilson, *Move to the rhythm: Oscillations in the subthalamic nucleus-external globus pallidus network*, Trends Neurosci **25** (2002), no. 10, 525–531.

[BNF+01] E. N. Brown, D. P. Nguyen, L. M. Frank, M. A. Wilson, and V. Solo, *An analysis of neural receptive field plasticity by point process adaptive filtering*, Pro Nat Acad Sci USA **98** (2001), no. 21, 12261–12266.

[BOM+01] P. Brown, A. Oliviero, P. Mazzone, A. Insola, P. Tonali, and V. Di Lazzaro, *Dopamine dependency of oscillations between subthalamic nucleus and pallidum in Parkinson's disease*, J Neurosci **21** (2001), no. 3, 1033–1038.

[Bon48] K. F. Bonhoeffer, *Activation of passive iron as a model for the excitation of nerve*, J Gen Physio **32** (1948), no. 1, 69–91.

[BP07] A. Bressan and B. Piccoli, *Introduction to the mathematical theory of control*, American Institute of Mathematical Sciences, 2007.

[BP98] G. Q. Bi and M. M. Poo, *Synaptic modifications in cultured hippocampal neurons: Dependence on spike timing, synaptic strength, and postsynaptic cell type*, J Neurosci **18** (1998 Dec), no. 24, 10464–10472.

[BPSV82] R. Benzi, G. Parisi, A. Sutera, and A. Vulpiani, *Stochastic resonance in climatic change*, Tellus **34** (1982), 10–16.

[BRK04] A. E. Brockwell, A. L. Rojas, and R. E. Kass, *Recursive Bayesian decoding of motor cortical signals by particle filtering*, J Neurophysiol **91** (2004), no. 4, 1899–1907.

[BSV81] R. Benzi, A. Sutera, and A. Vulpiani, *The mechanism of statistic resonance*, J Phys A: Math Gen **14** (1981), 453–457.

[BTSS04] M. Bazhenov, I. Timofeev, M. Steriade, and T. J. Sejnowski, *Potassium model for slow (2-3 hz) in vivo neocortical paroxysmal oscillations*, J Neurophysiol **92** (2004), no. 2, 1116–1132.

[BUZ86] G. Buzsáki, *Hippocampal sharp waves: Their origin and significance*, Brain Res **398** (1986 Nov), no. 2, 242–252.

[BV73] G. B. Benedek and F. Villars, *Physics, with illustrative examples from medicine and biology*, Addison-Wesley series in physics, Addison-Wesley Pub. Co, Reading, Mass., 1973.

[BW03] W. Bao and J. Y. Wu, *Propagating wave and irregular dynamics: Spatiotemporal patterns of cholinergic theta oscillations in neocortex in vitro*, J Neurophysiol **90** (2003), no. 1, 333–341.

[BW99] M. D. Bevan and C. J. Wilson, *Mechanisms underlying spontaneous oscillation and rhythmic firing in rat subthalamic neurons*, J Neurosci **19** (1999), no. 17, 7617–7628.

[BWFB05] R. Barbieri, M. A. Wilson, L. M. Frank, and E. N. Brown, *An analysis of hippocampal spatio-temporal representations using a Bayesian algorithm for neural spike train decoding*, IEEE Trans Neural Syst Rehabil Eng **13** (2005), no. 2, 131–136.

[Cvi88] P. Cvitanović, *Invariant measurement of strange sets in terms of cycles*, Phys Rev Lett **61** (1988), no. 24, 2729–2732.

[Caj09] S. R. y Cajal, *Histologie du système nerveux de l'homme et des vértébrés*, A. Maloine, Paris, 1909.

[Caj09] ———, *Revista trimestral micrografica, tomo iv*, Moya, Madrid, 1899.

[Cas00] J. L. Casti, *Five more golden rules: Knots, codes, chaos, and other great theories of 20th century mathematics*, Wiley, New York, 2000.

[CC39] K. S. Cole and H. J. Curtis, *Electric impedance of the squid giant axon during activity*, J Gen Physiol **22** (1939), no. 5, 649–670.

[CD08] N. Caporale and Y. Dan, *Spike timing-dependent plasticity: A Hebbian learning rule*, Annu Rev Neurosci **31** (2008), 25–46.

[CFVN50] J. G. Charney, R. Fjortoft, and J. Von Neumann, *Numerical integration of the barotropic vorticity equation*, Tellus **2** (1950), no. 4, 237–254.

[CH93] M. C. Cross and P. C. Hohenberg, *Pattern formation outside of equilibrium*, Rev Mod Phys **65** (1993), no. 3, 851–1112.

[Cha51] J. G. Charney, *Dynamic forecasting by numerical process, in compendium of meteorology* (T. F. Malone, ed.), American Meteorological Society, Boston, 1951.

[CHN88] C. Y. Chan, J. Hounsgaard, and C. Nicholson, *Effects of electric fields on transmembrane potential and excitability of turtle cerebellar purkinje cells in vitro*, J Physiol **402** (1988Aug), 751–771.

[CHO+09] M. Cornick, B. Hunt, E. Ott, H. Kurtuldu, and M. F. Schatz, *State and parameter estimation of spatiotemporally chaotic systems illustrated by an application to Rayleigh-Bénard convection*, Chaos: An Interdisciplinary J Nonlinear Sci **19** (2009), 013108.

[CLM57] A. Carlsson, M. Lindqvist, and T. Magnusson, *3, 4-dihydroxyphenylalanine and 5-hydroxytryptophan as reserpine antagonists*, Nature **180** (1957), 1200.

[CMO06] N. Chub, G. Z. Mentis, and M. J. O'Donovan, *Chloride-sensitive meq fluorescence in chick embryo motoneurons following manipulations of chloride and during spontaneous network activity*, J Neurophysiol **95** (2006), no. 1, 323–330.

[CN86] C. Y. Chan and C. Nicholson, *Modulation by applied electric fields of Purkinje and stellate cell activity in the isolated turtle cerebellum*, J Physiol **371** (1986 Feb), 89–114.

[Col49] K. S. Cole, *Dynamic electrical characteristics of the squid axon membrane*, Arch Sci Physiol **3** (1949), 253–258.

[Col68] K. S. Cole, *Membranes, ions, and impulses: A chapter of classical biophysics*, Vol. 1, University of California Press, Berkeley, 1968.

[Cow10] J. Cowan. A personal account of the development of large-scale brain activity from 1945 onward. in press 2010.

[CS03] R. K. Cloues and W. A. Sather, *Afterhyperpolarization regulates firing rate in neurons of the suprachiasmatic nucleus*, J Neurosci **23** (2003), no. 5, 1593–1604.

[CSG08] N. Chernyy, S. J. Schiff, and B. J. Gluckman, *Multi-taper transfer function estimation for stimulation artifact removal from neural recordings*, Conf Proc IEEE Eng Med Biol Soc **2008** (2008), 2772–2776.

[CUZ+09] J. R. Cressman, G. Ullah, J. Ziburkus, S. J. Schiff, and E. Barreto, *The influence of sodium and potassium dynamics on excitability, seizures, and the stability of persistent states: I. single neuron dynamics*, J Comput Neurosci **26** (2009), no. 2, 159–170.

[Dal01] P. Dayan and L. F. Abbott, *Theoretical neuroscience: Computational and mathematical modeling of neural systems*, MIT Press, Cambridge, Mass., 2001.

[Dal35] H. Dale, *Pharmacology and nerve-endings (Walter Ernest Dixon Memorial Lecture):(Section of Therapeutics and Pharmacology)*, Proc Roy Soc Med **28** (1935), no. 3, 319–332.

[DBNW07] A. D. Dorval 2nd, J. Bettencourt, T. I. Netoff and J. A. White, *Hybrid neuronal network studies under dynamic clamp*, Methods Mol Biol **403** (2007), 219–231.

[DDP+06] M. Dapretto, M. S. Davies, J. H. Pfeifer, A. A. Scott, M. Sigman, S. Y. Bookheimer, and M. Iacoboni, *Understanding emotions in others: Mirror neuron dysfunction in children with autism spectrum disorders*, Nat Neurosci **9** (2006 Jan), no. 1, 28–30.

[DeL71] M. R. DeLong, *Activity of pallidal neurons during movement*, J Neurophysiol **34** (1971), no. 3, 414–427.

[Den91] D. C. Dennett, *Consciousness explained*, 1st ed, Little, Brown and Co., Boston, 1991.

[DHM09] P. Danzl, J. Hespanha, and J. Moehlis, *Event-based minimum-time control of oscillatory neuron models*, Biological Cybernetics **101** (2009), no. 5, 387–399.

[Dic02] *Oxford English dictionary*, Oxford University Press, Oxford, 2002.

[dPFF+92] G. di Pellegrino, L. Fadiga, L. Fogassi, V. Gallese and G. Rizzolatti, *Understanding motor events: A neurophysiological study*, Exp Brain Res **91** (1992), no. 1, 176–180.

[DR05] G. Deco and E. T. Rolls, *Attention, short-term memory, and action selection: A unifying theory*, Prog Neurobiol **76** (2005Jul) no. 4, 236–256.

[DRFS01] A. Destexhe, M. Rudolph, J. M. Fellous, and T. J. Sejnowski, *Fluctuating synaptic conductances recreate in vivo-like activity in neocortical neurons*, Neuroscience **107** (2001), no. 1, 13–24.

[DRS90] W. L. Ditto, S. N. Rauseo, and M. L. Spano, *Experimental control of chaos*, Phys Rev Lett **65** (1990), no. 26, 3211–3214.

[DSBK$^+$06] G. Deuschl, C. Schade-Brittinger, P. Krack, J. Volkmann, H. Schäfer, K. Bötzel, et al. *A randomized trial of deep-brain stimulation for Parkinson's disease*, N Engl J Med **355** (2006), no. 9, 896–908.

[DTW06] G. S. Duane, J. J. Tribbia and J. B. Weiss, *Synchronicity in predictive modelling: A new view of data assimilation*, Nonlinear Proces in Geophys **13** (2006), no. 6, 601–612.

[DWC09] B. Deng, J. Wang, and Y. Che, *A combined method to estimate parameters of neuron from a heavily noise-corrupted time series of active potential*, Chaos: An Interdisciplinary Journal of Nonlinear Science **19** (2009), 015105.

[DWM$^+$79] G. C. Davis, A. C. Williams, S. P. Markey, M. H. Ebert, E. D. Caine, C. M. Reichert, and I. J. Kopin, *Chronic parkinsonism secondary to intravenous injection of meperidine analogues*, Psychiatry Res **1** (1979Dec), no. 3, 249–254.

[DY06] C. M. Danforth and J. A. Yorke, *Making forecasts for chaotic physical processes*, Physical Review Letters **96** (2006), 144102.

[DYK07] G. S. Duane, D. Yu, and L. Kocarev, *Identical synchronization, with translation invariance, implies parameter estimation*, Physics Letters A **371** (2007), no. 5-6, 416–420.

[DZM88] R. A. DeCarlo, S. H. Zak and G. P. Matthews, *Variable structure control of nonlinear multivariable systems: A tutorial*, Proceed IEEE **76** (1988), no. 3, 212–232.

[EA03] C. Eliasmith and C. H. Anderson, *Neural engineering: Computation, representation, and dynamics in neurobiological systems*, MIT Press, Cambridge, Mass., 2003.

[EBE$^+$07] A. Ergün, R. Barbieri, U. T. Eden, M. A. Wilson and E. N. Brown, *Construction of point process adaptive filter algorithms for neural systems using sequential Monte Carlo methods*, IEEE Trans Biomed Eng **54** (2007), no. 3, 419–428.

[Ecc53] J. C. Eccles, *The neurophysiological basis of mind: The principles of neurophysiology*, The Waynflete lectures, vol. 1952, Clarendon Press, Oxford, 1953.

[ECL$^+$08] A. Eusebio, C. C. Chen, C. S. Lu, S. T. Lee, C. H. Tsai, P. Limousin, et al., *Effects of low-frequency stimulation of the subthalamic nucleus on movement in Parkinson's disease*, Exp Neurol **209** (2008), no. 1, 125–130.

[ET10] B. Ermentrout and D. H. Terman, *Mathematical foundations of neuroscience*, Springer, New York, 2010.

[Eve94] G. Evensen, *Sequential data assimilation with a nonlinear quasi-geostrophic model using Monte Carlo methods to forecast error statistics*, J Geophysi Res **99** (1994), 10143–10162.

[EvL00] G. Evensen and P. J. van Leeuwen, *An ensemble Kalman smoother for nonlinear dynamics*, Monthly Weather Rev **128** (2000), no. 6, 1852–1867.

[FA71] J. M. Fuster and G. E. Alexander, *Neuron activity related to short-term memory*, Science **173** (1971), no. 3997, 652–654.

[FAD$^+$11] D. R. Freestone, P. Aram, M. Dewar, K. Scerri, D. B. Grayden, and V. Kadirkamanathan, *A data-driven framework for neural field modeling*, Neuroimage **56** (2011), no. 3, 1043–1058.

[FAK05] M. Feychting, A. Ahlbom and L. Kheifets, *Emf and health*, Annu Rev Public Health **26** (2005), 165–89.

[Far08] M. J. Farah, *Neuroethics and the problem of other minds: Implications of neuroscience for the moral status of brain-damaged patients and nonhuman animals*, Neuroethics **1** (2008), no. 1, 9–18.

[FBG98] I. A. Fleidervish, A. M. Binshtok and M. J. Gutnick, *Functionally distinct NMDA receptors mediate horizontal connectivity within layer 4 of mouse barrel cortex*, Neuron **21** (1998), 1055–1065.

[FBGR89] S. Funahashi, C. J. Bruce and P. S. Goldman-Rakic, *Mnemonic coding of visual space in the monkey's dorsolateral prefrontal cortex*, J Neurophysiol **61** (1989Feb), no. 2, 331–249.

[FC96] B. R. Frieden and W. J. Cocke, *Foundation for Fisher-information-based derivations of physical laws*, Phys Rev E **54** (1996), no. 1, 257–260.

[Fey65] R. P. Feynman, *The character of physical law*, The Messenger lectures, 1964, MIT Press, Cambridge, Mass., 1965.

[FGR$^+$07] X.-J. Feng, B. Greenwald, H. Rabitz, E. Shea-Brown, and R. Kosut, *Toward closed-loop optimization of deep brain stimulation for Parkinson's disease: Concepts and lessons from a computational model*, J Neural Eng **4** (2007b), no. 2, L14–21.

[FGS03] J. T. Francis, B. J. Gluckman, and S. J. Schiff, *Sensitivity of neurons to weak electric fields*, J Neurosci **23** (2003), no. 19, 7255–7261.

[FH56] B. Frankenhaeuser and A. L. Hodgkin, *The after-effects of impulses in the giant nerve fibres of loligo*, J Physiol **131** (1956), no. 2, 341–376.

[Fis34] R. A. Fisher, *Probability likelihood and quantity of information in the logic of uncertain inference*, Proceed R Soc Lond A, Containing Papers of a Mathematical and Physical Character (1934), 1–8.

[Fis98] M. E. Fisher, *Renormalization group theory: Its basis and formulation in statistical physics*, Rev Mod Phys **70** (1998), no. 2, 653–681.

[Fit60] R. Fitzhugh, *Thresholds and plateaus in the Hodgkin-Huxley nerve equations*, J Gen Physiol **43** (1960), no. 5, 867–896.

[Fit61] ———, *Impulses and physiological states in theoretical models of nerve membrane*, Biophysical J **1** (1961), no. 6, 445–466.

[Fit65] R. Fitzhugh, *A kinetic model of the conductance changes in nerve membrane*, J Cell Comp Physiol **66** (1965), 111–118.

[FJ00] G. T. Finnerty and J. G. Jefferys, *9-16 hz oscillation precedes secondary generalization of seizures in the rat tetanus toxin model of epilepsy*, J Neurophysiol **83** (2000 Apr), no. 4, 2217–2226.

[Flu97] B. Flury, *A first course in multivariate statistics*, Springer, New York, 1997.

[For90] D. Forster, *Hydrodynamic fluctuations, broken symmetry, and correlation functions*, Addison-Wesley, Advanced Book Program, Redwood City, Calif., 1990.

[FPP76] R. S. Fisher, T. A. Pedley, and D. A. Prince, *Kinetics of potassium movement in normal cortex*, Brain Res **101** (1976Jan), no. 2, 223–237.

[Fri24] H. Fricke, *A mathematical treatment of the electric conductivity and capacity of disperse systems I. The electric conductivity of a suspension of homogeneous spheroids*, Phys Rev **24** (1924), no. 5, 575–587.

[Fri25] H. Fricke, *The electric capacity of suspensions of red corpuscles of a dog*, Phys Rev **26** (1925 Nov), no. 5, 682–687.

[FSBG+07] X. J. Feng, E. Shea-Brown, B. Greenwald, R. Kosut, and H. Rabitz, *Optimal deep brain stimulation of the subthalamic nucleus–a computational study*, J Comput Neurosci **23** (2007a), no. 3, 265–282.

[FSGS01] J. T. Francis, P. So, B. J. Gluckman, and S. J. Schiff, *Differentiability implies continuity in neuronal dynamics*, Physica D: Nonlinear Phenomena **148** (2001), no. 1–2, 175–181.

[FTIKT04] Y. Fujiwara-Tsukamoto, Y. Isomura, K. Kaneda, and M. Takada, *Synaptic interactions between pyramidal cells and interneurone subtypes during seizure-like activity in the rat hippocampus*, J Physiol **557** (2004 Jun), no. Pt 3, 961–979.

[FTSB08] F. Frohlich, I. Timofeev, T. J. Sejnowski, and M. Bazhenov, *Extracellular potassium dynamics and epileptogenesis. in: Computational neuroscience in epilepsy* (Ivan Soltesz and Kevin Staley, eds.), Academic Press, New York, 2008.

[Fus95] J. M. Fuster, *Memory in the cerebral cortex: An empirical approach to neural networks in the human and nonhuman primate*, MIT Press, Cambridge, Mass., 1995.

[FW02] S. A. Factor and W. J. Weiner, *Parkinson's disease: Diagnosis and clinical management*, Demos, New York, 2002.

[Gai00] P. C. Gailey, Electrical signal detection and noise in systems with long-range coherence, in *Self-organized biological dynamics and nonlinear control.* (J. Wallenczek, ed.), Cambridge University Press, Cambridge, UK, 2000.

[Gal91] L. Galvani, *De viribus electricitatis in motu musculari: Commentarius*, Tip. Istituto delle Scienze, Bologna, 1791.

[Gau09] C. F. Gauss, *Theoria motvs corporvm coelestivm in sectionibvs conicis solem ambientivm*, Svmtibvs F. Perthes et I. H. Besser, Hambvrgi, 1809.

[GBC99] J. R. Gibson, M. Beierlein, and B. W. Connors, *Two networks of electrically coupled inhibitory neurons in neocortex*, Nature **402** (1999 Nov), no. 6757, 75–79.

[GBD00] R. S. Ghai, M. Bikson, and D. M. Durand, *Effects of applied electric fields on low-calcium epileptiform activity in the CA1 region of rat hippocampal slices*, J Neurophysiol **84** (2000Jul), no. 1, 274–280.

[GD57] C. F. Gauss and C. H. Davis, *Theory of the motion of the heavenly bodies moving about the sun in conic sections: A translation of Gauss's "theoria motus." with an appendix*, Little, Brown & Co., Boston, 1857.

[GDR+05] T. Gloveli, T. Dugladze, H. G. Rotstein, R. D. Traub, H. Monyer, U. Heinemann, M. A. Whittington, and N. J. Kopell, *Orthogonal arrangement of rhythm-generating microcircuits in the hippocampus*, Proc Natl Acad Sci USA **102** (2005), no. 37, 13295–13300.

[Gel74] A. Gelb, *Applied optimal estimation*, MIT Press, Cambridge, Mass., 1974.

[GFK+94] J. P. Gossard, M. K. Floeter, Y. Kawai, R. E. Burke, T. Chang, and S. J. Schiff, *Fluctuations of excitability in the monosynaptic reflex pathway to lumbar motoneurons in the cat*, J Neurophysiol **72** (1994), no. 3, 1227–1239.

[GK02] W. Gerstner and W. M. Kistler, *Spiking neuron models: Single neurons, populations, plasticity*, Cambridge University Press, Cambridge, UK, 2002.

[GKCM82] A. P. Georgopoulos, J. F. Kalaska, R. Caminiti, and J. T. Massey, *On the relations between the direction of two-dimensional arm movements and cell discharge in primate motor cortex*, J Neurosci **2** (1982), no. 11, 1527–1537.

[GLC+01] B. S. Gutkin, C. R. Laing, C. L. Colby, C. C. Chow, and G. B. Ermentrout, *Turning on and off with excitation: The role of spike-timing asynchrony and synchrony in sustained neural activity*, J Comput Neurosci **11** (Oct 2001 Sep), no. 2, 121–134.

[GM06] J.-M. Goaillard and E. Marder, *Dynamic clamp analyses of cardiac, endocrine, and neural function*, Physiology (Bethesda) **21** (2006 Jun), 197–207.

[GMN83] A. R. Gardner-Medwin and C. Nicholson, *Changes of extracellular potassium activity induced by electric current through brain tissue in the rat*, J Physiol **335** (1983 Feb), 375–392.

[GNN+96a] B. J. Gluckman, E. J. Neel, T. I. Netoff, W. L. Ditto, M. L. Spano, and S. J. Schiff, *Electric field suppression of epileptiform activity in hippocampal slices*, J Neurophysiol **76** (1996aDec), no. 6, 4202–4205.

[GNN+96b] B. J. Gluckman, T. I. Netoff, E. J. Neel, W. L. Ditto, M. L. Spano, and S. J. Schiff, *Stochastic resonance in a neuronal network from mammalian brain*, Phys Rev Lett **77** (1996b Nov), no. 19, 4098–4101.

[GNWS01] B. J. Gluckman, H. Nguyen, S. L. Weinstein, and S. J. Schiff, *Adaptive electric field control of epileptic seizures*, J Neurosci **21** (2001), no. 2, 590–600.

[Gol43] D. E. Goldman, *Potential, impedance, and rectification in membranes*, J Gen Physiol **27** (1943), no. 1, 37–60.

[Goo00] D. Goodstein, *In defense of Robert Andrews Millikan*, Engineering and Sci **4** (2000), 30–38.

[Gou81] S. J. Gould, *The mismeasure of man*, 1st ed, Norton, New York, 1981.

[Gow01] W. R. Gowers, *A manual of diseases of the nervous system v. 2, 1901*, 2nd ed., Vol. II, P. Blakiston's Sons & Co., Philadelphia, 1901.

[GR03] L. A. Geddes and R. A. Roeder, *De forest and the first electrosurgical unit*, IEEE Engine Med Biol Mag **22** (2003), no. 1, 84–87.

[GR95] P. S. Goldman-Rakic, *Cellular basis of working memory*, Neuron **14** (1995 Mar) no. 3, 477–485.

[GRM+08] Y. Guo, J. E. Rubin, C. C. McIntyre, J. L. Vitek, and D. Terman, *Thalamocortical relay fidelity varies across subthalamic nucleus deep brain stimulation protocols in a data-driven computational model*, J Neurophysiol **99** (2008), no. 3, 1477–1492.

[GS02] R. W. Guillery and S. M. Sherman, *The thalamus as a monitor of motor outputs*, Philos Trans R Soc Lond B Biol Sci **357** (2002), no. 1428, 1809–1821.

[GSAV+09] J. C. Gutiérrez, F. J. Seijo, M. A. Alvarez Vega, F. Fernández González, B. Lozano Aragoneses, and M. Blázquez, *Therapeutic extradural cortical stimulation for Parkinson's disease: Report of six cases and review of the literature*, Clin Neurol Neurosurg **111** (2009), no. 8, 703–707.

[GSDW92] A. Garfinkel, M. L. Spano, W. L. Ditto, and J. N. Weiss, *Controlling cardiac chaos*, Science **257** (1992), no. 5074, 1230.

[GSK86] A. P. Georgopoulos, A. B. Schwartz, and R. E. Kettner, *Neuronal population coding of movement direction*, Science **233** (1986), no. 4771, 1416–1419.

[GSL99] S. V. Girman, Y. Sauvé, and R. D. Lund, *Receptive field properties of single neurons in rat primary visual cortex*, J Neurophysiol **82** (1999), no. 1, 301–311.

[GWM00] A. Gupta, Y. Wang, and H. Markram, *Organizing principles for a diversity of gabaergic interneurons and synapses in the neocortex*, Science **287** (2000 Jan), no. 5451, 273–278.

[HA76] J. E. Hoke and R. A. Anthes, *The initialization of numerical models by a dynamic-initialization technique*, Monthly Weather Rev **104** (1976), no. 12, 1551–1556.

[Han82] A. Hannay, *Kierkegaard*, Routledge and K. Paul, London, 1982.

[Har96] P. Harder, *Functional semantics: A theory of meaning, structure, and tense in English*, Vol. 87, Mouton de Gruyter, Berlin, 1996.

[HB05] N. E. Hallworth and M. D. Bevan, *Globus pallidus neurons dynamically regulate the activity pattern of subthalamic nucleus neurons through the frequency-dependent activation of postsynaptic gabaa and gabab receptors*, J Neurosci **25** (2005), no. 27, 6304–6315.

[HDHM06] B. Haider, A. Duque, A. R. Hasenstaub and D. A. McCormick, *Neocortical network activity in vivo is generated through a dynamic balance of excitation and inhibition*, J Neurosci **26** (2006), no. 17, 4535–4545.

[Heb49] D. O. Hebb, *The organization of behavior: A neuropsychological theory*, Wiley, New York, 1949.

[HEO+03] T. Hashimoto, C. M. Elder, M. S. Okun, S. K. Patrick, and J. L. Vitek, *Stimulation of the subthalamic nucleus changes the firing pattern of pallidal neurons*, J Neurosci **23** (2003), no. 5, 1916–1923.

[HGJ+00] U. Heinemann, S. Gabriel, R. Jauch, K. Schulze, A. Kivi, A. Eilers, et al., *Alterations of glial cell function in temporal lobe epilepsy*, Epilepsia **41 Suppl 6** (2000), S185–189.

[HH39] A. L. Hodgkin and A. F. Huxley, *Action potentials recorded from inside a nerve fibre*, Nature **144** (1939), 710–711.

[HH52a] ———, *Currents carried by sodium and potassium ions through the membrane of the giant axon of loligo*, J Physiol **116** (1952a), no. 4, 449–472.

[HH52b] ———, *The components of membrane conductance in the giant axon of loligo*, J Physiol **116** (1952b), no. 4, 473–496.

[HH52c] ———, *The dual effect of membrane potential on sodium conductance in the giant axon of loligo*, J Physiol **116** (1952c), no. 4, 497–506.

[HH52d] ———, *A quantitative description of membrane current and its application to conduction and excitation in nerve*, J Physiol **117** (1952d), no. 4, 500–544.

[HHK52] A. L. Hodgkin, A. F. Huxley, and B. Katz, *Measurement of current-voltage relations in the membrane of the giant axon of loligo*, J Physiol **116** (1952), no. 4, 424–448.

[HK49] A. L. Hodgkin and B. Katz, *The effect of sodium ions on the electrical activity of the giant axon of the squid*, J Physiol **108** (1949), no. 1, 37.

[HKS07] B. R. Hunt, E. J. Kostelich, and I. Szunyogh, *Efficient data assimilation for spatiotemporal chaos: A local ensemble transform Kalman filter*, Physica D: Nonlinear Phenomena **230** (2007), nos. 1–2, 112–126.

[HLB96] P. Holmes, J. L. Lumley, and G. Berkooz, *Turbulence, coherent structures, dynamical systems, and symmetry*, Cambridge University Press, Cambridge, 1996.

[HLG77] U. Heinemann, H. D. Lux, and M. J. Gutnick, *Extracellular free calcium and potassium during paroxsmal activity in the cerebral cortex of the cat*, Exp Brain Res **27** (1977 Mar) no. 3–4, 237–243.

[HM10] S. Hawking and L. Mlodinow, *The Grand Design*, Bantam Books, New York, 2010.

[HP05] C. Q. Howe and D. Purves, *Perceiving geometry: Geometrical illusions explained by natural scene statistics*, Springer, New York, 2005.

[HP09] Q. J. M. Huys and L. Paninski, *Smoothing of, and parameter estimation from, noisy biophysical recordings*, PLoS Comput Biol **5** (2009), no. 5, e1000379.

[HSF+06] L. R. Hochberg, M. D. Serruya, G. M. Friehs, J. A. Mukand, M. Saleh, A. H. Caplan, et al., *Neuronal ensemble control of prosthetic devices by a human with tetraplegia*, Nature **442** (2006), no. 7099, 164–171.

[HSG06] M. D. Humphries, R. D. Stewart, and K. N. Gurney, *A physiologically plausible model of action selection and oscillatory activity in the basal ganglia*, J Neurosci **26** (2006), 12921–12942.

[HSH+00] S. Hinterkeuser, W. Schröder, G. Hager, G. Seifert, I. Blümcke, C. E. Elger, et al., *Astrocytes in the hippocampus of patients with temporal lobe epilepsy display changes in potassium conductances*, Eur J Neurosci **12** (2000Jun), no. 6, 2087–2096.

[HSW$^+$91] L. J. Hirsch, S. S. Spencer, P. D. Williamson, D. D. Spencer, and R. H. Mattson, *Comparison of bitemporal and unitemporal epilepsy defined by depth electroencephalography*, Ann Neurol **30** (1991 Sep), no. 3, 340–346.

[HTY$^+$04] X. Huang, W. C. Troy, Q. Yang, H. Ma, C. R. Laing, S. J. Schiff, and J. Y. Wu, *Spiral waves in disinhibited mammalian neocortex*, J Neurosci **24** (2004), no. 44, 9897–9902.

[Hua06] D. Huang, *Adaptive-feedback control algorithm*, Physical Review E **73** (2006), no. 6, 66204.

[Hum07] D. Hume, *An enquiry concerning human understanding*, Open Court Publishing Co., Chicago, 1907.

[HWDL02] M. Hodaie, R. A. Wennberg, J. O. Dostrovsky, and A. M. Lozano, *Chronic anterior thalamus stimulation for intractable epilepsy*, Epilepsia **43** (2002), no. 6, 603–608.

[Iac09] M. Iacoboni, *Imitation, empathy, and mirror neurons*, Annu Rev Psychol **60** (2009), 653–670.

[ID06] M. Iacoboni and M. Dapretto, *The mirror neuron system and the consequences of its dysfunction*, Nat Rev Neurosci **7** (2006 Dec) no. 12, 942–951.

[Izh07] E. M. Izhikevich, *Dynamical systems in neuroscience: The geometry of excitability and bursting*, MIT Press, Cambridge, Mass., 2007.

[Jam90] W. James, *The principles of psychology*, H. Holt and Co., New York, 1890.

[JDBF03] J. G. R. Jefferys, J. Deans, M. Bikson, and J. Fox, *Effects of weak electric fields on the activity of neurons and neuronal networks*, Radiat Prot Dosimetry **106** (2003), no. 4, 321–323.

[Jef81] J. G. Jefferys, *Influence of electric fields on the excitability of granule cells in guinea-pig hippocampal slices*, J Physiol **319** (1981), 143–152.

[JH82] J. G. Jefferys and H. L. Haas, *Synchronized bursting of ca1 hippocampal pyramidal cells in the absence of synaptic transmission*, Nature **300** (1982 Dec) no. 5891, 448–450.

[JMK95] P. Jung and G. Mayer-Kress, *Spatiotemporal stochastic resonance in excitable media*, Phys Rev Lett **74** (1995), no. 11, 2130–2133.

[JP55a] C. G. Jung and W. Pauli, *Synchronicity: An acausal connecting principle, in the interpretation of nature and the psyche:*, Vol. 51, Pantheon Books, New York, 1955a.

[JP55b] ———, *The influence of archetypal ideas on the scientific theories of Kepler, in the interpretation of nature and the psyche:*, Vol. 51, Pantheon Books, New York, 1955b.

[JU97a] S. J. Julier and J. K. Uhlmann, *A new extension of the Kalman filter to nonlinear systems*, SPIE **3068** (1997a), 182–193.

[JU97b] ———, *A consistent, debiased method for converting between polar and Cartesian coordinate systems*, SPIE **3086** (1997b), 110–121.

[JUDW00] S. Julier, J. Uhlmann, and H. F. Durrant-Whyte, *A new method for the nonlinear transformation of means and covariances in filters and estimators*, Automatic Control, IEEE Trans **45** (2000), no. 3, 477–482.

[Kac66] M. Kac, *Can one hear the shape of a drum?*, Amer Math Monthly **73** (1966), no. 4, 1–23.

[Kal60] R. E. Kalman, *A new approach to linear filtering and prediction problems*, Trans ASME–J Basic Engine **82, Series D** (1960), 35–45.

[Kal03] E. Kalnay, *Atmospheric modeling, data assimilation, and predictability*, Cambridge University Press, New York, 2003.

[KBC99] D. Kernell, R. Bakels, and J. C. Copray, *Discharge properties of motoneurones: How are they matched to the properties and use of their muscle units?*, J Physiol Paris **93** (1999), no. 1–2, 87–96.

[KBM95] U. Kim, T. Bal, and D. A. McCormick, *Spindle waves are propagating synchronized oscillations in the ferret lgnd in vitro*, J Neurophysiol **74** (1995), no. 3, 1301–1323.

[KBVB$^+$03] P. Krack, A. Batir, N. Van Blercom, S. Chabardes, V. Fraix, C. Ardouin, et al., *Five-year follow-up of bilateral stimulation of the subthalamic nucleus in advanced Parkinson's disease*, N Engl J Med **349** (2003), no. 20, 1925–1934.

[KF70] A. N. Kolmogorov and S. V. Fomin, *Introductory real analysis*, Rev. English ed., Dover Publications, New York, 1970.

[KFF07] J. M. Kilner, K. J. Friston, and C. D. Frith, *Predictive coding: An account of the mirror neuron system*, Cogn Process **8** (2007), no. 3, 159–166.

[KHY84] A. Konnerth, U. Heinemann, and Y. Yaari, *Slow transmission of neural activity in hippocampal area ca1 in absence of active chemical synapses*, Nature **307** (1984), no. 5946, 69–71.

[Kie69] S. A. Kierkegaard, *Af Søren Kierkegaards efterladte papirer, 1833–1843*, C. A. Reitzel, Kjøbenhavn, 1869.

[Kle05] F. C. Klebaner, *Introduction to stochastic calculus with applications*, 2nd ed., Imperial College Press, London and Hackensack, N.J., 2005.

[KP08] J. E. Kulkarni and L. Paninski, *State-space decoding of goal-directed movements*, IEEE Signal Processing Magazine **25** (2008), no. 1, 78–86.

[KPLA99] W. C. Koller, R. Pahwa, K. E. Lyons, and A. Albanese, *Surgical treatment of Parkinson's disease*, J Neurol Sci **167** (1999), no. 1, 1–10.

[Kow21] J. Kowarschki, *Die diathermie*, Springer Verlag, Berlin, 1921.

[KS40] B. Katz and O. H. Schmitt, *Electric interaction between two adjacent nerve fibres*, J Physiol **97** (1940Feb), no. 4, 471–488.

[KS90] M. Kirby and L. Sirovich, *Application of the Karhunen-Loeve procedure for the characterization of human faces*, IEEE Trans Pattern Anal Machine Intell **12** (1990), no. 1, 103–108.

[KS98] C. Koch and I. Segev, *Methods in neuronal modeling: From ions to networks*, 2nd ed, MIT Press, Cambridge, Mass., 1998.

[KSK00] A. Krawiecki, A. Sukiennicki, and R. A. Kosiński, *Stochastic resonance and noise-enhanced order with spatiotemporal periodic signal*, Phys Rev E **62** (2000), no. 6, 7683–7689.

[KJOA07] M. L. Kringelbach, N. Jenkinson, S. L. F. Owen, and T. Z. Aziz, *Translational principles of deep brain stimulation*, Nat Rev Neurosci **8** (2007), no. 8, 623–635.

[KWBT71] W. Kuschinsky, M. Wahl, O. Bosse, and K. Thurau, *The dependency of the pial arterial and arteriolar resistance on the perivascular H + and K + concentrations. A micropuncture study*, Eur Neurol **6** (1971), no. 1, 92–95.

[KWS00] H. Kager, W. J. Wadman, and G. G. Somjen, *Simulated seizures and spreading depression in a neuron model incorporating interstitial space and ion concentrations*, J Neurophysiol **84** (2000 Jul), no. 1, 495–512.

[KWS02] ———, *Conditions for the triggering of spreading depression studied with computer simulations*, J Neurophysiol **88** (2002 Nov), no. 5, 2700–2712.

[KWS07] ———, *Seizure-like afterdischarges simulated in a model neuron*, J Comput Neurosci **22** (2007 Apr), no. 2, 105–128.

[KWS+09] R. Kirov, C. Weiss, H. R. Siebner, J. Born, and L. Marshall, *Slow oscillation electrical brain stimulation during waking promotes EEG theta activity and memory encoding*, Proc Natl Acad Sci USA **106** (2009 Sep), no. 36, 15460–15465.

[KXX+05] N. Kang, J. Xu, Q. Xu, M. Nedergaard, and J. Kang, *Astrocytic glutamate release-induced transient depolarization and epileptiform discharges in hippocampal ca1 pyramidal neurons*, J Neurophysiol **94** (2005 Dec), no. 6, 4121–4130.

[LA09] C. Letellier and L. A. Aguirre, *Symbolic observability coefficients for univariate and multivariate analysis*, Physical Review E **79** (2009), no. 6, 066210.

[LA98] E. N. Lorenz and K. A. Emanuel, *Optimal sites for supplementary optimal sites for supplementary weather observations: Simulation with a small model.*, J Atmos Sci **55** (1998), 399–414.

[Lai05] C. R. Laing, *Spiral waves in nonlocal equations*, SIAM J Appl Dynamical Sys **4** (2005), no. 3, 588–606.

[LBTI83] J. W. Langston, P. Ballard, J. W. Tetrud, and I. Irwin, *Chronic parkinsonism in humans due to a product of meperidine-analog synthesis*, Science **219** (1983), no. 4587, 979–980.

[Lau91] P. Lauger, *Electrogenic ion pumps*, Distinguished lecture series of the Society of General Physiologists. 5 (1991).

[Len03] P. Lennie, *The cost of cortical computation*, Curr Biol **13** (2003), no. 6, 493–497.

[Lew75] G. H. Lewes, *Problems of life and mind, 2 vols*, Kegan Paul, Trench, Turbner, and Co., London, 1875.

[LGT09] A. M. Lozano, P. L. Gildenberg, and R. R. Tasker (eds.), *Textbook of stereotactic and functional neurosurgery*, 2nd ed, Springer, Berlin, 2009.

[LHJ95] X. B. Liu, C. N. Honda, and E. G. Jones, *Distribution of four types of synapse on physiologically identified relay neurons in the ventral posterior thalamic nucleus of the cat*, J Comp Neurol **352** (1995), no. 1, 69–91.

[LHR09] A. J. Lees, J. Hardy, and T. Revesz, *Parkinson's disease*, Lancet **373** (2009), no. 9680, 2055–2066.

[LJ06] H. C. Lai and L. Y. Jan, *The distribution and targeting of neuronal voltage-gated ion channels*, Nature Rev Neurosci **7** (2006), no. 7, 548–562.

[LKC⁺02] X. Leinekugel, R. Khazipov, R. Cannon, H. Hirase, Y. Ben-Ari, and G. Buzsáki, *Correlated bursts of activity in the neonatal hippocampus in vivo*, Science **296** (2002 Jun), no. 5575, 2049–2052.

[LL05] K. Lehnertz and B. Litt, *The first international collaborative workshop on seizure prediction: Summary and data description*, Clin Neurophysiol **116** (2005 Mar), no. 3, 493–505.

[LL25] G. W. Leibniz and R. Latta, *The monadology and other philosophical writings*, Oxford University Press, H. Milford, London, 1925.

[Llo43] D. P. C. Lloyd, *Reflex action in relation to pattern and peripheral source of afferent stimulation*, J Neurophysiol **6** (1943), no. 2, 111–119.

[LMD⁺95] J. F. Lindner, B. K. Meadows, W. L. Ditto, M. E. Inchiosa, and A. R. Bulsara, *Array enhanced stochastic resonance and spatiotemporal synchronization*, Phys Rev Lett **75** (1995), no. 1, 3–6.

[LMMP59] J. Y. Lettvin, H. R. Maturana, W. S. McCulloch, and W. H. Pitts, *What the frog's eye tells the frog's brain*, Proc Ire **47** (1959), no. 11, 1940–1959.

[LOH⁺09] Z. Li, J. E. O'Doherty, T. L. Hanson, M. A. Lebedev, C. S. Henriquez, and M. A. L. Nicolelis, *Unscented Kalman filter for brain-machine interfaces*, PLoS One **4** (2009), no. 7, e6243.

[Lor63] E. N. Lorenz, *Deterministic nonperiodic flow*, J Atmospheric Sci **20** (1963), 130–141.

[LP95] J. W. Langston and J. Palfreman, *The case of the frozen addicts*, 1st ed, Pantheon Books, New York, 1995.

[LS09] E. V. Lubenov and A. G. Siapas, *Hippocampal theta oscillations are travelling waves*, Nature **459** (2009), no. 7246, 534–539.

[LT03] C. R. Laing and W. C. Troy, *Two-bump solutions of Amari-type models of neuronal pattern formation*, Physica D: Nonlinear Phenomena **178** (2003), no. 3-4, 190–218.

[LTGE02] C. R. Laing, W. C. Troy, B. Gutkin, and G. B. Ermentrout, *Multiple bumps in a neuronal model of working memory*, SIAM J Appl Math (2002), 62–97.

[LY75] T. Y. Li and J. A. Yorke, *Period three implies chaos*, Ameri Math Monthly **82** (1975), no. 10, 985–992.

[Man82] B. B. Mandelbrot, *The fractal geometry of nature*, W. H. Freeman, San Francisco, 1982.

[Mar06] H. Markram, *The blue brain project*, Nat Rev Neurosci **7** (2006), no. 2, 153–160.

[Mar49] G. Marmont, *Studies on the axon membrane*, J Cell Comp Physiol **34** (1949), 351–382.

[May82] P. S. Maybeck, *Stochastic models, estimation and control*, Vol. 141, Academic Press, New York, 1982.

[MB08] P. Mitra and H. Bokil, *Observed brain dynamics*, Oxford University Press, Oxford, 2008.

[MB97] D. A. McCormick and T. Bal, *Sleep and arousal: Thalamocortical mechanisms*, Annu Rev Neurosci **20** (1997), 185–215.

[MBJ05] D. R. Merrill, M. Bikson, and J. G. R. Jefferys, *Electrical stimulation of excitable tissue: Design of efficacious and safe protocols*, J Neurosci Methods **141** (2005 Feb), no. 2, 171–198.

[MBRW03] P. Miller, C. D. Brody, R. Romo, and X. J. Wang, *A recurrent network model of somatosensory parametric working memory in the prefrontal cortex*, Cereb Cortex **13** (2003 Nov), no. 11, 1208–1218.

[MC93] J. G. Milton and J. D. Cowan, *Spiral waves in integrate-and-fire neural networks*, Adv Neu Infor Proc Sys (1993), 1001–1001.

[McB94] C. J. McBain, *Hippocampal inhibitory neuron activity in the elevated potassium model of epilepsy*, J Neurophysiol **72** (1994 Dec), no. 6, 2853–2863.

[McC45] W. S. McCulloch, *A heterarchy of values determined by the topology of nervous nets*, Bull Math Biol **7** (1945), no. 2, 89–93.

[McG71] J. D. McGervey, *Introduction to modern physics*, Academic Press, New York, 1971.

[MCNR01] R. D. Meyer, S. F. Cogan, T. H. Nguyen, and R. D. Rauh, *Electrodeposited iridium oxide for neural stimulation and recording electrodes*, IEEE Trans Neural Syst Rehabil Eng **9** (2001 Mar), no. 1, 2–11.

[MDD+08] C. Moreau, L. Defebvre, A. Destée, S. Bleuse, F. Clement, J. L. Blatt, et al., *STN-DBS frequency effects on freezing of gait in advanced Parkinson disease*, Neurology **71** (2008), no. 2, 80–84.

[MED96] E. K. Miller, C. A. Erickson, and R. Desimone, *Neural mechanisms of visual working memory in prefrontal cortex of the macaque*, J Neurosci **16** (1996 Aug), no. 16, 5154–5167.

[MEW82] J. McCulloch, L. Edvinsson, and P. Watt, *Comparison of the effects of potassium and pH on the calibre of cerebral veins and arteries*, Pflugers Arch **393** (1982 Mar), no. 1, 95–98.

[MFL89] V. Montana, D. L. Farkas, and L. M. Loew, *Dual-wavelength ratiometric fluorescence measurements of membrane potential*, Biochemistry **28** (1989 May), no. 11, 4536–4539.

[MFP74] W. J. Moody, K. J. Futamachi, and D. A. Prince, *Extracellular potassium activity during epileptogenesis*, Exp Neurol **42** (1974 Feb) no. 2, 248–263.

[MG77] M. C. MacKey and L. Glass, *Oscillation and chaos in physiological control systems*, Science **197** (1977), no. 4300, 287–289.

[MGST04] C. C. McIntyre, W. M. Grill, D. L. Sherman, and N. V. Thakor, *Cellular effects of deep brain stimulation: Model-based analysis of activation and inhibition*, J Neurophysiol **91** (2004), no. 4, 1457.

[MHMB06] L. Marshall, H. Helgadóttir, M. Mölle and J. Born, *Boosting slow oscillations during sleep potentiates memory*, Nature **444** (2006), no. 7119, 610–613.

[MKR+05] F. Mormann, T. Kreuz, C. Rieke, R. G. Andrzejak, A. Kraskov, P. David, et al., *On the predictability of epileptic seizures*, Clin Neurophysiol **116** (2005 Mar), no. 3, 569–587.

[MM08] E. O. Mann and I. Mody, *The multifaceted role of inhibition in epilepsy: Seizure-genesis through excessive GABAergic inhibition in autosomal dominant nocturnal frontal lobe epilepsy*, Curr Opin Neurol **21** (2008), no. 2, 155–160.

[MMJ00] M. Magnin, A. Morel, and D. Jeanmonod, *Single-unit analysis of the pallidum, thalamus and subthalamic nucleus in parkinsonian patients*, Neuroscience **96** (2000), no. 3, 549–564.

[Moh91a] R. R. Mohler, *Nonlinear systems: Dynamics and control*, Vol. 1, Prentice Hall, Englewood Cliffs, N.J., 1991.

[Moh91b] R. R. Mohler, *Nonlinear systems: Applications to bilinear control*, Vol. 2, Prentice-Hall, Englewood Cliffs, N.J., 1991.

[Moo59] J. W. Moore, *Excitation of the squid axon membrane in isosmotic potassium chloride*, Nature **183** (1959), no. 4656, 265–266.

[MP43] W. S. McCulloch and W. Pitts, *A logical calculus of the ideas immanent in nervous activity*, Bull Math Biol **5** (1943), no. 4, 115–133.

[MP99] P. P. Mitra and B. Pesaran, *Analysis of dynamic brain imaging data*, Biophys J **76** (1999), no. 2, 691–708.

[MPIQ09] J. Martinez, C. Pedreira, M. J. Ison, and R. Q. Quiroga, *Realistic simulations of extracellular recordings*, J Neurosci Methods **184** (2009), 285–293.

[MR74] J. W. Moore and F. Ramon, *On numerical integration of the Hodgkin and Huxley equations for a membrane action potential.*, J Theor Biol **45** (1974), no. 1, 249–273.

[MS04] P. E. McSharry and L. A. Smith, *Consistent nonlinear dynamics: Identifying model inadequacy*, Physica D: Nonlinear Phenomena **192** (2004), 1–22.

[MS95] Z. F. Mainen and T. J. Sejnowski, *Reliability of spike timing in neocortical neurons*, Science **268** (1995), no. 5216, 1503–1506.

[MS99a] P. E. McSharry and L. A. Smith, *Better nonlinear models from noisy data: Attractors with maximum likelihood*, Phys Rev Lett **83** (1999), 4285–4288.

[MS99b] D. W. Moran and A. B. Schwartz, *Motor cortical representation of speed and direction during reaching*, J Neurophysiol **82** (1999), no. 5, 2676–2692.

[MSF+10] K. J. Miller, G. Schalk, E. E. Fetz, M. den Nijs, J. G. Ojemann, and R. P. N. Rao, *Cortical activity during motor execution, motor imagery, and imagery-based online feedback*, Proc Natl Acad Sci USA **107** (2010 Mar), no. 9, 4430–4435.

[MSS82] J. W. Moore, N. Stockbridge, and S. J. Schiff, *On the squid axon membrane's response to sequential voltage and current clamps*, Biophysi J **40** (1982), no. 3, 259–262.

[MSY07] D. A. McCormick, Y. Shu, and Y. Yu, *Hodgkin and Huxley model—still standing?*, Nature **445** (2007), E1–E2.

[MTD90] C. J. McBain, S. F. Traynelis, and R. Dingledine, *Regional variation of extracellular space in the hippocampus*, Science **249** (1990 Aug), no. 4969, 674–677.

[MTRW⁺04] H. Markram, M. Toledo-Rodriguez, Y. Wang, A. Gupta, G. Silberberg, and C. Wu, *Interneurons of the neocortical inhibitory system*, Nat Rev Neurosci **5** (2004Oct), no. 10, 793–807.

[Mur03] J. D. Murray, *Mathematical biology: Spatial models and biomedical applications*, Springer Verlag, Berlin, 2003.

[Mur98] K. P. Murphy, *Switching Kalman filters*, University of California, Berkeley, 1998.

[MVC⁺10] M. M. Monti, A. Vanhaudenhuyse, M. R. Coleman, M. Boly, J. D. Pickard, L. Tshibanda, et al., *Willful modulation of brain activity in disorders of consciousness*, N Engl J Med **362** (2010 Feb), no. 7, 579–589.

[MW84] R. Miles and R. K. Wong, *Unitary inhibitory synaptic potentials in the guinea-pig hippocampus in vitro*, J Physiol **356** (1984), no. 1, 97–113.

[MW86] ———, *Excitatory synaptic interactions between CA3 neurones in the guinea-pig hippocampus*, J Physiol **373** (1986), no. 1, 397–418.

[MWC65] J. Monod, J. Wyman, and J. P. Changeux, *On the nature of allosteric transitions: A plausible model*, J Mol Biol **12** (1965), 88–118.

[NAW⁺91] H. T. Nagamoto, L. E. Adler, M. C. Waldo, J. Griffith, and R. Freedman, *Gating of auditory response in schizophrenics and normal controls. effects of recording site and stimulation interval on the p50 wave*, Schizophr Res **4** (1991 Jan to Feb), no. 1, 31–40.

[NAY62] J. Nagumo, S. Arimoto, and S. Yoshizawa, *An active pulse transmission line simulating nerve axon*, Proc IRE **50** (1962), no. 10, 2061–2070.

[NCPS06] T. I. Netoff, T. Carroll, L. M. Pecora, and S. J. Schiff, *Detecting coupling in the presence of noise an nonlinearity*, Handbook of time series analysis. Recent theoretical developments and applications. Wiley-Vch Weinheim, 2006.

[NH07] P. Nachev and M. Husain, *Comment on "detecting awareness in the vegetative state,"* Science **315** (2007 Mar), no. 5816, 1221; author reply 1221.

[NJ03] S. Nadkarni and P. Jung, *Spontaneous oscillations of dressed neurons: A new mechanism for epilepsy?*, Phys Rev Lett **91** (2003 Dec), no. 26 Pt 1, 268101.

[NPS04] T. I. Netoff, L. M. Pecora, and S. J. Schiff, *Analytical coupling detection in the presence of noise and nonlinearity*, Phys Rev E **69** (2004), no. 1, 17201.

[NRB04] P. C. Nelson, M. Radosavljević, and S. Bromberg, *Biological physics: Energy, information, life*, W.H. Freeman and Co., New York, 2004.

[NS06] P. L. Nunez and R. Srinivasan, *Electric fields of the brain: The neurophysics of EEG*, 2nd ed., Oxford University Press, Oxford, 2006.

[OCB⁺06] A. M. Owen, M. R. Coleman, M. Boly, M. H. Davis, S. Laureys, and J. D. Pickard, *Detecting awareness in the vegetative state*, Science **313** (2006Sep), no. 5792, 1402.

[OD71] J. O'Keefe and J. Dostrovsky, *The hippocampus as a spatial map. Preliminary evidence from unit activity in the freely-moving rat*, Brain Res **34** (1971), no. 1, 171–175.

[OGY90] E. Ott, C. Grebogi, and J. A. Yorke, *Controlling chaos*, Phys Rev Lett **64** (1990), no. 11, 1196–1199.

[OHM⁺05] L. M. Oberman, E. M. Hubbard, J. P. McCleery, E. L. Altschuler, V. S. Ramachandran, and J. A. Pineda, *EEG evidence for mirror neuron dysfunction in autism spectrum disorders*, Brain Res Cogn Brain Res **24** (2005 Jul), no. 2, 190–198.

[OHS⁺04] E. Ott, B. R. Hunt, I. Szunyogh, A. V. Zimin, E. J. Kostelich, M. Corazza, et al., *Estimating the state of large spatio-temporally chaotic systems*, Physics Lett A **330** (2004), no. 5, 365–370.

[OMRO⁺08] J. A. Obeso, C. Marin, C. Rodriguez-Oroz, J. Blesa, B. Benitez-Temiño, J. Mena-Segovia, et al., *The basal ganglia in Parkinson's disease: Current concepts and unexplained observations*, Ann Neurol **64 Suppl 2** (2008), S30–46.

[OR93] J. O'Keefe and M. L. Recce, *Phase relationship between hippocampal place units and the EEG theta rhythm*, Hippocampus **3** (1993), no. 3, 317–330.

[OS94] E. Ott and J. C. Sommerer, *Blowout bifurcations: The occurence of riddled basins and on-off intermittency phys*, Rev Lett **188** (1994), 39–47.

[OSB+10] S. Ozen, A. Sirota, M. A. Belluscio, C. A. Anastassiou, C. Koch, G. Buzsáki, *Transcranial electric stimulation entrains cortical neuronal populations in rats*, J Neurosci **30** (2010), 11476–85.

[PAF+] L. Paninski, Y. Ahmadian, D. G. Ferreira, S. Koyama, K. Rahnama Rad, M. Vidne, J. et al., *A new look at state-space models for neural data*, J Comput Neurosci **29** (2010), 107–126.

[PAM04] A. A. Prinz, L. F. Abbott, and E. Marder, *The dynamic clamp comes of age*, Trends Neurosci **27** (2004 Apr), no. 4, 218–224.

[Par17] J. Parkinson, *An essay on the shaking palsy*, Sherwood, Neely, and Jones, London, 1817.

[PBG+05] E. H. Park, E. Barreto, B. J. Gluckman, S. J. Schiff, and P. So, *A model of the effects of applied electric fields on neuronal synchronization*, J Comput Neurosci **19** (2005), no. 1, 53–70.

[PBL+94] V. Parpura, T. A. Basarsky, F. Liu, K. Jeftinija, S. Jeftinija, and P. G. Haydon, *Glutamate-mediated astrocyte-neuron signalling*, Nature **369** (1994 Jun), no. 6483, 744–747.

[PC90] L. M. Pecora and T. L. Carroll, *Synchronization in chaotic systems*, Phys Rev Lett **64** (1990), no. 8, 821–824.

[PCH95] L. M. Pecora, T. L. Carroll, and J. F. Heagy, *Statistics for mathematical properties of maps between time series embeddings*, Phys Rev E **52** (1995), no. 4, 3420–3439.

[PCP+97] J. C. Prechtl, L. B. Cohen, B. Pesaran, P. P. Mitra, and D. Kleinfeld, *Visual stimuli induce waves of electrical activity in turtle cortex*, Proc Natl Acad Sci **94** (1997), 7621–7626.

[PD06] E.-H. Park and D. M. Durand, *Role of potassium lateral diffusion in non-synaptic epilepsy: A computational study*, J Theor Biol **238** (2006 Feb), no. 3, 666–682.

[PE01a] D. J. Pinto and G. B. Ermentrout, *Spatially structured activity in synaptically coupled neuronal networks: I. Traveling fronts and pulses*, SIAM J Appl Math (2001), 206–225.

[PE01b] ———, *Spatially structured activity in synaptically coupled neuronal networks: II. Lateral inhibition and standing pulses*, SIAM J Appl Math (2001), 226–243.

[Pen94] R. Penrose, *Shadows of the mind: A search for the missing science of consciousness*, Oxford University Press, Oxford, 1994.

[PFHD04] L. Paninski, M. R. Fellows, N. G. Hatsopoulos, and J. P. Donoghue, *Spatiotemporal tuning of motor cortical neurons for hand position and velocity*, J Neurophysiol **91** (2004), no. 1, 515–532.

[PH00] V. Parpura and P. G. Haydon, *Physiological astrocytic calcium levels stimulate glutamate release to modulate adjacent neurons*, Proc Natl Acad Sci USA **97** (2000 Jul), no. 15, 8629–8634.

[PHK02] S. Panda, J. B. Hogenesch, and S. A. Kay, *Circadian rhythms from flies to human*, Nature **417** (2002), no. 6886, 329–335.

[PHK+01] D. J. Patil, B. R. Hunt, E. Kalnay, J. A. Yorke, and E. Ott, *Local low dimensionality of atmospheric dynamics*, Phys Rev Lett **86** (2001), no. 26, 5878–5881.

[Pic97] M. Piccolino, *Luigi Galvani and animal electricity: Two centuries after the foundation of electrophysiology*, Trends Neurosci **20** (1997 Oct), no. 10, 443–448.

[PK99] D. Plenz and S. T. Kital, *A basal ganglia pacemaker formed by the subthalamic nucleus and external globus pallidus*, Nature **400** (1999), no. 6745, 677–682.

[PN87] O. B. Paulson and E. A. Newman, *Does the release of potassium from astrocyte endfeet regulate cerebral blood flow?*, Science **237** (1987 Aug), no. 4817, 896–898.

[PP07] J. B. Posner and F. Plum, *Plum and Posner's diagnosis of stupor and coma*, 4th ed., Vol. 71, Oxford University Press, Oxford, 2007.

[PR94] P. F. Pinsky and J. Rinzel, *Intrinsic and network rhythmogenesis in a reduced traub model for CA3 neurons*, J Comput Neurosci **1** (1994 Jun) no. 1–2, 39–60.

[Pre07] W. H. Press, *Numerical recipes: The art of scientific computing*, 3rd ed., Cambridge University Press, Cambridge, 2007.

[PRSC08] M. Pirini, L. Rocchi, M. Sensi, and L. Chiari, *A computational modelling approach to investigate different targets in deep brain stimulation for Parkinson's disease*, J Comput Neurosci **26** (2008), 91–107.

[PM65] D. P. Purpura and J. G. McMurtry, *Intracellular Activities and Evoked Potential Changes During Polarization of Motor Cortex*, J Neurophysiol **28** (1965Jan), 166–85.

[PM66] D. P. Purpura and A. Malliani, *Spike generation and propagation initiated in dendrites by transhippocampal polarization*, Brian Res **1** (1966), no. 4, 403–6.

[PVC99] J. L. Perez-Velazquez and P. L. Carlen, *Synchronization of GABAergic interneuronal networks during seizure-like activity in the rat horizontal hippocampal slice*, Eur J Neurosci **11** (1999 Nov), no. 11, 4110–4118.

[PW93] D. B. Percival and A. T. Walden, *Spectral analysis for physical applications: Multitaper and conventional univariate techniques*, Cambridge University Press, Cambridge, 1993.

[Pyr96] K. Pyragas, *Weak and strong synchronization of chaos*, Phys Rev E **54** (1996), no. 5, 4508–4511.

[Qia99] Y. Qian, *Periodic orbits: A novel language for neuronal dynamics*, Adv Mech [Chinese] **29** (1999), 121–133.

[Ral55a] W. Rall, *A statistical theory of monosynaptic Input-output relations*, J Cell Compar Physiol **46** (1955), no. 3, 373–411.

[Ral55b] ———, *Experimental monosynaptic input-output relations in the mammalian spinal cord*, J Cell Compar Physiol **46** (1955), no. 3, 413–437.

[RAM98] G. Rainer, W. F. Asaad, and E. K. Miller, *Memory fields of neurons in the primate prefrontal cortex*, Proc Natl Acad Sci USA **95** (1998 Dec), no. 25, 15008–15013.

[Ran75] J. B. Ranck, Jr. *Which elements are excited in electrical stimulation of mammalian central nervous system: A review*, Brain Res **98** (1975 Nov), no. 3, 417–440.

[RB97] R. P. Rao and D. H. Ballard, *Dynamic model of visual recognition predicts neural response properties in the visual cortex*, Neural Comput **9** (1997), no. 4, 721–763.

[RBHL99] R. Romo, C. D. Brody, A. Hernández, and L. Lemus, *Neuronal correlates of parametric working memory in the prefrontal cortex*, Nature **399** (1999 Jun), no. 6735, 470–473.

[RFDC09] G. Rizzolatti, M. Fabbri-Destro, and L. Cattaneo, *Mirror neurons and their clinical relevance*, Nat Clin Pract Neurol **5** (2009 Jan), no. 1, 24–34.

[RFMF+01] A. Raz, V. Frechter-Mazar, A. Feingold, M. Abeles, E. Vaadia, and H. Bergman, *Activity of pallidal and striatal tonically active neurons is correlated in MPTP-treated monkeys but not in normal monkeys*, J Neurosci **21** (2001), no. 3, RC128.

[RFPSS05] J. E. Ruppert-Felsot, O. Praud, E. Sharon, and H. L. Swinney, *Extraction of coherent structures in a rotating turbulent flow experiment*, Phys Rev E **72** (2005), no. 1, 16311.

[RG05] F. Ramón and W. Gronenberg, *Electrical potentials indicate stimulus expectancy in the brains of ants and bees*, Cell Mol Neurobiol **25** (2005 Mar), no. 2, 313–327.

[RGDA97] A. Riehle, S. Grün, M. Diesmann, and A. Aertsen, *Spike synchronization and rate modulation differentially involved in motor cortical function*, Science **278** (1997 Dec), no. 5345, 1950–1953.

[RGW+03] K. A. Richardson, B. J. Gluckman, S. L. Weinstein, C. E. Glosch, J. B. Moon, R. P. Gwinn, et al., *In vivo modulation of hippocampal epileptiform activity with radial electric fields*, Epilepsia **44** (2003 Jun), no. 6, 768–777.

[Rin85] J. Rinzel, *Excitation dynamics: Insights from simplified membrane models*, Fed Proc **44** (1985), no. 15, 2944–2946.

[Rin90] J. Rinzel, *Discussion: Electrical excitability of cells, theory and experiment: Review of the Hodgkin-Huxley foundation and an update*, Bulletin of Mathematical Biology **52** (1990) no. 1, 3–23.

[RLJ85] P. A. Rutecki, F. J. Lebeda, and D. Johnston, *Epileptiform activity induced by changes in extracellular potassium in hippocampus*, J Neurophysiol **54** (1985 Nov), no. 5, 1363–1374.

[Rob57] J. D. Robertson, *New observations on the ultrastructure of the membranes of frog peripheral nerve fibers*, J Biophys Biochem Cytol **3** (1957), no. 6, 1043–1048.

[Rop10] A. H. Ropper, *Cogito ergo sum by MRI*, N Engl J Med **362** (2010 Feb), no. 7, 648–649.

[RRH06] D. Rubino, K. A. Robbins, and N. G. Hatsopoulos, *Propagating waves mediate information transfer in the motor cortex*, Nat Neurosci **9** (2006), no. 12, 1549–1557.

[RRS00] C. B. Ransom, B. R. Ransom, and H. Sontheimer, *Activity-dependent extracellular K+ accumulation in rat optic nerve: The role of glial and axonal na+ pumps*, J Physiol **522 Pt 3** (2000Feb), 427–442.

[RSG05] K. A. Richardson, S. J. Schiff, and B. J. Gluckman, *Control of traveling waves in the mammalian cortex*, Phys Rev Lett **94** (2005), no. 2, 28103.

[RSTA95] N. F. Rulkov, M. M. Sushchik, L. S. Tsimring, and H. D. I. Abarbanel, *Generalized synchronization of chaos in directionally coupled chaotic systems*, Phys Rev E **51** (1995), no. 2, 980–994.

[RT04] J. E. Rubin and D. Terman, *High frequency stimulation of the subthalamic nucleus eliminates pathological thalamic rhythmicity in a computational model*, J Comput Neurosci **16** (2004), no. 3, 211–235.

[Rus27] W. A. Rushton, *The effect upon the threshold for nervous excitation of the length of nerve exposed, and the angle between current and nerve*, J Physiol **63** (1927 Sep), no. 4, 357–377.

[SB03] A. C. Smith and E. N. Brown, *Estimating a state-space model from point process observations*, Neural Comput **15** (2003), no. 5, 965–991.

[SB07] L. Srinivasan and E. N. Brown, *A state-space framework for movement control to dynamic goals through brain-driven interfaces*, IEEE Trans Biomed Eng **54** (2007), no. 3, 526–535.

[SBIJ08] E. T. Spiller, A. Budhiraja, K. Ide, and C. K. R. T. Jones, *Modified particle filter methods for assimilating Lagrangian data into a point-vortex model*, Physica D: Nonlinear Phenomena **237** (2008), 1498–1506.

[SBJ⁺02] P. So, E. Barreto, K. Josić, E. Sander, and S. J. Schiff, *Limits to the experimental detection of nonlinear synchrony*, Phys Rev E **65** (2002), no. 4, 46225.

[SBOM00] W. J. Song, Y. Baba, T. Otsuka, and F. Murakami, *Characterization of Ca(2+) channels in rat subthalamic nucleus neurons*, J Neurophysiol **84** (2000), no. 5, 2630–2637.

[SBP07] S. Schütz-Bosbach and W. Prinz, *Prospective coding in event representation*, Cogn Process **8** (2007), no. 2, 93–102.

[Sch10] S. J. Schiff, *Towards model-based control of Parkinson's disease*, Phil Trans R Soc Lond A **368** (2010) 2269–2308.

[Sch44a] E. Scharrer, *The blood vessels of the nervous tissue*, Q Rev Biol **19** (1944), no. 4, 308–318.

[Sch44b] E. Schrödinger, *What is life?: The physical aspect of the living cell*, Cambridge University Press, Cambridge, 1944.

[Sco03] A. Scott, *Nonlinear science: Emergence and dynamics of coherent structures*, 2nd ed., Vol. 8, Oxford University Press, Oxford, 2003.

[Sco95] ———, *Stairway to the mind: The controversial new science of consciousness*, Copernicus, New York, 1995.

[SCP⁺07] S. Sunderam, N. Chernyy, N. Peixoto, J. P. Mason, S. L. Weinstein, S. J. Schiff, and B. J. Gluckman, *Improved sleep-wake and behavior discrimination using mems accelerometers*, J Neurosci Methods **163** (2007 Jul), no. 2, 373–383.

[SCP⁺09] ———, *Seizure entrainment with polarizing low-frequency electric fields in a chronic animal epilepsy model*, J Neural Eng **6** (2009 Aug), no. 4, 046009.

[SDW02] S. J. Schiff, B. K. Dunagan, and R. M. Worth, *Failure of single-unit neuronal activity to differentiate globus pallidus internus and externus in Parkinson disease*, J Neurosurg **97** (2002), no. 1, 119–128.

[Sea80] J. Searle, *Minds, brains, and programs*, Behav Brain Sci **3** (1980), 417–457.

[SEJP09] A. H. V. Schapira, M. Emre, P. Jenner, and W. Poewe, *Levodopa in the treatment of Parkinson's disease*, Eur J Neurol **16** (2009), no. 9, 982–989.

[SEMB07] L. Srinivasan, U. T. Eden, S. K. Mitter, and E. N. Brown, *General-purpose filter design for neural prosthetic devices*, J Neurophysiol **98** (2007), no. 4, 2456–2475.

[Seu09] H. S. Seung, *Reading the book of memory: Sparse sampling versus dense mapping of connectomes*, Neuron **62** (2009), no. 1, 17–29.

[SFN⁺98] P. So, J. T. Francis, T. I. Netoff, B. J. Gluckman, and S. J. Schiff, *Periodic orbits: A new language for neuronal dynamics*, Biophysi J **74** (1998), no. 6, 2776–2785.

[SFS98] E. Schneidman, B. Freedman, and I. Segev, *Ion channel stochasticity may be critical in determining the reliability and precision of spike timing*, Neural Comput **10** (1998), no. 7, 1679–1703.

[SGY97] T. Sauer, C. Grebogi, and J. A. Yorke, *How long do numerical chaotic solutions remain valid?*, Phys. Rev. Lett. **79** (1997 Jul), no. 1, 59–62.

[Sha19] E. H. Shackleton, *South: The story of Shackleton's last expedition, 1914–1917*, W. Heinemann, London, 1919.

[She06] C. S. Sherrington, *The integrative action of the nervous system*, C. Scribner's Sons, New York, 1906.

[SHM03] Y. Shu, A. Hasenstaub, and D. A. McCormick, *Turning on and off recurrent balanced cortical activity*, Nature **423** (2003), no. 6937, 288–293.

[SHP+02] M. D. Serruya, N. G. Hatsopoulos, L. Paninski, M. R. Fellows, and J. P. Donoghue, *Instant neural control of a movement signal*, Nature **416** (2002), no. 6877, 141–142.

[SHW07] S. J. Schiff, X. Huang, and J.-Y. Wu, *Dynamical evolution of spatiotemporal patterns in mammalian middle cortex*, Phys Rev Lett **98** (2007), no. 17, 178102.

[Sim06a] D. Simon, *Optimal state estimation: Kalman, H and nonlinear approaches*, Wiley-Interscience, Hoboken, N.J., 2006.

[Sim06b] ———, *Using nonlinear Kalman filtering to estimate signals*, Embedded Systems Design **19** (2006), no. 7, 38–57.

[Sir03] L. Sirovich, *A pattern analysis of the second Rehnquist US Supreme Court*, Proc Nat Acad Sci **100** (2003), no. 13, 7432–7437.

[Sir87] ———, *Turbulence and the dynamics of coherent structures*. I- Coherent structures., Q Appl Math **45** (1987), 561–571.

[Sir89] ———, *Chaotic dynamics of coherent structures*, Physica D **37** (1989), no. 1-3, 126–145.

[SJ07] R. D. Saunders and J. G. R. Jefferys, *A neurobiological basis for elf guidelines*, Health Phys **92** (2007 Jun), no. 6, 596–603.

[SJD+94] S. J. Schiff, K. Jerger, D. H. Duong, T. Chang, M. L. Spano, and W. L. Ditto, *Controlling chaos in the brain*, Nature **370** (1994), no. 6491, 615–620.

[SK05] P. Somogyi and T. Klausberger, *Defined types of cortical interneurone structure space and spike timing in the hippocampus*, J Physiol **562** (2005 Jan), no. Pt 1, 9–26.

[SK87] L. Sirovich and M. Kirby, *Low-dimensional procedure for the characterization of human faces*, J Opt Soc Am A **4**, (1987), 519–524.

[SKJI06] H. Salman, L. Kuznetsov, C. Jones, and K. Ide, *A method for assimilating Lagrangian data into a shallow-water-equation ocean model*, Monthly Weather Rev **134** (2006), no. 4, 1081–1101.

[SKM+07] G. Schalk, J. Kubánek, K. J. Miller, N. R. Anderson, E. C. Leuthardt, J. G. Ojemann, et al., *Decoding two-dimensional movement trajectories using electrocorticographic signals in humans*, J Neural Eng **4** (2007), no. 3, 264–275.

[SKW08] G. G. Somjen, H. Kager, and W. J. Wadman, *Computer simulations of neuron-glia interactions mediated by ion flux*, J Comput Neurosci **25** (2008 Oct), no. 2, 349–365.

[SLBY98] K. J. Staley, M. Longacher, J. S. Bains, and A. Yee, *Presynaptic modulation of CA3 network activity*, Nat Neurosci **1** (1998), no. 3, 201–209.

[SMH05a] T. R. Smith, J. Moehlis, and P. Holmes, *Low-dimensional modelling of turbulence using the proper orthogonal decomposition: A tutorial*, Nonlinear Dynamics **41** (2005), no. 1, 275–307.

[SMH05b] T. R. Smith, J. Moehlis, and P. Holmes, *Low-dimensional models for turbulent plane Couette flow in a minimal flow unit*, J Fluid Mech **538**, (2005), 71–110.

[Sod07] T. Soderstrom, *Errors-in-variables methods in system identification*, Automatica **43** (2007), no. 6, 939–958.

[Som04] G. G. Somjen, *Ions in the brain: Normal function, seizures, and stroke*, Oxford University Press, Oxford, 2004.

[Sor70] H. W. Sorensen, *Least-squares estimation: From Gauss to Kalman*, IEEE Spectrum **7** (1970), 63–68.

[SOS06] S. Shoham, D. H. O'Connor, and R. Segev, *How silent Is the brain: is there a "dark matter" problem in neuroscience?*, J Comp Physiol A **192** (2006), no. 8, 777–784.

[SOS+96] P. So, E. Ott, S. J. Schiff, D. T. Kaplan, T. Sauer, and C. Grebogi, *Detecting unstable periodic orbits in chaotic experimental data*, Phys Rev Lett **76** (1996), no. 25, 4705–4708.

[SOS+97] P. So, E. Ott, T. Sauer, B. J. Gluckman, C. Grebogi, and S. J. Schiff, *Extracting unstable periodic orbits from chaotic time series data*, Phys Rev E **55** (1997), no. 5, 5398–5417.

[SP74] A. Schopenhauer and E. F. J. Payne, *Parerga and paralipomena: Short philosophical essays*, Clarendon Press, Oxford, 1974.

[SR99] D. M. Senseman and K. A. Robbins, *Modal behavior of cortical neural networks during visual processing*, J Neurosci **19** (1999), no. 10, RC3, 1–7.

[SRSRSL99] M. L. Steyn-Ross, D. A. Steyn-Ross, J. W. Sleigh, and D. T. J. Liley, *Theoretical electroencephalogram stationary spectrum for a white-noise-driven cortex: Evidence for a general anesthetic-induced phase transition*, Phys Rev E **60** (1999), no. 6, 7299–7311.

[SS08] S. J. Schiff and T. Sauer, *Kalman filter control of a model of spatiotemporal cortical dynamics*, J Neural Eng **5** (2008), no. 1, 1–8.

[SS09] T. D. Sauer and S. J. Schiff, *Data assimilation for heterogeneous networks: The consensus set*, Phys Rev E **79** (2009), no. 5 Pt 1, 051909.

[SS85] S. J. Schiff and G. G. Somjen, *Hyperexcitability following moderate hypoxia in hippocampal tissue slices*, Brain Res **337** (1985 Jul), no. 2, 337–340.

[SS87] ———, *The effect of graded hypoxia on the hippocampal slice: An in vitro model of the ischemic penumbra*, Stroke **18** (1987 Jan–Feb), no. 1, 30–37.

[SSC+96] S. J. Schiff, P. So, T. Chang, K. R. Burke, and T. D, Sauer, *Detecting dynamical interdependence and generalized synchrony through mutual prediction in a neural ensemble*, Phys Rev E **54** (1996), no. 6, 6708–6724.

[SSKV02] A. Sitz, U. Schwarz, J. Kurths, and H. U. Voss, *Estimation of parameters and unobserved components for nonlinear systems from noisy time series*, Phys. Rev. E **66** (2002), no. 1, 016210.

[SSKW05] S. J. Schiff, T. Sauer, R. Kumar, and S. L. Weinstein, *Neuronal spatiotemporal pattern discrimination: The dynamical evolution of seizures*, Neuroimage **28** (2005), no. 4, 1043–1055.

[Sta72] H. E. Stanley, *Biomedical physics and biomaterials science*, MIT Press, Cambridge, Mass., 1972.

[Sta87] ———, *Introduction to phase transitions and critical phenomena*, Oxford University Press, New York, 1987.

[Ste93] G. W. Stewart, *On the early history of the singular value decomposition*, SIAM Rev (1993), 551–566.

[Str06] G. Strang, *Linear algebra and its applications*, 4th ed, Thomson, Brooks/Cole, Belmont, Calif., 2006.

[Str86] ———, *Introduction to applied mathematics*, Wellesley-Cambridge Press, Wellesley, Mass., 1986.

[SVM00] M. V. Sanchez-Vives and D. A. McCormick, *Cellular and network mechanisms of rhythmic recurrent activity in neocortex*, Nat Neurosci **3** (2000Oct), no. 10, 1027–1034.

[SW49] C. E. Shannon and W. Weaver, *The mathematical theory of communication*, University of Illinois Press, Urbana, 1949.

[SWSB07] A. C. Smith, S. Wirth, W. A. Suzuki, and E. N. Brown, *Bayesian analysis of interleaved learning and response bias in behavioral experiments*, J Neurophysiol **97** (2007), no. 3, 2516–2524.

[Tal07] N. Taleb, *The black swan: The impact of the highly improbable*, 1st ed., Random House, New York, 2007.

[Tas98] R. R. Tasker, *Deep brain stimulation is preferable to thalamotomy for tremor suppression*, Surg Neurol **49** (1998), no. 2, 145–153; discussion 153–154.

[TAT+05] G.-F. Tian, H. Azmi, T. Takano, Q. Xu, W. Peng, J. Lin, et al., *An astrocytic basis of epilepsy*, Nat Med **11** (2005 Sep), no. 9, 973–981.

[TB56] C. A. Terzuolo and T. H. Bullock, *Measurement of imposed voltage gradient adequate to modulate neuronal firing*, Proc Natl Acad Sci USA **42** (1956 Sep), no. 9, 687–694.

[TBF05] S. Thrun, W. Burgard, and D. Fox, *Probabilistic robotics*, MIT Press, Cambridge, Mass., 2005.

[TC98] A. E. Telfeian and B. W. Connors, *Layer-specific pathways for the horizontal propagation of epileptiform discharges in neocortex*, Epilepsia **39** (1998), no. 7, 700–708.

[TD82] C. P. Taylor and F. E. Dudek, *Synchronous neural afterdischarges in rat hippocampal slices without active chemical synapses*, Science **218** (1982 Nov), no. 4574, 810–812.

[TD88] S. F. Traynelis and R. Dingledine, *Potassium-induced spontaneous electrographic seizures in the rat hippocampal slice*, J Neurophysiol **59** (1988 Jan), no. 1, 259–276.

[TDSK85] R. D. Traub, F. E. Dudek, R. W. Snow, and W. D. Knowles, *Computer simulations indicate that electrical field effects contribute to the shape of the epileptiform field potential*, Neuroscience **15** (1985b Aug), no. 4, 947–958.

[TDTK85] R. D. Traub, F. E. Dudek, C. P. Taylor, and W. D. Knowles, *Simulation of hippocampal afterdischarges synchronized by electrical interactions*, Neuroscience **14** (1985a Apr), no. 4, 1033–1038.

[Tho82] D. J. Thomson, *Spectrum estimation and harmonic analysis*, Proc IEEE **70** (1982), no. 9, 1055–1096.

[Tho86] S. M. Thompson, *Relations between chord and slope conductances and equivalent electromotive forces*, Am J Physiol **250** (1986), no. 2 Pt 1, C333–339.

[TM91] R. D. Traub and R. Miles, *Neuronal networks of the hippocampus*, Cambridge University Press, Cambridge, 1991.

[TN86] D. Tranchina and C. Nicholson, *A model for the polarization of neurons by extrinsically applied electric fields*, Biophys J **50** (1986 Dec), no. 6, 1139–1156.

[TP07] Z. Toth and M. Peña, *Data assimilation and numerical forecasting with imperfect models: The mapping paradigm*, Physica D: Nonlin Phenomena **230** (2007), no. 1–2, 146–158.

[TRD⁺07] A. B. L. Tort, H. G. Rotstein, T. Dugladze, T. Gloveli, and N. J. Kopell, *On the formation of gamma-coherent cell assemblies by oriens lacunosum-moleculare interneurons in the hippocampus*, Proc Natl Acad Sci USA **104** (2007 Aug), no. 33, 13490–13495.

[TRYW02] D. Terman, J. E. Rubin, A. C. Yew, and C. J. Wilson, *Activity patterns in a model for the sub-thalamopallidal network of the basal ganglia*, J Neurosci **22** (2002), no. 7, 2963–2976.

[TTS02] D. M. Taylor, S. I. H. Tillery, and A. B. Schwartz, *Direct cortical control of 3d neuroprosthetic devices*, Science **296** (2002), no. 5574, 1829–1832.

[Tur50] A. M. Turing, *Computing machinery and intelligence*, Mind **59** (1950), no. 236, 433–460.

[TW99] D. Teets and K. Whitehead, *The discovery of ceres: How Gauss became famous*, Mathematics Mag **72** (1999), no. 2, 83–93.

[UCJBS09] G. Ullah, J. R. Cressman Jr., E. Barreto, and S. J. Schiff, *The influence of sodium and potassium dynamics on excitability, seizures, and the stability of persistent states: II. Network and glial dynamics*, J Comput Neurosci **26** (2009a), no. 2, 171–183.

[US09] G. Ullah and S. J. Schiff, *Tracking and control of neuronal Hodgkin-Huxley dynamics*, Phys Rev E **79** (2009b), no. 4, 40901.

[US10] G. Ullah and S. J. Schiff, *Assimilating seizure dynamics*, PLoS Comput Biol **6** (2010), 1000776.

[Utk77] V. Utkin, *Variable structure systems with sliding modes*, IEEE Trans on Automatic Control **22** (1977), no. 2, 212–222.

[VBF⁺03] J. L. Vitek, R. A. E. Bakay, A. Freeman, M. Evatt, J. Green, W. McDonald, et al., *Randomized trial of pallidotomy versus medical therapy for Parkinson's disease*, Ann Neurol **53** (2003), no. 5, 558–569.

[VdP26] B. Van der Pol, *LXXXVIII. On "relaxation-oscillations,"* Phil Mag Series 7 **2** (1926), no. 11, 978–992.

[vdPVdM27] B. van der Pol and J. Van der Mark, *Frequency demultiplication*, Nature **120** (1927), no. 3019, 363–364.

[VGPN03] E. Von Goldammer, J. Paul, and J. Newbury, *Heterarchy-hierarchy. Two complementary categories of description*, 2003. Available at <http://www.vordenker.de/hetarchy/a_hetarchy-e.pdf>.

[VTA06] G. Vassort, K. Talavera, and J. L. Alvarez, *Role of t-type Ca2+ channels in the heart*, Cell Calcium **40** (2006), no. 2, 205–220.

[VTK04] H. U. Voss, J. Timmer, and J. Kurths, *Nonlinear dynamical system identification from uncertain and indirect measurements*, Interl J Bifur Chaos **14** (2004), no. 6, 1905–1933.

[Wan03] X.-J. Wang, *Persistent neural activity: Experiments and theory*, Cereb Cortex **13** (2003 Nov), no. 11, 1123.

[WB07] C. K. Wikle and L. M. Berliner, *A Bayesian tutorial for data assimilation*, Physica D: Nonlin Phenomena **230** (2007), no. 1-2, 1–16.

[WC72] H. R. Wilson and J. D. Cowan, *Excitatory and inhibitory interactions in localized populations of model neurons*, Biophysical J **12** (1972), no. 1, 1–24.

[WC73] ———, *A mathematical theory of the functional dynamics of cortical and thalamic nervous tissue*, Biol Cybern **13** (1973), no. 2, 55–80.

[WD03] T. Wichmann and M. R. DeLong, *Pathophysiology of parkinson's disease: The MPTP primate model of the human disorder*, Ann NY Acad Sci **991** (2003), 199–213.

[WDK03] D. M. Wolpert, K. Doya, and M. Kawato, *A unifying computational framework for motor control and social interaction*, Philos Trans R Soc Lond B **358** (2003 Mar), no. 1431, 593–602.

[WFS⁺09] F. M. Weaver, K. Follett, M. Stern, K. Hur, C. Harris, W. J. Marks Jr., et al., *Bilateral deep brain stimulation vs best medical therapy for patients with advanced Parkinson disease: A randomized controlled trial*, JAMA **301** (2009), no. 1, 63–73.

[WGB⁺06] W. Wu, Y. Gao, E. Bienenstock, J. P. Donoghue, and M. J. Black, *Bayesian population decoding of motor cortical activity using a Kalman filter*, Neural Comput **18** (2006), no. 1, 80–118.

[WGD92] E. N. Warman, W. M. Grill, and D. Durand, *Modeling the effects of electric fields on nerve fibers: Determination of excitation thresholds*, IEEE Trans Biomed Eng **39** (1992), no. 12, 1244–1254.

[Wie48] N. Wiener, *Cybernetics*, John Wiley, New York, 1948.

[Wil99] H. R. Wilson, *Spikes, decisions, and actions: The dynamical foundations of neuroscience*, Oxford University Press, Oxford, 1999.

[WM04] J. R. Wolpaw and D. J. McFarland, *Control of a two-dimensional movement signal by a noninvasive brain-computer interface in humans*, Proc Natl Acad Sci USA **101** (2004), no. 51, 17849–17854.

[WM95] K. Wiesenfeld and F. Moss, *Stochastic resonance and the benefits of noise: From ice ages to crayfish and squids*, Nature **373** (1995), no. 6509, 33–36.

[WMVS03] J. R. Wolpaw, D. J. McFarland, T. M. Vaughan, and G. Schalk, *The Wadsworth Center brain-computer interface (BCI) research and development program*, IEEE Trans Neural Syst Rehabil Eng **11** (2003), no. 2, 204–207.

[Woo06] A. W. Wood, *How dangerous are mobile phones, transmission masts, and electricity pylons?*, Arch Dis Child **91** (2006 Apr), no. 4, 361–366.

[WS93] I. P. Weiss and S. J. Schiff, *Reflex variability in selective dorsal rhizotomy*, J Neurosurg **79** (1993), no. 3, 346–353.

[WVAA98] J. C. Weaver, T. E. Vaughan, R. K. Adair, and R. D. Astumian, *Theoretical limits on the threshold for the response of long cells to weak extremely low frequency electric fields due to ionic and molecular flux rectification*, Biophys J **75** (1998 Nov), no. 5, 2251–2254.

[XHTW07] W. Xu, X. Huang, K. Takagaki, and J.-Y. Wu, *Compression and reflection of compression and reflection of visually evoked cortical waves*, Neuron **55** (2007), 119–129.

[YBL⁺06] S. C. Yang, D. Baker, H. Li, K. Cordes, M. Huff, G. Nagpal, et al., *Data assimilation as synchronization of truth and model: Experiments with the three-variable Lorenz system*, J Atmos Sci **63** (2006), no. 9, 2340–2354.

[YCX⁺01] X. Yu, G. Chen, Y. Xia, Y. Song, and Z. Cao, *An invariant-manifold-based method for chaos control*, IEEE Trans Circuits Systems I: Fundamental Theory and Applications **48** (2001), no. 8, 930–937.

[YKH83] Y. Yaari, A. Konnerth, and U. Heinemann, *Spontaneous epileptiform activity of ca1 hippocampal neurons in low extracellular calcium solutions*, Exp Brain Res **51** (1983), no. 1, 153–156.

[You36] J. Z. Young, *The structure of nerve fibres in cephalopods and crustacea*, Proc R Soc Lond B (1936), 319–337.

[YUO99] K. D. Young, V. I. Utkin, and A. Ozguner, *A control engineer's guide to sliding mode control. Control systems technology*, IEEE Trans **7** (1999), no. 3, 328–342.

[ZCBS06] J. Ziburkus, J. R. Cressman, E. Barreto, and S. J. Schiff, *Interneuron and pyramidal cell interplay during in vitro seizure-like events*, J Neurophysiol **95** (2006), no. 6, 3948–3954.

[ZDG96] K. Zhou, J. C. Doyle, and K. Glover, *Robust and optimal control*, Prentice Hall, Englewood Cliffs, N.J., 1996.

[ZRC07] S. Zenker, J. Rubin, and G. Clermont, *From inverse problems in mathematical physiology to quantitative differential diagnoses*, PLoS Comput Biol **3** (2007 Nov), no. 11, e204.

Index

Printed in the United States
by Baker & Taylor Publisher Services